VOLCANIC ACTIVITY AND CLIMATE

**Studies in
Geophysical Optics and Remote Sensing
Series Editor: Adarsh Deepak
Published Volumes and Volumes in Preparation**

Action Plan for Remote Sensing Applications for Rice Production
Adarsh Deepak (Ed.)

Advances in Remote Sensing Retrieval Methods
Adarsh Deepak, Henry E. Fleming, and Moustafa T. Chahine (Eds.)

Aerosols and Climate
Peter V. Hobbs and M. Patrick McCormick (Eds.)

Aerosols and Their Climatic Effects
Hermann E. Gerber and Adarsh Deepak (Eds.)

A Lunar-Based Analytical Laboratory
Charles W. Gehrke, Mitchell K. Hobish, Robert W. Zumwalt, Michel Prost, and Jean Desgrès (Eds.)

A Lunar-Based Chemical Analysis Laboratory
Cyril Ponnamperuma and Charles W. Gehrke (Eds.)

Applications of Remote Sensing for Rice Production
Adarsh Deepak and K.R. Rao (Eds.)

Atmospheric Aerosols: Global Climatology and Radiative Characteristics
Guillaume A. d'Almeida, Peter Koepke, and Eric P. Shettle

Atmospheric Aerosols: Their Formation, Optical Properties, and Effects
Adarsh Deepak (Ed.)

Atmospheric Radiative Transfer
Jacqueline Lenoble

Hygroscopic Aerosols
Lothar H. Ruhnke and Adarsh Deepak (Eds.)

Geographic Information Systems in Government
Bruce K. Opitz (Ed.)

IRS '84: Current Problems in Atmospheric Radiation
Giorgio Fiocco (Ed.)

IRS '88: Current Problems in Atmospheric Radiation
J. Lenoble and J.P. Geleyn (Eds.)

IRS '92: Current Problems in Atmospheric Radiation
S. Keevallik and O. Kärner (Eds.)

IRS '96: Current Problems in Atmospheric Radiation
William Smith and Knut Stamnes (Eds.)

Light Absorption by Aerosol Particles
Hermann E. Gerber and Edward E. Hindman (Eds.)

Microwave Remote Sensing of the Earth System
Alain Chedin (Ed.)

Nucleation and Atmospheric Aerosols
N. Fukuta and P.E. Wagner (Eds.)

Optical Properties of Clouds
Peter V. Hobbs and Adarsh Deepak (Eds.)

Ozone in the Atmosphere
Rumen D. Bojkov and Peter Fabian (Eds.)

Polar and Arctic Lows
Paul F. Twitchell, Erik A. Rasmussen, and Kenneth L. Davidson (Eds.)

Polarization and Intensity of Light in the Atmosphere
Kinsell L. Coulson

Prebiological Self Organization of Matter
Cyril Ponnamperuma and Frederick R. Eirich (Eds.)

Proceedings in Atmospheric Electricity
Lothar H. Ruhnke and John Latham (Eds.)

Ozone in the Atmosphere
R.D. Bojkov and P. Fabian (Eds.)

Radiative Transfer in Scattering and Absorbing Atmospheres: Standard Computational Procedures
Jacqueline Lenoble (Ed.)

Remote Sensing Calibration Systems: An Introduction
Hsi Chen

RSRM '87: Advances in Remote Sensing Retrieval Methods
Adarsh Deepak, Henry E. Fleming, and John S. Theon (Eds.)

Spectral Line Shapes, Volume 4
R.J. Exton (Ed.)

Tropical Rainfall Measurements
John S. Theon and Nobuyoshi Fugono (Eds.)

The Global Role of Tropical Rainfall
J.S. Theon, T. Matsuno, T. Sakata, and N. Fugono (Eds.)

Volcanic Activity and Climate
Kirill Ya. Kondratyev and Ignacio Galindo

VOLCANIC ACTIVITY AND CLIMATE

K. Ya. Kondratyev
Nansen International Environmental and Remote Sensing Centre
St. Petersburg, Russia

and

I. Galindo
Universidad de Colima
Colima, Mexico

A. DEEPAK Publishing **1997**
A Division of Science and Technology Corporation
Hampton, Virginia USA

Copyright © 1997 by *A. DEEPAK Publishing*
All rights reserved.
No part of this publication may be reproduced or transmitted in any form or by any means, electronic or mechanical, including photocopy, recording, or any information storage and retrieval system, without permission in writing from the publisher.

A. DEEPAK Publishing
A Division of Science and Technology Corporation
101 Research Drive
Hampton, Virginia 23666-1340 USA

Library of Congress Cataloging-in-Publication Data

Kondrat´ev, K. I͡A. (Kirill I͡Akovlevich)
 Volcanic activity and climate / K. Ya. Kondratyev and I. Galindo.
  p. cm.
 Includes bibliographical references and index.
 ISBN 0-937194-37-9 (hardcover)
 1. Weather—Effect of volcanic eruptions on. 2. Climatic changes—Environmental aspects. I. Galindo, I. (Ignacio) II. Title.
QC981.8.V65K66   1997
551.5—dc21
                                                                            97-9780
                                                                             CIP

Printed in the United States of America

CONTENTS

PREFACE vii

INTRODUCTION 1
   1. Sulfate Aerosols 2
   2. Modeling Aerosol Properties 5
   3. Climate Impact of Aerosols 8
      Tropospheric Aerosols 8
      Stratospheric Aerosols 14
   4. Conclusions 21
   References 22

CHAPTER 1 GLOBAL CLIMATE DIAGNOSTIC DATA AND NUMERICAL
            MODELING RESULTS 29
   1.1 Conventional Observations 29
   1.2 Cause and Effect Studies 39
   1.3 Satellite Observations 44
   1.4 Recognition of Anthropogenic Signals 56
   1.5 Numerical Climate Modeling Results 66
   References 80

CHAPTER 2 SOME PAST ERUPTIONS: VOLCANIC AEROSOLS AND
            CLIMATOLOGICAL EVIDENCE 86
   2.1 Development of Scientific Views 94
   2.2 Prominent Eruptions During Past 200 Years 97
   2.3 Volcanic Aerosols 118
      2.3.1 General Background 118
      2.3.2 Comments on the Stratospheric Ions 121
      2.3.3 Aerosol Measurements after Some Eruptions ... 124
      2.3.4 Comparative Data with Other Planets 134
          Modeling the Optical Properties of an
             Evolving Atmosphere 138
          Climatic Implications of the Volcanic Activity
             in the Process of the Earth's Evolution. 144

   The Climatic Impact of Volcanic Activity in the
     Evolution on Mars 151
 References 156

CHAPTER 3 RECENT MAJOR EXPLOSIVE VOLCANIC ERUPTIONS 166
 3.1 The Mount St. Helens Volcano Eruption 166
 3.2 The El Chichón Volcano Eruption 187
 3.3 The Mount Pinatubo Volcano Eruption 213
   3.3.1 Basic Information Concerning the
     Mount Pinatubo Eruption 213
   3.3.2 Satellite Observations 216
   3.3.3 Aircraft Observations 221
   3.3.4 Balloon Observations 225
   3.3.5 Remote Sensing Observations from the Surface ... 228
   3.3.6 Climatic Impact 233
   3.3.7 Impact on the Ozone Layer 238
 References 239

CHAPTER 4 PROPERTIES OF AEROSOLS — RADIATIVE EFFECTS ... 249
 4.1 *In situ* Measurements 250
 4.2 Remote Sounding of Stratospheric Aerosol Properties 258
 References 271

CHAPTER 5 ASSESSMENT OF THE EFFECT OF VOLCANIC ERUPTIONS
   ON CLIMATE 278
 5.1 Observational Data and Radiation Budget 278
 5.2 Numerical Modeling of the Climatic Impact of
   Stratospheric Aerosols 307
 5.3 Effects of Volcanic Eruptions on Atmospheric Ozone 331
   5.3.1 Influence on Ozone Retrievals 332
   5.3.2 Volcanic Impact on Ozone Concentration and
     Relevant Climatic Implications 335
 5.4 Conclusion 345
 References 346

SUMMARY 354
GLOSSARY OF ABBREVIATIONS AND ACRONYMS 361
SUBJECT INDEX 365

PREFACE

The principal aim of this book is to discuss the effects of explosive volcanic eruptions on climate in all their complexity, based on observational (with special emphasis on satellite remote sensing data) and numerical modeling results. Because the topic of climate change may be of interest to many different types of readers, the authors have presented the material in what we hope will be an easily understood manner. The book has been written both to introduce and to provide a comprehensive summary of the subject to researchers and advanced students working in the fields of climatology, atmospheric aerosols, volcanology, global warming, environmental sciences, and meteorology, as well as to specialists in global climate change.

Despite the long history of efforts to study the impacts of explosive volcanic eruptions on global climate and the environment, the results are, as yet, inconclusive. This is primarily because climate change is the result of a complex combination of many highly interactive processes. For instance, climate cooling was expected at the Earth's surface after the El Chichón eruption, but because the anomalous 1982/83 El Niño event induced heating, the situation drastically changed. The only possibility of identifying the "volcanic climatic signal" is to study global climate change in all its complexity on the basis of both observations and simulation modeling. In the Introduction, the authors provide an overview of the complexity of climate change due to aerosol variability. Chapter 1 then follows with a general survey of global climate diagnostics based mainly on the analysis of observational data as well as some numerical modeling results.

The rest of the book is devoted to the subject of climate change itself as it discusses the history of relevant studies and volcanic aerosol properties in Chapter 2 and the Mount St. Helens, El Chichón and Pinatubo explosive volcanic eruptions in Chapter 3. Subsequently, Chapter 4 presents fundamental aerosol optical properties and their radiative effects and Chapter 5 provides a more in-depth assessment of volcanic climatic impacts.

A part of the WMO monograph "Volcanoes and Climate" by K. Ya. Kondratyev (WCP-54, 1985) has been used in the book. We wish to express our most sincere gratitude to Drs. Rumen D. Bojkov and Byron W Boville, who were editors of the monograph.

The authors acknowledge the assistance received from Ms. Myriam Cruz for her excellent computer work.

Part of this work was supported with Grants FOMES93, FOMES95 and CONACYT 1701-T9209.

In detail, we summarize the approach the book takes as follows:

After a certain overemphasis of the importance of anthropogenic inputs on global climate (due mainly to the atmospheric greenhouse effect), the time has come to more objectively assess the role of natural climate variations due to internal variability of the climate system as well as such external impacts as volcanic eruptions and solar activity. It is a paradox that atmospheric aerosols, which significantly affect climate, have only recently been extensively studied. The new importance given to aerosols is evidenced by the expanded number of studies that propose assessing tropospheric sulfate aerosols and how they, in part, compensate for greenhouse warming. Important new efforts have also been undertaken to study the climatic consequences of explosive volcanic eruptions and the development of adequate observational means which include, among others, various kinds of remote sensing techniques.

Studies of volcanic eruptions are of great interest to atmospheric scientists seeking a greater understanding of the nature of both paleoclimatic and present-day climate changes. Such eruptions have been important from the very early stages of the Earth's evolution and have significantly contributed to the formation of the atmosphere and the oceans. Studies of volcanic eruption product diffusion also contribute to a better understanding of tropospheric and stratospheric dynamics and their interaction with changing radiative regimes.

The primary mechanism by which volcanic activity influences the climate system is through changes in the stratospheric aerosol content and its impact on radiative flux divergence. Volcanic aerosol origin can be classified into two main categories according to their composition and mechanism of formation or injection. Ash particles are directly injected into the atmosphere during the eruption, whereas sulfate particles are largely generated *in situ* from the gas phase through not yet properly understood gas-to-particle conversion mechanisms. Because of competing growth and transport processes, the aerosol population is expected to change as a function of space and time. Volcanic aerosols scatter solar radiation and, thus, modify both the amount of energy reaching the ground and backscattered into space. Aerosols also absorb and re-emit infrared radiation, thus modifying the planetary thermal radiation field. Radiation absorption and subsequent heat exchanges and reemissions lead to *in situ* atmospheric warming or cooling. The net effect is generally one of surface cooling, whereas the net stratospheric effect is one of warming, due to absorption of both solar and infrared radiation. Stratospheric heating is also affected by the albedo of the underlying atmosphere and surface.

As a rule, the identification of volcanic climate signals is not a simple problem because of the complexity of the climatic system. It has so happened, for instance, that the two most powerful recent eruptions of the El Chichón and

Pinatubo volcanoes coincided with significant El Niño/Southern Oscillation (ENSO) events.

Several factors determine the climatic significance of a volcanic eruption. The explosive strength of an eruption affects the height of injection and, in particular, whether or not the dust and gas enters the stratosphere, where residence times are much longer than in the lower atmosphere. The dust (which is largely silicate ash) ejected by the volcano affects the energy balance in the initial stages of volcanic cloud development. The amount of sulfur is, however, considered to be a more critical measure of climatic significance. The initial injection of sulfur gases is rapidly followed by their transformation into sulfate aerosols (secondary aerosol production), which have both a greater effect on the energy balance and a longer residence time than the particulate injection.

Recent studies indicate that the products of most volcanic activity are confined to the troposphere and are usually returned to the surface by sedimentation or rain-out within a few months. However, the products of strong explosive eruptions reach the stratosphere and may remain there a year or longer. As they undergo chemical transformations in the stratosphere, their effects on the Earth's radiation budget changes. The effect of changing aerosol properties on radiative transfer processes is generally considered the most important factor in studies related to the role volcanic eruptions have on climate change and, hence, is the major focus of this book.

The most important question related to aerosol climatic impacts is whether aerosols actually warm or cool the Earth. This debate is primarily due to a lack of aerosol optical properties observational data and to over-simplifications which result when only one type of particle with prescribed properties is considered. However, in nature, after volcanic eruptions, there are aerosols with various optical characteristics and they are characterized by inhomogeneous spatial distributions. It is also very important to separately consider two aerosol layers: one in the stratosphere and one in the troposphere. Stratospheric aerosols, 20 or more kilometers above the Earth's surface, absorb solar radiation and infrared radiation emitted from the surface and lower atmospheric layers. Stratospheric aerosols have a warming effect on the surface because they emit infrared radiation toward the lower atmosphere. This effect, however, is offset by aerosols which absorb solar energy or backscatter sunlight into space. Whether stratospheric aerosols have a net surface warming or cooling effect depends mainly on particle size. Because these particles are generally rather small (0.1 µm) H_2SO_4 droplets, they usually cool the surface, although they actually warm the atmospheric layer in which they are found. The tropospheric aerosol layer cools the surface by backscattering sunlight while itself warming as it absorbs sunlight. Whether aerosols will have a net warming or cooling effect depends upon the absorption to scattering ratio. This ratio, in turn, depends on tropospheric aerosol composition, which is quite variable, as well as on surface albedo and solar zenith angle.

Tropospheric aerosols primarily affect climate due to their interaction with clouds. Meanwhile, stratospheric aerosols affect the ozone layer and may represent a significant coupling mechanism between stratospheric aerosols and ozone.

Onset of eruptive process of Popocatepetl, 21 December 1994. (I. Galindo)

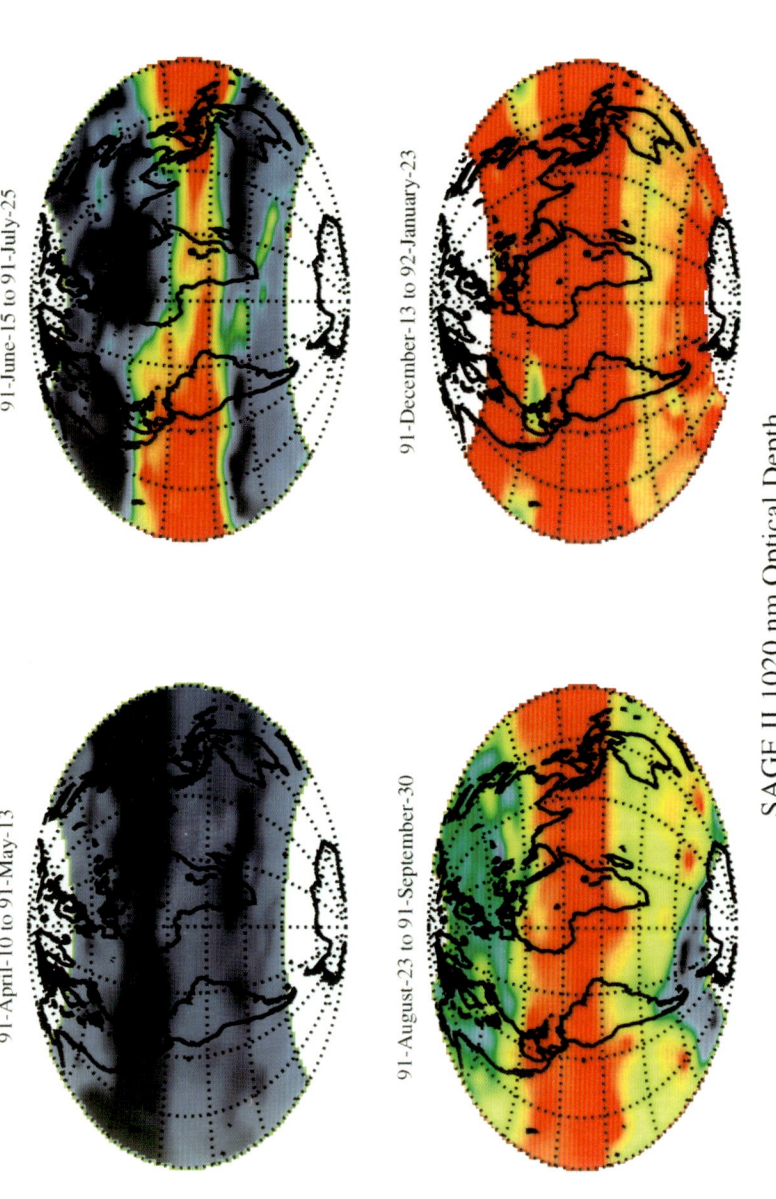

INTRODUCTION

The problem of climate and its global change due to both natural and anthropogenic causes is, no doubt, one of the central issues confronting present science (Bolle, 1982; Jennings, 1993; Kondratyev and Cracknell, 1996; Kondratyev and Grassl, 1996; Newell, 1984). Among the most important factors affecting climate are atmospheric aerosols. These aerosols are, in fact, the product of complicated physical and chemical processes which take place in the environment as well as the direct injection of dust, ash or smoke particles into the atmosphere. Some of these processes are caused by growing anthropogenic loading on the environment. Man's impact on atmospheric aerosols is manifested through changes in the intensity of formation or disintegration and in the physical and chemical properties of aerosol particles (Ackerman and Chang, 1992; d'Almeida et al., 1991; Deepak, 1982; Gerber, 1982; Gerber and Deepak, 1984; Hobbs and McCormick, 1988; Husar et al., 1991; Jennings, 1993; Kondratyev, 1972, 1973, 1976, 1978, 1990, 1992a,b,c, 1993). There are three primary manners in which aerosols affect climate:

1. Direct aerosol effects on the radiation budget of the surface-atmosphere system through a redistribution of shortwave solar radiation and infrared thermal emissions due to scattering and absorption by non uniformly distributed atmospheric aerosol particles (Coakley et al., 1983; Galindo, 1965, 1975, 1978, 1984, 1992a,b; Galindo and Muhlia, 1970; Galindo et al.,1975; Grassl, 1973, 1974a,b, 1975, 1992; Kondratyev, 1980, 1983, 1985, 1988, 1991, 1992a; 1992b; 1992c; Zuev and Kabanov, 1987).

2. The crucial role of aerosol particles in atmospheric water phase transformations and cloud formation, both of which are important mechanisms that substantially affect atmospheric energetics to a greater extent than direct effects (Deepak and Vali, 1991; Feigelson, 1981; Gerber and Deepak, 1984; Hegg, 1991; Ivlev, 1982; Jones et al., 1994; Kondratyev, 1972, 1991). The role of aerosol forming gas-to-particle reactions (condensation nuclei) related to natural and anthropogenic atmospheric emissions of compounds such as dimethylsulfide and sulfur dioxide is of great concern.

3. Heterogeneous chemical processes, particularly the disintegration of ozone molecules on aerosol particle surfaces, considerably affect atmospheric gas composition changes (Bach, 1976; Finlayson-Pitts and Pitts, 1986; IGBP, 1988; Kondratyev, 1989, 1990).

Complex field aerosol-radiation experiments undertaken in various global regions as part of the Complete Atmospheric Energetics Experiment (CAENEX) and Global Atmospheric Aerosol Radiation Experiment (GAAREX) programs (Kondratyev, 1972, 1978, 1991; Kondratyev and Prokofyev, 1984; Kondratyev and Ter Markaryants, 1976) reveal that aerosols absorb substantial amounts of solar radiation and greatly affect the atmospheric radiation budget. Urgent problems include:

(i) The impact of dust outbreaks from arid regions on the atmospheric radiative regime over both continents and oceans;
(ii) The transport of aerosols to high latitudes leading to the formation of arctic haze and extended stratified cloudiness in the Arctic (Marchuk et al, 1986; Kondratyev, 1992a,b; Kondratyev and Johannessen, 1993). Sporadic volcanic eruptions play a special role, contributing to these aerosol perturbations.

An extremely complicated problem is the impact of radiation on the formation and evolution of the atmospheric aerosol field. The relationships between the radiative, convective and turbulent heat-and-mass exchange in the processes of aerosol generation, formation of its chemical composition, size distribution, and aerosol optical properties needs to be further studied (Bolle, 1982; Hobbs and McCormick, 1988; Husar et al., 1991; Kolomeev and Sorokovnikova, 1991; Kondratyev, 1991; SMIC, 1971). In this context, numerical modeling of aerosol formation and global distribution needs to be improved.

1. Sulfate Aerosols

Sulfate aerosols have recently caused great concern (Bates et al., 1992; Charlson et al., 1987; Fried et al., 1992). Numerical modeling has allowed assessment of changes in tropospheric sulfate aerosol distribution and the deposition of nonmarine sulfate aerosol on underlying surfaces which occurred during the industrial revolution (Langner and Rodhe, 1991; Langner et al., 1992). The 10-layer model (up to the 100 hPa level) was utilized over a 10° latitude by 10° longitudinal grid with the prescribed monthly meteorological parameter means.

Concentrations of dimethylsulfide (DMS), sulfur dioxide (SO_2), and nonmarine sulfate aerosol (SO_4^{2-}) precursors were carefully analyzed. The final transformation product of all these gaseous components is assumed to be SO_4^{2-}. For comparison purposes, estimates of these concentrations have been made with prescribed emissions before the industrial revolution and at present (Table 1).

Table 1. Total global aerosol emissions. After Rodhe and Langner (1993). [Printed with kind permission from Kluwer Academic Publishers.]

Sources	Emission (TgS/year)
1. Industrial	(95% SO_2; 5% SO_4^{2-}); 70 (6).
2. Biomass burning	(SO_2) 2.5 (1.4).
3. DMS emissions by the ocean:	25 (14.2).
4. Volcanic SO_2	8.5 (2.7).
5. Emissions from soils and vegetation	(DMS, H_2S): 1 (0.4).

The first parentheses contain the composition of Northern Hemisphere emissions. The second parentheses contain Southern Hemisphere contributions.

Numerical global distribution modeling maps of sulfate concentrations in the lower 500 m layer of the atmosphere reveal the effects of anthropogenic emissions in the Northern Hemisphere (NH). These modelings reveal a maxima of more than 1 ppb in European regions, North America, and China. A SO_4^{2-} lifetime of about 5 days determines the long range transport of sulfate aerosol which is observed east of North America, China and eastern and southeastern parts of Europe. In the Southern Hemisphere (SH) the anthropogenic impact can be observed in Africa, South America, and Australia. When there is an absence of anthropogenic emissions, maximum sulfate concentrations occur in the tropical belt, where natural DMS emissions are most intensive. Calculations of pre-industrial and present sulfate concentrations in the 0.5–1.5 km layer, where, principally, lower-level stratus clouds are formed, led to the conclusion that sulfate concentrations increased by two orders of magnitude in the most polluted European regions. Furthermore, the increased NH concentrations reached 51% in July and 0.8% in January, respectively.

Both wet and dry sulfate aerosol deposition has intensified: there is more than a 58% increase over the NH and more than a 16% increase over the SH. Since only about half of the SO_2 is deposited onto the surface, a large percentage of the remaining SO_2 is oxidized in cloud droplets. Only 6% of total anthropogenic emissions (72.2 TgS/year) is transformed into aerosol particle condensation nuclei (CN). This greatly affects the possibility of indirect SO_2

emissions having impacts on cloud microphysical and optical parameters. Reliable quantitative estimates are, as of now, impossible. A similar study concerning dimethylsulfide reveals the existence of gas-to-particle oceanic DMS reactions participating in the formation of sulfate aerosol. Dimethylsulfide production depends primarily on oceanic bioproductivity dynamics. An assessment has been made of possible sulfate aerosol climatic impacts under these conditions (Rodhe and Langner, 1993).

Numerical modeling (global 3-D model MOGUNTIA) of the transport and transformation of aerosols and minor gaseous atmospheric components reveals the contribution of ocean-emitted DMS to the global sulfur cycle (Rodhe and Langner, 1993). Since DMS oxidation gives off SO_2 and subsequent gas-to-particle reactions which form sulfate aerosol are proportional to the concentration of hydroxyl radicals (OH), the initial task consists in simulating the OH concentration global field considering possible reactions for the CH_4-CO-NO_x-OH-O_3 system. The degree of SO_2 oxidation to cloud droplet sulfate is assumed to be proportional to their respective concentrations, wet and dry deposition of SO_2 and sulfate.

Table 2 shows sulfur compound emissions (TgS/year) in both hemispheres.

Table 2. Northern Hemisphere and Southern Hemisphere sulfur compound natural and anthropogenic emissions (TgS/yr). After Rodhe and Langner (1993). [Printed with kind permission from Kluwer Academic Publishers.]

Emission Source	Global emissions	NH	SH
Industry (SO_2)	70 (62–80)	64	6
Biomass burning (SO_2)	2.5 (2.3–2.8)	1.1	1.4
Oceans (DMS)	16 (8–51)	6.9	9.1
Volcanoes (SO_2)	8.5 (7.4–9.4)	5.8	2.7
Soils and vegetation (DMS, H_2S)	1 (0.2–4.3)	0.6	0.4
Total	107.0	82.3	24.7

Ranges are in parentheses.

Assessments of natural ocean DMS emissions are rather uncertain and can lead to inaccurately determined total emissions. Oceanic DMS emissions are characterized by strong spatial variability and annual change, especially in high and middle latitudes. Naturally, such a variability is also typical of the DMS concentrations near the surface, where they reach maxima in summer over high latitudes ocean surfaces and in the equatorial belt throughout the entire year. On the other hand, minima are found over continents, where biogenic DMS emissions are less substantial. The calculated data, on the whole, concur with available observational data.

Estimates of the global distribution of sulfate aerosol cooling impacts due to DMS emissions revealed maxima (up to 1 W/m^2) over ocean equatorial belts. Another vast region with values greater than 0.5 W/m^2, stretching from the equatorial Atlantic, across North Africa, towards the northern Indian Ocean, was also found. The strong effects that prevail in these areas are primarily due to prevailing clear skies which favor a prolonged sulfate aerosol lifetime.

Bearing in mind that the average climatic impact of NH anthropogenic sulfate aerosols was -1.1 W/m^2, and the positive greenhouse effect was about 2.5 W/m^2, it becomes clear that the effects of DMS are rather substantial. To obtain more reliable estimates of the role of DMS, further studies are needed to determine:

(i)   DMS flux across the ocean-atmosphere interface;
(ii) Atmospheric DMS oxidation;
(iii) SO_2 oxidation processes within cloud droplets and in aerosol particles themselves;
(iv) Sulfate aerosol washout by precipitation.

The impacts of aerosols on climate as well as their chemical composition and particle size distribution, both of which determine aerosol radiative characteristics, represent areas that require considerable future study.

A number of surveys on atmospheric aerosols and their radiative properties have been published. However, most of the results are not accurate enough because aerosol property data which form the basis for radiative characteristic calculations are fragmentary and nonrepresentative. The estimates for anthropogenic aerosols, mostly sulfates, are more reliable than for natural aerosols—especially in not easily accessible ocean and continental regions. This fact tends to affect assessment reliability concerning aerosol variability and climate, although there is no doubt that man's activities do impact the sulfur, carbon and nitrogen cycles.

2. Modeling Aerosol Properties

Variability of aerosol particle optical properties and size distribution—even within a certain climatic zone, is now well-known (Ivlev and Andreev, 1986; Jaenicke, 1987; Kondratyev, 1978; Kondratyev et al., 1983; Meszáros, 1981). Consequently, undertaking climatological calculations and assessments with the use of directly measured aerosol characteristics is not always possible. On the other hand, some aerosol characteristics are relatively stable and are subject to certain laws pertaining to particle size distribution and chemical composition (d'Almeida et al., 1991; Jaenicke, 1987; Jennings, 1993; Kondratyev, 1991; Meszáros, 1981; Stowe et al., 1990). The difficult problem of substantiating representative aerosol models is of fundamental importance.

The difficulties of modeling aerosol particle characteristics and properties are related to the number of parameters that influence their variability and the absence of reliable parameter selection criteria. Besides, aerosol spatial and temporal variability is determined by the complicated processes of generation, transformation and evolution of an ensemble of aerosol particles as well as their propagation from the source and particle removal. An isolated consideration of these processes is not adequate, since all of these processes are interrelated.

The problem of aerosol modeling can be defined by two major characteristics:

1. Aerosol global spatial distribution (the field of the aerosol mass concentration) which describes the average characteristics of generation, propagation, and particle removal from the atmosphere (Kondratyev, 1976, 1980; Kondratyev et al., 1983).

2. Local properties of aerosols resulting from the effect of different-origin atmospheric aerosol transformational processes (Galindo, 1992a; 1992b; Galindo and Bravo, 1975; Galindo et al., 1996; Goetz et al., 1991; Grassl, 1974a,b; Ivlev and Andreev, 1986; Meszáros, 1981).

Particle generation processes, first of all, determine aerosol physical and chemical properties as well as their concentration, and, therefore, subsequent transformational processes concerning the size distribution spectra and the rate of atmospheric particle removal.

The following types of natural aerosol particles can be identified:
(i) Sea spray disintegration and evaporation products;
(ii) Mineral dust that is wind-driven into the atmosphere;
(iii) Volcanic eruption products that are either directly ejected into the atmosphere or result from gas-to-particle conversion;
(iv) Particles of biogenic origin that are directly ejected into the atmosphere as well as those resulting from the condensation of volatile organic compounds and chemical reactions such as those which occur between terpenes;
(v) Smoke from biomass burning on land;
(vi) Products of natural gas-to-particle conversion such as sulfates that are formed from ocean surface sulfur compounds.

The principal types of anthropogenic aerosols are particles that result from industrial emissions which include, for example, soot, smoke, cement, road dust, etc., as well as products of gas-to-particle conversion. Other aerosol substances result from heterogeneous chemical reactions and, in particular, photocatalytic reactions.

The long lifetime of aerosol particles promotes their transport, mixing and interaction with each other and with various gaseous compounds, causing particle physical and chemical properties to be smoothed down, subsequently forming background aerosols.

Background aerosol characteristics also suffer substantial changes in different atmospheric layers. The weak mixing of tropospheric and stratospheric aerosols causes stratospheric particles to substantially differ in both composition and size distribution from tropospheric aerosols. The troposphere can contain regions with different-type particles, especially in view of the relatively short lifetime of tropospheric particles. This is especially true in the surface layer, where the presence of aerosols from certain specific sources can markedly dominate the intensity of the others.

The processes of atmospheric aerosol transport and removal are determined by both the meteorological factors and the particular aerosol properties. These processes depend on meteorological conditions and include organized convective and advective particle transport and mixing due to turbulent diffusion. The individual properties of particles determine, in particular, their removal from the atmosphere as sedimentation on the obstacles.

Particle atmospheric transport and removal as well as their size (mass) and chemical composition are influenced by coagulation and condensation growth as well as heterogeneous reactions. These processes are responsible for different particle sizes and determine the value of the complex refractive index and particle shape as well as their optical properties (d'Almeida et al., 1991; Gerstell et al., 1995; Ivlev and Andreev, 1986; Patterson et al., 1983; Pudykiewicz and Dastoor, 1995).

The most substantial gas-to-particle reactions affecting aerosol particle formation include:

(i) Reactions between sulfur dioxide and hydroxyl-radicals which, in the presence of water molecules, eventually lead to the formation of easily condensible sulfuric acid;

(ii) Hydrocarbon compound reactions with ozone and hydroxyl radicals which subsequently form primary nitrogen oxides such as peroxyacetylnitrate.

Aerosol system formation and evolution depend on periodic surface atmospheric solar radiation variations. The processes of atmospheric heating near the surface, water evaporation and aerosol chemical reactions have a clear-cut diurnal change. This can result in variations in aerosol characteristics which, in some cases, are stronger than the annual change. Strong daily anthropogenic aerosol variations have been documented (Galindo, 1965; Galindo and Muhlia, 1970).

Aerosol matter cycles are closely related to atmospheric hydrological processes. On the one hand, clouds and precipitation play an important role in the formation, transformation and removal of aerosol particles from the atmosphere, and on the other hand, aerosols are condensation nuclei whose physicochemical properties determine cloud microphysical processes. It is not accidental that both water vapor molecules and aerosol particles have approximately equal lifetimes. Therefore, more in-depth knowledge about aerosols and cloud interactions is needed to better understand aerosol formation processes and their temporal and spatial variability.

3. Climatic Impacts of Aerosols

The main problem concerning aerosol climatic impact modeling consists of determining their influence on longwave and shortwave atmospheric radiation transfer. Climate change can take place due to aerosol concentration regional and global scale variations, as well as size distribution and chemical composition variability (Bach, 1976; Galindo, 1992a,b; Grassl, 1973, 1974a,b; 1975, 1982; Kondratyev, 1980, 1985, 1988; Kondratyev and Cracknell, 1996; Kondratyev and Grassl, 1993, 1996). Related to this is the possible climatic impact of both tropospheric and stratospheric soot particles resulting from atmospheric nuclear explosions (Kondratyev, 1988). During recent years, scientists have assessed the climatic implications of the Kuwait oil field fires (Cahalan, 1992, Cofer et al., 1992; Hudson and Clarke, 1992; Parungo et al., 1992).

Tropospheric Aerosols

Numerical modeling needs to consider aerosol climatic effects based on adequate and realistic models that can assess climate sensitivity to various atmospheric aerosol characteristics (Coakley et al., 1983; Hansen et al., 1992; Hobbs and McCormick, 1988; Jennings, 1993; Kawamata et al., 1992; Kondratyev, 1991; Lacis et al., 1992; Penner et al., 1993; Pollack, 1983; Schlesinger et al., 1993; SMIC, 1971).

Future studies should particularly concentrate on the following:

(i)   The impact of aerosols on regional and global processes;

(ii) The identification of different aerosol types and their properties in order to better determine their climatic implications;
(iii) Assessments of potential impacts of aerosol and minor gaseous components such as H_2O, O_3, CO_2, CO, etc., clouds and surface albedo on both regional and global climate.

Models for aerosol formation, transformation and transport dynamics and their interaction with climatic processes need to be developed. Numerical experiments on climate sensitivity to aerosol effects should be accomplished to reveal aerosol impacts on the radiation budget of the surface-atmosphere system, atmospheric heat flux divergence, heat balance, and the climatic system. Radiative-convective equilibrium models can provide approximate assessments of aerosol climatic effects. These models have revealed strong aerosol effects on the vertical temperature profile when the aerosol optical depth is large. Conversely, these same models show a weak effect if aerosol optical depth is less than 0.5. Depending on surface albedo, aerosol optical properties and their spatial distribution, cooling and warming of up to several degrees can take place (Charlson et al., 1987; Deepak, 1982; Hobbs and McCormick, 1988; Jennings, 1993; Kondratyev, 1980; 1983, 1985, 1992a,b,c; Penner et al., 1993; Pollack, 1983).

Aerosol effects on surface temperatures using the zonal model of atmospheric general circulation (AGCM) suggest that a several-degrees temperature change is possible. Initial calculations for desert regions using the 3-D AGCM as well as specific aerosol characteristics reveal that:

(i) Surface temperatures can decrease by up to 2.5 °C, provided solar radiation is exclusively absorbed by aerosols;
(ii) A surface temperature increase of up to 3.5 °C is possible, provided aerosol thermal radiation transfer impacts (greenhouse effect) are taken into account;
(iii) There is increased atmospheric stability near the Earth's surface and increased free-atmosphere instability in the upper part of a dust-loaded atmosphere;
(iv) A marked temperature field transformation occurs in the dust-loaded part of the atmosphere in western Europe, Asia and tropical Africa, in the specific case of Saharan desert dust outbreaks.

Assessments of tropospheric background aerosol effects on the global radiative regime demonstrates that aerosols can cause a shortwave radiation budget decrease of as much as 5.0 W/m² at the surface and 3.5 W/m² at the top of the atmosphere. In other words, it causes approximately the same, but opposite sign, radiative disturbances as a CO_2 doubling. Naturally, aerosol effects on the thermal emission field should also be taken into account. In contrast to global comparative homogeneity of radiative disturbances due to CO_2, specific features of aerosols that affect the radiative regime include:

  (i)   spatial inhomogeneity,
  (ii) variability of aerosol content,
  (iii) surface albedo changes,
  (iv) sun elevation variations.

Results from surface air temperature (SAT) calculations, using the AGCM and an energy-balance climate model, reveal serious differences: energy-balance model calculations can differ by as much as an order of magnitude. This is primarily because the energy balance model considers only albedo feedback, characterizes mean annual conditions and provides a state of equilibrium which, in the case of AGCM, strongly depends on the prescribed sea surface temperature (SST).

An analysis of calculated zonally averaged temperature fields shows that heating, due to aerosol shortwave radiation absorption, is compensated for in the troposphere (first of all, near the surface) due to nonradiative factors. Therefore temperature field does not substantially change. In the stratosphere, heating is partly compensated for by intensified longwave cooling.

The geographical distribution of aerosol climatic effects has two characteristic features. One of these characteristics is decreased solar global radiation due to desert aerosols. For example, solar global radiation decreases can reach 60 W/m^2 in the Sahara and on the Arabian Peninsula. The second feature refers to outgoing longwave radiation in a belt extending from the northeastern coast of Brazil to Texas. Aerosol concentrations significantly decrease in this belt, causing an abnormal increase of outgoing radiation in the center of this zone.

Numerical modeling used to assess various aerosol-type contributions to climate change reveal the nonlinear nature of their interaction: the sum of various aerosol-type effects is not equal to the effect of the sum of their contributions. The strongest spatial homogeneity of aerosol climatic effects is observed in the case of stratospheric aerosols. Stratospheric aerosols cause some suppression of Hadley circulation as well as slowing down tropical and subtropical tropospheric zonal transport.

Increased sulfur dioxide emissions and the subsequent formation of atmospheric sulfate aerosol during the second half of this century has caused scientists to attempt to assess relevant aerosol climatic effects. These effects are manifested by increased surface-atmosphere system albedo due to the following processes:

(i) Aerosol overloading increases backscattering;
(ii) Increased cloud albedo results from greater condensation nuclei concentration, which leads to an increase in small droplet concentration.
(iii) Anthropogenic aerosols should have the greatest climatic effects in the summer when there is maximum insolation in regions with concentrated industry (Galindo, 1965, 1992a; Galindo et al., 1996).

Under the assumption that direct and indirect aerosol effects are proportional to atmospheric aerosol content, Engardt and Rodhe (1993) undertook an analysis of summer and winter SAT trends for different NH regions. They considered monthly mean SAT anomalies using monthly mean SAT data compiled 1950 and 1979. Table 3 illustrates SAT variations (ΔT) 1971 to 1990, as compared to 1926 to 1945 during the winter (December-January-February) and summer (June-July-August) seasons.

Table 3. Different regional Northern Hemisphere temperature trends. After Engardt and Rodhe (1993). [© American Geophysical Union]

Aerosol content increase $B_{SO_4^{2-}}$, mg/m²	ΔT, °C summer (JJA)	ΔT, °C winter (DJF)	Percent of NH surface
>10	-0.33 ±0.39	0.62 ±0.70	4
<10	0.05 ±0.13	0.11 ±0.24	96
>8	-0.13 ±0.26	0.88 ±0.69	12
<8	0.06 ±0.13	0.01 ±0.21	88
>6	-0.09 ±0.23	0.62 ±0.60	21
<6	0.07 ±0.13	-0.02 ±0.22	79

Numbers in the right-hand columns represent standard deviations. $B_{SO_4^{2-}}$ represents the sulfate aerosol content increase in an air column (mg/m²).

In regions with $B_{SO_4^{2-}}$ pollution levels greater than 6, 8, and 10 mg/m² $B_{SO_4^{2-}}$, a SAT summertime decrease of 0.2–0.4 °C takes place, when compared with other parts of the NH. A strong warming of 0.5–0.9 °C takes place in the winter. These results agree with observations made in highly polluted regions (e.g., Galindo, 1992a).

Global distribution maps of ΔT show that in summer a warming takes place in most of the SH and in many regions of the NH. However, there is a cooling in vast continental regions of the NH. The most substantial cooling trend in polluted regions was recorded between 1960 and 1970. These coolings coincide with a period in which substantial sulfur compound emissions were released into the atmosphere. Fortunately, since 1970, there has been a gradual decrease of these sulfur compound emissions.

Despite the difficulty in interpreting some of the results, the following conclusions can be drawn:

(i) Summer cooling in heavily polluted regions qualitatively agrees with sulfate aerosol effects. Other factors, however, may also be significant—especially increased cloudiness in these regions;
(ii) Winter warming in polluted regions can be interpreted as a manifestation of the greenhouse effect of pollution;
(iii) Summer cooling trends in the North Atlantic could have been influenced by either atmospheric pollution or other unrelated causes. Because the statistical significance of the results is relatively low, they should only be considered as preliminary.

Further studies of sulfate climatic impact have been accomplished by Charlson and Wigley, 1994; Chylek et al., 1995; Kondratyev et al., 1996.

Sensitivity studies of global mean SAT variations due to CO_2 doubling (ΔT_{2X}) include the effects of two external forcings, prescribed as radiation budget variations of the surface-troposphere system. These results show a nonlinear relationship between surface temperature and sulfate increases, which depend on CO_2 and anthropogenic sulfate aerosol content variations. The most likely temperature range of between 1.39 °C and 3.00 °C refers to 1978 sulfate estimates (Schlesinger et al., 1993). The values of ΔT_{2X} were calculated using the energy-balance model for the atmosphere-ocean system in a 40-layer ocean. This system was divided into polar regions, where bottom waters form, and nonpolar regions, where upwelling and diffusion are also taken into account. Calculations, which began in 1765, were reported as SAT deviations from the mean between 1861 and 1990. The scenarios of greenhouse gas (GHG) dynamics have been prescribed due to the Intergovernmental Panel of Experts on Climate Change (IPCC) Report and the 1978 aerosol-induced radiative disturbance. $\Delta FSO_4^{2-}(1978) = -1.1 \pm 0.9$ W/m²; this can vary from -0.2 to -2.0 W/m².

Since values of this effect can only be roughly estimated, numerical modeling has been undertaken using ΔFSO_4^{2-} as the prescribed variable. For calculation purposes, ΔFSO_4^{2-} was considered under both clear and cloudy skies:

$$\Delta FSO_4^{2-}(1978) = \Delta FSO_4^{2-}{}_{,clear}(1978) + \Delta FSO_4^{2-}{}_{,cloud}(1978). \tag{1}$$

Calculation results of ΔT_{2X} were compared to observational data between 1861 and 1990. Temporal variations of the aerosol effect are described by the following equation:

$$\Delta FSO_4^{2-}(t) = \Delta FSO_4^{2-}(1978) \, [QSO_2 - S(t)] / [QSO_2 - S(1978)] \tag{2}$$

where $QSO_2 - S(t)$ represents atmospheric sulfur emissions in the form of SO_2, from data obtained between 1860 and 1977. Approximate estimates have led to the following relationship:

$$\Delta FSO_4^{2-}{}_{,clear} = -8.2 \times 10^{-3} (QSO_2 - S) \text{ [Tg/year]}. \quad (3)$$

Accounting for uncertain input parameters and emission levels:

$$\Delta FSO_4^{2-}{}_{,clear}(1978) = -0.58 \pm 0.42 \text{ W/m}^2. \quad (4)$$

Cloud component estimates of ΔFSO_4^{2-} are less reliable because they must be studied in the context of indirect sulfate aerosol effects as condensation nuclei (CN) on cloud microphysical and radiative characteristics. Anthropogenic CN advection, equal to 10^{19} s^{-1}, from the continents to the oceans is assumed to be taking place. This process increases nucleation 20% over naturally occurring levels when marine stratus clouds form. This, in turn, leads to a tropospheric radiation budget decrease of about 1 W/m² in the Northern Hemisphere. This effect, however, is considered absent in the Southern Hemisphere.

$\Delta FSO_4^{2-}{}_{,cloud}(1978) = -0.5 \pm 0.5$ W/m² and total effects are equal to what has been provided above (-1.1±0.9 W/m²). Calculations of SAT anomalies from 1765 ($\Delta T = 0$) to 1990 were made with various $\Delta FSO_4^{2-}(1978)$ and ΔT_{2X} values.

Calculation results can be approximated using the following regression equation:

$$1/\Delta T_{2X} = 0.7042 + 0.4584x - 0.1005x^2 - 0.06847x^3 \quad (5)$$

where $x = \Delta FSO_4^{2-}(1978)$.

Thus, ΔT_{2X} varies from 1.39 °C for $x = 0$ (without the aerosol effect) to 5.92 °C for the nominal value of $\Delta FSO_4^{2-}(1978) = -1.1$ W/m². Apparently, the most likely interval will be 1.39 °C $\leq \Delta T_{2X} \leq$ 3.00 °C. In other words, $\Delta T_{2X} = 2.2 \pm 0.8$ °C corresponds to the interval:

$$0 \geq \Delta FSO_4^{2-}(1978) \geq 0.74 \text{ W/m}^2. \quad (6)$$

To reduce estimated uncertainties of $\Delta FSO_4^{2-}{}_{,clear}$, research must obtain more reliable data on the efficiency of aerosol scattering calculated per unit mass, atmospheric SO_4^{2-} lifetime and aerosol backscattering. The problem of improving $\Delta FSO_4^{2-}{}_{,cloud}$ estimate reliability is much more complicated because the probable values are very small.

Stratospheric Aerosols

Volcanic eruptions can have significant climatic impacts because they inject aerosols into the stratosphere (Kondratyev, 1989). In order to substantiate some of these climatic effects, Lacis et al. (1992) calculated the global-mean vertical temperature profile using a 1-D radiative-convective model with a cloud cover of 50% and vertical temperature lapse rate of 6.5 °C/km. Under these conditions, they were able to obtain the following parameters: extinction coefficient Q_{ext}, single-scattering albedo $\bar{\omega}$, and the coefficient of asymmetry of phase function $\cos\theta$ in the 0.3-30 µm region for droplets of a 75% water solution of H_2SO_4, as well as spectral reflection, transmission and absorption with an optical depth τ at the wavelength 0.55 µm equal to 0.1 and 1.0.

Size distribution was prescribed for stratospheric aerosols in May and October 1982, 1.5 and 6.5 months, respectively, after the El Chichón eruption. Many aerosols contained considerable amounts of large particles with ≥ 1 µm radii, which determined the eruptive cloud's enhanced thermal infrared radiation absorption.

In order to adequately assess stratospheric aerosol climatic forcing, it is necessary to determine changes of both the radiation budget ΔF_{net} and its ΔF_{solar} and ΔF_{IR} components for the surface-troposphere system. Calculations of ΔF_{solar} and ΔF_{IR} as a function of the effective radius of particles r_{eff} have shown that IR forcing is strongly related to particle radii, increasing with radius growth within the interval 0.5–3.5 µm. The albedo for shortwave radiation, however, is almost independent of particle radii, provided it is greater than the effective wavelength (i.e., 0.5 µm). Aerosols with a critical particle radius of $r_c \geq 2$ µm can cause strong tropospheric heating due to the attenuation of longwave radiative cooling.

It turns out that r_c weakly depends on particle composition and shape. Although small aerosol particle size ($r_{eff} < 0.05$ µm) also determines warming, it has weak effects because of the relatively small amount of such aerosols. The width of the particle size distribution which is characterized by v_{eff} markedly affects ΔF_{net}. Therefore, it can be reliably concluded that the variability of net radiative forcing of the troposphere principally depends on r_{eff} and v_{eff}. Furthermore, aerosol size distribution can be approximated by a γ distribution. With low values τ ($\tau < 0.2$) the simplest approximations are justified:

$$\Delta F_{net}(W/m^2) \sim 30\,\tau; \quad \Delta T\,(°C) \sim 9\,\tau. \tag{7}$$

The presence of admixtures in H_2SO_4 aerosol droplets causes changes in the single-scattering albedo. Even for $\bar{\omega}_c = 0.99$, stratospheric cooling due to backscattering is 50% less as compared to $\bar{\omega}_c = 1$ and stratospheric temperatures substantially increase. This, however, does not have substantial tropospheric impacts. The principal conclusion derived from the above data is that radiative forcing of stratospheric aerosols on the troposphere can be simulated taking only stratospheric aerosol optical depth into account. The exception to this is if particle size exceeds 1 μm when an enhancement of the large particle-induced greenhouse warming can be compared with or even exceed the albedo effect of cooling, provided $r_{eff} > 2$ μm.

The minimal combination of aerosol optical parameters must be first determined in order to appropriately assess climate changes. The various aerosol optical parameters that affect climate must first be determined. Present calculations are usually based on the following optical characteristics:

(i) The volume coefficient of attenuation (extinction) σ_e i.e., the availability of the data on the vertical $\sigma_e(z)$ profile makes the calculation of the atmospheric optical depth τ_a possible;
(ii) the volume coefficient of scattering σ_{sc}, or, alternatively, the single-scattering albedo $\bar{\omega}_a = \sigma_{sc}/\sigma_e$;
(iii) the volume phase function, which is approximately simulated by introducing the coefficient g of asymmetry of scattering function.

The representative characteristic of aerosol effects on climate is the ratio of the coefficients of aerosol radiation attenuation in the longwave ($\lambda = 10$ μm) and shortwave ($\lambda = 0.55$ μm) spectral regions. Some parameters for different aerosol types are shown in Table 4.

Table 4. Principal characteristics of aerosol models. After Lacis et al. (1992). [© American Geophysical Union]

Aerosol origin	Wavelength, μm				Horizontal variations	Vertical structure
	0.55		10.0			
	τ	$\bar{\omega}$	τ	$\bar{\omega}$		
Desert	0.54	0.891	0.0052	0.44	+	ED in ABL SH = 3 km
Stratosphere	0.045	1.000	0.0049	0.06	-	Homogeneous
Ocean	0.033	0.988	0.0083	0.76	+	ED in ABL SH = 1 km
Continents	0.031	0.890	0.0030	0.44	+	ED in ABL SH = 1 km
Troposphere	0.030	0.890	0.0029	0.44	-	Homogeneous
Cities	0.008	0.647	0.0003	0.15	+	ED in ABL SH = 1 km
Volcanoes	0.007	0.943	0.0002	0.15	-	Homogeneous

ED represents the exponential decrease; ABL – atmospheric boundary layer; SH – scale height.

To obtain reliable data on aerosol optical characteristics, it is necessary to have either complete information on the distribution and chemical composition of aerosol particles: number density, size distribution, complex refractive index and shape, or a statistically substantiated optical measurement database.

Although, during recent years, a number of studies have considered atmospheric aerosol optical properties (Bolle, 1982; Deepak, 1982; Feigelson, 1981; Hobbs and McCormick, 1988; Husar et al., 1991; Ivlev and Andreev, 1986; Kondratyev, 1973, 1978, Kondratyev et al., 1988; Stowe et al, 1990; Zakharov et al., 1990; Zuev and Krekov, 1986), available information is still rather incomplete.

Reliable *in situ* aerosol size distribution and chemical composition measurements cannot, however, be considered representative for large spatial and temporal scales. Therefore, recent ground, aircraft, balloon, and satellite observational results should form the basis for climatological aerosol models, combining complex modeling and available observational data. The development of both theoretical and empirical models which can explain aerosol particle formation, evolution, and removal from the atmosphere, as well as various models to determine their size distribution and optical characteristics, are necessary. The Atmospheric Radiation Measurement (ARM) program, which recently began in the United States, with broad-based international cooperation, was specifically established to more precisely answer some of these questions.

Models employing relatively few parameters, which are based on both the physics of aerosol formation processes and reliable observational data, are extremely important. Further studies to substantiate models based on relative humidity $H(\%)$ and meteorological visibility S_m (km) can also significantly contribute to our knowledge.

Careful analysis of the coupling between the El Niño and Southern Oscillation (ENSO) phenomena is important in the context of identifying climate signals due to volcanic aerosols. Portman and Gutzler (1996) studied surface air temperature changes in the United States to determine the impact of both volcanic eruptions and ENSO, using observational data from 1219 meteorological stations. The time series analysis was made separating certain time periods for various eruptions (Table 5).

The intensity of these eruptions has been characterized using various parameters such as the dust veil index, sulfate content, etc. The criteria for determining ENSO, Southern Oscillation Index (SOI) and SST anomalies in the eastern sector of the tropical Pacific have also been chosen.

Table 6 contains the list of observed ENSO events, shown in brackets. S-SM-very strong-strong (moderate) ENSO.

Table 5. Volcanic Eruptions.

Year	Volcano	Latitude	Month
1902	Mount Pelée	15° N	May
	Soufrière	13° N	May
	Santa Maria	15° N	October
1907	Ksudach	52° N	March
1912	Katmai	58° N	June
1963	Agung	8° S	March
1982	El Chichón	17° N	April

Table 6. El Niño/Southern Oscillation phenomena (years).

A.- Warm Phase (VS-S)				
1911	1917	1925	1932	1940
1957	1972	[1982]		

B.- Warm Phase (M)				
[1902]	1904	[1907]	1910	1914
1923	1930	[1939]	1943	1951
1953	1965	1969	1976	

C.- Cold Phase				
[1903]	1906	[1908]	1916	1920
1924	1928	1938	1942	1949
1954	[1964]	1970		

A technique of superimposed epochs has been used to analyze observational data using the time series for various eruptions. Single station mean monthly values have been calculated as deviations from 6-year averages calculated for every calendar month, filtering out annual variations and long-term trends.

The analysis of five eruptions indicates that SAT anomalies increased with latitude. After an eruption, stable negative anomalies of about 1–2 °C occurred in the northern and central belts of the United States. During the next

6 months, however, positive anomalies were observed in the eastern and northern parts of the country, whereas negative anomalies prevailed in the western sector of the central and southern parts of the USA. Subsequently, during the second year after an eruption, negative anomalies were prevalent again over the entire United States. These negative temperature anomalies reached up to -2 °C in the northern belt.

The consideration of seven ENSO cases (VS-S) has led to the discovery of a number of significant anomalies, including positive (negative), with the interval of 6–17 months after the onset of ENSO in northern (southern) belts (such situations have also been observed earlier). All volcanic eruptions considered (except Agung in 1963) took place during warm ENSOs of at least moderate intensity.

Christy (1993) has analyzed the combined impact of the Pinatubo eruption and the 1991–1993 ENSO event at tropospheric and lower stratospheric layers, using the National Oceanic and Atmospheric Administration (NOAA) satellite microwave sounding unit (MSU) data for channels N2R, to determine mean tropospheric temperatures at 0–10 km altitudes, and N4 to determine lower stratospheric temperatures in the 100–50 hPa layer. The time series analysis of data for global and zonal average lower stratospheric temperatures from 1979 to 1992 distinctly revealed climatic impacts from the following volcanic eruptions: Nyamuragira, December 1981; El Chichón, April 1982; and Pinatubo, June 1991.

It is important to consider the two primary effects of volcanic sulfur acid aerosol:

1. Absorption of the upward longwave (thermal) radiation flux, leading to stratospheric warming;
2. Transformation of shortwave radiation fluxes, whose characteristic feature is an almost complete mutual compensation between direct solar radiation attenuation (about 25%) and increased downward flux of scattered radiation.

Global radiation decrease did not exceed 3% during the year following powerful eruptions, which led to certain tropospheric and surface cooling. In both hemispheres, the lower stratospheres of both subtropical belts reacted most rapidly to the combined effects of the Nyamuragira-El Chichón (N-CH) eruptions, which occurred in December 1981 and March-April 1982, respectively. They also reacted similarly to the June 1991 eruption of Mount Pinatubo. However, relevant warming was not zonally symmetric: the combined effects of the N-CH eruptions were stronger in the Northern Hemisphere, whereas Pinatubo's effects were stronger in the Southern Hemisphere.

In the year following the N-CH eruptions, global lower stratosphere temperature (LST) increased by 1.0 °C. However, during the subsequent 18 months, it decreased at the rate of 0.0025 °C/day. Within the equatorial belt, LST increased by approximately 0.01 °C/day for the N-CH eruptions and 0.015 °C/day for Pinatubo. Relatively stable mean global LSTs from 1984 to 1990 were about 0.5 °C lower than those before the eruptions. Related to this, some scientists believe that volcanic aerosol stimulates ozone "wash-out," which subsequently leads to stratospheric cooling.

Time series of LST substantially differed in the cases considered here. Rather rapid warming (0.013 °C/day) after the Pinatubo eruption (June-September 1991) reached 1.3 °C and was five times stronger than warming after the N-CH eruptions, but the rate of subsequent cooling was approximately equal.

Analysis of tropospheric temperature (TT) data has also convincingly revealed ENSO event and volcanic eruption impacts. For instance, ENSO 1982–83 produced global tropospheric warming of approximately 0.5 °C greater than registered background levels. This tropospheric warming had increased by 0.6 °C by the end of 1986 and remained stable until September 1988. This, however, contrasts with the onset of the cold ENSO phase in 1989, which led to a TT drop of 0.6 °C. The impact of ENSO 1991–92 was insignificant.

It is possible that the decreasing TT trend between 1982 and 1986 was enhanced by Nyamuragira-El Chichón impacts. If one ignores possible ENSO 1982-83 contributions, the average global TT decrease during the 4–5 year period was 0.3 °C. During this period, the strongest anomaly that occurred was a rapid cooling in the Northern Hemisphere from June 1991 to August 1992, when the global mean TT decrease was 0.73 °C. This cooling was especially strong over land surfaces. For the entire 4-year MSU time series, TT minimums were observed in June and August 1992. It has been very difficult to perform a cause-and-effect analysis of observational data to assess ENSO event and volcanic eruption contributions because amplitudes of high frequency and interannual TT variability were approximately equal.

An unambiguous identification of the volcanic climatic signal has been possible using satellite ERBE (Earth Radiation Budget Experiment) data obtained from the June 1991 Pinatubo eruption (Minnis et al., 1993). For the first time, observational analyses show large-scale volcanic forcing (VF) along with strong cooling, which was observed immediately after the eruption. This cooling increased through September 1991 and, in the course of shortwave forcing, increased relative to longwave forcing. The primary effects of aerosols were to directly increase albedo over mostly clear areas and to both directly and indirectly contribute to an increased albedo in cloudy areas. Stratospheric aerosol optical depth increased by up to two orders of magnitude between 40° N and 40° S during the first 5 months after the eruption.

Minnis et al. (1993) have characterized the strength of VF by examining zonal monthly mean variations of reflected shortwave (SW) fluxes (M_{SW}) and outgoing longwave (LW) radiation (M_{LW}) relative to the 5-year means. The M_{SW} increase after the August 1991 Pinatubo eruption, which occurred between 5° N and 5° S, reached 10 W/m². Meanwhile, the LW volcanic signal was barely detectable over interannual variability noise. Therefore, a net cooling of almost 8 W/m², which is twice the value of any other monthly anomaly, occurred in August 1991.

This trend continued between 40° N and 40° S with a net VF of -4.3 W/m² for August 1991. Since the standard deviation (SD) for the 5-year period is 1.5 W/m², the anomaly is considered statistically significant at the 95% confidence level. Globally averaged VF, excluding volcanic impacts outside the 40° latitudinal belt, was 2.7 ± 1.0 W/m² during August and September 1991.

The amount and distribution of observed VF closely followed the spread of the volcanic aerosol cloud. Simple linear regressions may be utilized to obtain an approximate assessment:

$$\Delta M_{SW} = (17.1 \pm 5.6) \tau - 0.7,$$
$$\Delta M_{LW} = (-7.9 \pm 5.1) \tau + 1.9,$$
$$\Delta M_{NET} = (-11.5 \pm 4.4) \tau - 0.5,$$

where τ is the zonal mean optical depth. However, such correlations are correct only for an initial post-eruption period. During later stages, transformation of aerosol properties, and principally their size distribution, becomes an important factor.

Data show pronounced spatial inhomogeneity of VF distribution, even within a given latitudinal zone. In general, Pinatubo's effects were most evident over areas that tend to be cloud-free, especially over dark surfaces. This is why the apparent lack of a significant VF signal over the Sahara is not surprising.

A substantial penetration of volcanic aerosol into the troposphere, primarily in the vicinity of tall convective storms and in tropopause folds, took place a few months after the eruption. This is why lidar observations during August 1991 discovered significantly enhanced tropospheric aerosol loading as far as north as 40° N. Such a penetration created favorable conditions for an interaction between volcanic aerosol and clouds. Since volcanic aerosol particles can serve as cloud and ice nuclei, this could increase the number and decrease the effective radius of cloud particles, resulting in increased cloud albedo. Although such an indirect effect is most noticeable for deep clouds ($M_{LW} < 220$ W/m²), it may also occur for other cloud types.

Robock and Mao (1995) have further discussed identification of volcanic signal in surface temperature observation data.

VOLCANIC ACTIVITY AND CLIMATE

4. Conclusions

Further studies of atmospheric aerosols properties and their climatic impacts should:

1. Improve the standard radiative models of the atmosphere to more adequately simulate the characteristics of various aerosol types, including low humidity effects, their size distribution, vertical concentration profiles, etc.
2. Obtain more adequate observational information about the spatial and temporal variability of aerosols on global scales.
3. Develop techniques to identify anthropogenically induced aerosol content and composition trends on both regional and global scales, with special consideration given to monitoring sulfate dynamics and soot components.
4. Study the dependence of aerosol optical properties and their chemical composition and size distribution, accounting for the effects of varying air humidity.
5. Compare special techniques to better measure aerosol characeristics including remote sensing techniques to complement *in situ* measurements.
6. Include both observational and theoretical studies of internal aerosol particle structure, considering their multicomponent composition, as well as the role of aerosol particles as condensation nuclei.

Both tropospheric and stratospheric chemical processes are important climate-forming factors (Bach, 1976; Finlayson-Pitts and Pitts, 1986; Kondratyev, 1992c). Therefore, the problem of stratospheric ozone, which is closely related to stratospheric heterogeneous chemistry, has been intensively discussed (Kondratyev, 1980, 1989).

Apparently, the most important mechanisms affecting heterogeneous reactions related to the formation and disintegration of ozone molecules include:

(i) Changes in the stratospheric radiative regime due to aerosol particle scattering;
(ii) Heterogeneous reactions taking place with an adsorption of gases on aerosol particles (stratospheric aerosols, polar stratospheric clouds).

The following phenomena are related to heterogeneous tropospheric processes: formation of acid rains, atmospheric removal of various polluting gases, reduced visibility, fogs, clouds, and the special phenomenon of Arctic haze.

The effects of anthropogenic factors manifest themselves most as a change of cloud systems' physical and chemical properties: their optical characteristics and stability (Jennings, 1993; Feigelson, 1981; Hegg, 1991; Kondratyev, 1991; Marchuk et al., 1986).

All the atmospheric climate-forming processes are closely coupled, forming a chain in which there are direct and inverse, positive and negative feedbacks. Under certain circumstances, conditions appear that create an unstable equilibrium, and the connection between the links is such that coupling with one link, often energetically insignificant, brings forth more substantial changes in other links. In other words, a trigger mechanism is at work. Aerosols are the most likely candidates for such an atmospheric trigger mechanism. It is especially important to thoroughly study the physico-chemical processes that govern the dynamics of the atmosphere as a colloidal system. The urgency and complexity of these problems demonstrate the urgent need for greater international cooperation in this field to better study and understand global advertent and inadvertent impacts on weather and climate.

REFERENCES

Ackerman, S.A., and Chang, H., 1992: Radiative effects of airborne dust on regional energy budgets at the top of the atmosphere. *J. Appl. Meteorol.* **31**, 223-233.

d'Almeida, G.A., Koepke, P., and Shettle, E., 1991: *Atmospheric Aerosols: Global Climatology and Radiative Characteristics* A. Deepak Publ., Hampton, Virginia, 549 pp.

Bach, W., 1976: Global air pollution and climatic change. *Rev. Geophys. Space Phys.* **14**, 429-474.

Bates, T.S., Calhoun, J.A., and Quinn, P.K., 1992: Variations in the methanol-sulfonate to sulfate molar ratio in submicrometer marine aerosol particles over the South Pacific Ocean. *J. Geophys. Res.* **97, D9**, 9859-9866.

Bolle, H.-J., 1982: Aerosol research within the World Climate Research Programme. In *Light Absorption by Aerosol Particles*. Spectrum Press (A. Deepak Publishing), Hampton, Virginia, 1-51.

Cahalan, R.F., 1992: The Kuwait oil fires as seen by Landsat. *J. Geophys. Res.* **97, D13**, 14565-14572.

Charlson, R.J., Lovelock, J.S., Andreas, M.O., and Warren, S.G., 1987: Oceanic phytoplankton, atmospheric sulphur, cloud albedo and climate. *Nature* **326**, 655-661.

Charlson, R.J., and Wigley, T.M.L., 1994: Sulfate aerosol and climatic change. *Sci. American* **270**, 28-35.
Christy, J.R., 1993: Impacts on global temperatures due to Mt. Pinatubo and the 1991-92 ENSO. *Proc. Seventeenth Annual Climate Diagnostic Workshop.* NOAA, Washington, D.C., 13-17.
Chylek, P., Videen, G., Ngo. D., Pinnick, R.G., Klett, G.D. 1995: Effect of black carbon on the optical properties and climate forcing of sulfate aerosols. *J. Geophys. Res.*, **100**, 16325-16332.
Coakley, J.A, Jr., Cess, R.D., and Yurevich, P.B., 1983: The effect of tropospheric aerosols on the Earth's radiation budget: a parameterization for climate models. *J. Atmos. Sci.* **40**, 116-138.
Cofer, W.R. III, Stevens, R.K., Winstead, E.L., et al., 1992: Kuwait oil fires: Compositions of source smoke. *J. Geophys. Res.* **97, D13**, 14521-14526.
Deepak, A. (Ed.), 1982: *Atmospheric aerosols. Their Formation, Optical Properties and Effects.* Spectrum Press, Hampton, Virginia, 480 pp.
Deepak, A., and Vali, G. (Eds.), 1991: The International Global Aerosol Program (IGAP) Plan: Overview. A. Deepak Publ., Hampton, Virginia, 1, 17.
Engardt, M., and Rodhe, H., 1993: A comparison between patterns of temperature trends and sulfate aerosol pollution. *Geophys. Res. Lett.* **20** (2), 117-120.
Feigelson, E.M. (Ed)., 1981: *Radiation in a Cloudy Atmosphere.* Leningrad, Gidrometeoizdat, 280 pp. (in Russian).
Finlayson-Pitts, B.J., and Pitts J.N., 1986: *Atmospheric Chemistry. Fundamentals and Experimental Techniques.* 4, John Wiley and Sons, 1098 pp.
Fried, A., Hery, B., Regazzi, R.A., et al., 1992: Measurements of carbonyl sulfide in automotive emissions and an assessment of its importance to the global sulfur cycle. *J. Geophys. Res.* **97, D13**, 14621-14634.
Galindo, I., 1965: Turbidiometric estimations in Mexico City using the Volz sun photometer. *Pure Appl. Geophys.* **60**, 189-196.
Galindo I., 1975: Physikalische und mathematische Untersuchungen zur atmosphärischen Trübung. *Arch. Met. Geoph. Biokl., Ser. B* **23**, 225-254.
Galindo, I., 1978: On the presence of saharian aerosol at the Western part of the Atlantic Ocean. *Zeitschrift f. Meteorologie,* **Heft 6, Band 28**, 352-360.
Galindo, I., 1984: Anthropogenic aerosols and their regional scale climatic factors. In *Aerosols and Their Climatic Effects*, Gerber, H.E., and Deepak, A. (Eds.), A. Deepak Publ., Hampton, Virginia, 245-259.
Galindo, I., 1992a: Cambios climáticos regionales como componentes del cambio climático global. *Ciencia* **45**, 21-27.
Galindo, I., 1992b: Extinction of short-wave solar radiation due to El Chichon stratospheric aerosol. *Atmosfera* **5**, 253-268.

Galindo, I., and Muhlia, A., 1970: Contribution to the turbidity problem in Mexico City. *Arch. Met. Geoph. Biokl., Ser. B* **18**, 169-186.
Galindo, I., and Bravo, J.L., 1975: On the presence of a volcanic stratospheric dust stratum over a polluted atmosphere. *Geofis. Int.* **15**, 157-167.
Galindo, I., Muhlia, A., and Leyva, A., 1975: Atmospheric turbidity and sky radiation at maritime environments: Gulf of Mexico and Tropical Atlantic. *Beitr. Atmos. Phys.* **48**, 168-184.
Galindo, I., Kondratyev, K.Ya., and Zenteno, G., 1996: Determination of the atmospheric optical depth due to El Chichón stratospheric aerosol cloud in the polluted atmosphere of México City. *Atmósfera* **9,** 23-32.
Gerber, H.E., 1982: Nature of fog nuclei from measurements of relative humidity. *Idojárás* **86**, 175-180
Gerber, H.E., and Deepak, A., (Eds.), 1984: *Aerosols and Their Climatic Effects.* A. Deepak Publ., Hampton. Virginia, 297 pp.
Gerstell, M.F., Crisp., J., Crisp., D., 1995: Radiative forcing at the stratosphere by SO_2 gas, silicate ash, and H_2SO_4 aerosols shortly after the 1982 eruptions of El Chichón. *J. Climate* **8**, 1060-1070.
Goetz, G., Mészaros, S., and Vali, G., 1991: *Atmospheric Particles and Nuclei.* Budapest, Ak. Kiado, 274 pp.
Grassl, H., 1973: Aerosol influence on radiative cooling. *Tellus* **25**, 386-395.
Grassl, H., 1974a: Erwärmung und Abkühlung durch Atmosphärisches Aerosol. *Ann. Meteorol. Neue Folge* **9**, 55-59.
Grassl, H., 1974b: Atmospheric absorbers in the window region and their influence on radiative heating and cooling. *Proc. Int. Conf. on Structure, Composition and General Circulation of the Upper and Lower Atmospheres and Possible Anthropogenic Perturbations.* January 14-25, 1974. Toronto, Vol. 2, 1129-1142.
Grassl, H., 1975: Albedo reduction and radiative heating of clouds by absorbing aerosol particles. *Beitr. Phys. Atm.* **48**, (3), 199-210.
Grassl, H., 1992: The influence of aerosol particles on radiation parameters of clouds. *Idojárás* **86**, 60-75.
Hansen, J., Lacis, A., Ruedy, R., and Sato, M., 1992: Potential climate impact of Mount Pinatubo eruption. *Geophys. Res. Lett.* **19**, 215-218.
Hegg, D.A., (Ed.), 1991: Report of the Experts Meeting on Aerosol Physics and Chemistry. Hampton, Virginia. WMO/TD No. 439, GAW No. 74. WMO, Geneva, 23 pp.
Hobbs, P.V., and McCormick, M.P. (Eds.), 1988: *Aerosols and Climate*, A. Deepak Publ., Hampton, Virginia, 486 pp.
Hudson, J.G., and Clarke, A.D., 1992: Aerosol and cloud condensation nuclei measurements in the Kuwait plume. *J. Geophys. Res.* **97** (D13), 14533-14536.

Husar, R.B., Stowe, L.L., and Deepak, A. (Eds.), 1991: Report of the Experts Meeting on Global Aerosol Data System (GADS), Hampton, Virginia, GAW No. 73. WMO, Geneva 20 pp.
International Geosphere-Biosphere Program, 1988: A Study of Global Changes. A Plan for Action 1988. Report of Special Committee for JGBP for First Meeting of Scientific Advisory Council for JGBP, Stockholm, Sweden.
Ivlev, L.S., 1982: *Chemistry and Structure of Atmospheric Aerosols*. LGU Publ., Leningrad, 386 pp. (in Russian).
Ivlev, L.S., and Andreev, S.D., 1986: *Optical properties of atmospheric aerosols*. Leningrad, LGU Publ., 359 pp. (in Russian).
Jaenicke, R., 1987: Aerosol physics and chemistry. In *Landolt-Bornstein, New Series,* Vol. 43, Meteorology, Berlin Springer, 391-457.
Jennings, S.G. (Ed.), 1993: *Aerosol Effects on Climate*. Arizona, The University of Arizona Press, 305 pp.
Jones, A., Roberts, D.I., and Slingo, A., 1994: A climate model study of indirect radiative forcing of anthropogenic sulphate aerosols. *Nature* **370**, 450-453.
Kawamata, M., Yamada, Sh., Kudoh, T., and Takano K., 1992: Atmospheric temperature variation after the 1991 Mt. Pinatubo eruption. *J. Meteorol. Soc. Jap.* **70**, 1161-1166.
Kolomeev, M.P., and Sorokovnikova, O.S., 1991: Effect of stratospheric aerosol on the zonal mean parameters of atmospheric general circulation. *Izv. AN SSSR, FAO* **27**, 483-491 (in Russian)
Kondratyev, K.Ya., 1972: Complex Atmospheric Energetics Experiment. *WMO Bull.* **No. 12**, Geneva, 94 pp.
Kondratyev, K.Ya. (Ed.), 1973: *Effects of Aerosols on Radiation Transfer: Possible Climatic Consequences.* LGU Publ., Leningrad, 286 pp. (in Russian).
Kondratyev, K.Ya., 1976: Aerosol and Climate. *Trudy GGO.*- **Issue 381**, 3-66. (in Russian).
Kondratyev, K.Ya., (Ed.), 1978: *Atmospheric Aerosols and Their Effect on Radiation Transfer.*Gidrometeoizdat, Leningrad, 120 pp. (in Russian).
Kondratyev, K.Ya., 1980: *Radiative Factors of the Present Global Climate Change*. Gidrometeoizdat, Leningrad, 279 pp. (in Russian).
Kondratyev, K.Ya., 1983: *Earth Radiation Budget, Aerosol and Cloudiness* Progress in Sci. and Technol., Meteorology and Climatology, Vol. 10, Moscow, VINITI, 316 pp. (in Russian).
Kondratyev, K.Ya., 1985: *Volcanoes and Climate.* Progress in Sci. and Technol. *Meteorology and Climatology*, Vol. 14. Moscow. VINITI, 204 pp. (in Russian).
Kondratyev, K.Ya., 1988: *Climate Shocks: Natural and Anthropogenic*. J. Wiley and Sons, Chichester, 269 pp.

Kondratyev, K.Ya., 1989: *Global Ozone Dynamics.* Prog. in Sci. and Technol. Geomagnetism and Upper Atmosphere, Vol. 14, VINITI, Moscow, 212 pp. (in Russian).
Kondratyev, K.Ya., 1990: Atmospheric chemistry and climate. *Progress in Chemistry* **59** (10), 1587-1600 (in Russian).
Kondratyev, K.Ya., (Ed.), 1991: *Aerosols and Climate.* Gidrometeoizdat, Leningrad, 541 pp. (in Russian).
Kondratyev, K.Ya., 1992a: Aerosol-cloud-climate interaction. 1. Aerosol. *Optics of the Atmosphere and the Ocean* **5** (3), 317-335 (in Russian).
Kondratyev K.Ya., 1992b: Aerosol-cloud-climate interaction. 2. Clouds. *Optics of the Atmosphere and the Ocean* **5** (3), 324-335 (in Russian).
Kondratyev, K.Ya., 1992c: *Global Climate.* Gidrometeoizdat, St. Petersburg, 359 pp. (in Russian).
Kondratyev, K.Ya., 1993: Complex monitoring of the Mt. Pinatubo eruption consequences. *Studies of the Earth from Space* 1, 111-122. (in Russian).
Kondratyev, K.Ya., and Ter Markaryants, N.E. (Eds.), 1976: *Complete Radiation Experiment.* Gidrometeoizdat, Leningrad, 238 pp. (in Russian).
Kondratyev, K.Ya., Moskalenko, N.I., and Pozdnyakov D.V., 1983: *Atmospheric Aerosol.* Gidrometeoizdat, Leningrad, 224 pp. (in Russian).
Kondratyev, K.Ya. and Prokofyev, M.A., 1984: Atmospheric aerosols and their effect on climate. *Izv. AN SSSR, FAO.* **20**, (11), 1055-1064 (in Russian).
Kondratyev, K.Ya., and Johannessen., O.M., 1993: Arctic and Climate. PROPO, St. Petersburg, 141 pp. (in Russian).
Kondratyev, K.Ya., Johannessen., O.M., and Melentyev, V.V., 1996: *High Latitude Climate and Remote Sensing.* Wiley/PRAXIS, Chichester e.a. 248 pp.
Kondratyev, K.Ya., and Grassl, H., 1993: *Global Climate Change in the Context of Global Ecodynamics.* St. Petersburg, Center of Ecological Safety, 195 pp. (in Russian).
Kondratyev, K.Ya., and Grassl, H., 1996: *Global Climate Dynamics in the Context of Global Change.* Springer-Verlag, Berlin e.a. (in press).
Kondratyev, K.Ya. and Cracknell, A.P., 1996: Observing Global Climate Change. Taylor and Francis, London e.a., (in press).
Kondratyev, K.Ya, V.I., Binenko., and I.N. Melnikova, 1996: Solar radiation absorption in the visible spectral region in cloudy and cloudless atmospheres. *Meteorol. and Hidrol.*, **N2**, 31-45 (in Russian).
Lacis, A., Hansen, J., and Sato, M., 1992: Climate forcing by stratospheric aerosols. *Geophys. Res. Lett.* **19**, 1607-1610.
Langner, J., and Rodhe, M., 1991: A global three-dimensional model of the tropospheric sulfur cycle. *J. Atmos. Chem.* **13**, 225-263.
Langner, J., Rodhe, N., Crutzen, J., and Zimmerman, P., 1992: Anthropogenic influence on the distribution of tropospheric sulphate aerosol. *Nature* **358**, 712-715.

Marchuk, G.I., Kondratyev, K.Ya., Kozoderov, V.V, and Khvorostyanov, V.I., 1986: *Clouds and Climate*. Gidrometeoizdat, Leningrad, 512 pp. (in Russian).
Meszáros, E., 1981: *Atmospheric Chemistry. Fundamental Aspects.* Budapest, Akad. Kiado, 201 pp.
Minnis, P., Harrison, E.F., Stowe, L.L., Gibson, G.G., Denn, F.M., Dorlling, D.R., and Smith, W.L., Jr., 1993: Radiative climate forcing by the Mount Pinatubo eruption. *Science* **259**, 1411-1415.
Newell, R.E., 1984: Volcanism and climate. In *1985 Yearbook of Science and Technology*, McGraw Hill, New York, 206-225.
Parungo, P., Kopcewicz, B., Nagamoto, C., et al., 1992: Aerosol particles in the Kuwait oil fire plumes: Their morphology, size distribution, chemical composition, transport, and potential effect on climate. *J. Geophys. Res.* **97** (D14), 15867-15882.
Patterson, E.M., Pollard, C.O., and Galindo, I., 1983: Optical properties of the ash from "El Chichon" Volcano. *Geophys. Res. Lett.* **10**, 317-320.
Penner, J.B., Charlson, R.J., Hales, J. M., et al., 1993: Quantifying and minimizing uncertainty of climate forcing by anthropogenic aerosols. *Bull. Amer. Meteorol. Soc.* **75** (N3), 375-400.
Pollack, J.B., 1983: Aerosol, radiation, and climate. *Fifth Conf. on Atmos. Radiation.* Oct. 31–Nov. 4, 1983, Baltimore, Maryland. Amer. Meteorol. Soc., Boston, Massachusetts, 44-345.
Portman, D.A., and Gutzler, D.S., 1996: Explosive volcanic eruptions, the ENSO, and United States climate variability. *J. Climate* **(9)**, 17–33.
Pudykiewicz, J.A., and Dastoor, A.P., 1995: On numerical simulation of the global distribution of sulfate aerosol produced by a large volcanic eruption. *J. Climate*, **8,** 464-473.
Robock, A., and Mao, J., 1995: The volcanic signal in surface temperature observations. *J. Climate*, **8,** 1086-1103.
Rodhe, H., and Langner, J., 1993: Atmospheric concentration of DMS and its oxidation products estimated in a global 3-D model. In *Dimethylsulfide: Oceans, Atmosphere and Climate.* Proc. Int. Symp. Held in Belgirate, Italy, October 13–15, 1992. G. Restelli and G. Angeletti (Eds.) Reidel Publ., Dordrecht, 333–343.
Schlesinger, M.E., Jiang, K., and Charlson, R.J., 1993: Implications of anthropogenic atmospheric sulfate for the sensitivity of the climate system. In *Climate Change and Energy Policy*, L. Rosen and R. Glasser (Eds.). Amer. Inst. Phys., New York, N.Y., 75-108.
SMIC – Report of the Study of Man's Impact on Climate. *Inadvertent Climate Modification* (1971). MIT Press, Cambridge, Mass., 178 pp.
Stowe, L.L., Heitzenberger, R., and Deepak, A., 1990: Experts Meeting on Space Observations of Tropospheric Aerosols and Complimentary Measurements. WCRP-48, WMO/TD-No. 389, Geneva.

Zakharov, V.M., Kostko, O.K., and Khmelevtsov, S.S., 1990: *Lidars and Climate Studies.* Gidrometeoizdat, Leningrad, 320 pp. (in Russian).

Zuev, V.E. and Krekov, G.M., 1986: *Optical Models of the Atmosphere.* Gidrometeoizdat, Leningrad, 256 pp. (in Russian).

Zuev, V.E. and Kabanov, M.V., 1987: *Optics of Atmospheric Aerosol.* Gidrometeoizdat, Leningrad, 254 pp. (in Russian).

CHAPTER 1

GLOBAL CLIMATE DIAGNOSTICS DATA AND NUMERICAL MODELING RESULTS

Climate change is a result of numerous interactive climatic forcings. Therefore, to identify the "volcanic signal" it is necessary to analyze global climatic observational data to understand what it can and cannot tell us (Karl et al., 1989, 1993; Kondratyev, 1992; Madden et al., 1993; Plantico et al., 1990; Richards, 1992; Woodward and Gray, 1993).

1.1 Conventional Observations

A thorough analysis of SAT and SST observations confirm earlier suspicions that, from the end of the 19th century to the present, the global mean SAT has risen by about 0.5 °C. To support this conclusion, Jones and Briffa (1992) undertook a detailed analysis of information needed to provide a homogeneous data set. They then considered specific features of the spatial and temporal variability of SATs during this century, particularly SAT variations between 1981 and 1990, and compared these data to 1931–1940 data.

The existing analysis of SAT observations is confined to the period after 1850. During 1850–1990, the observational network gradually expanded so that by 1920, the only regions not covered by regular meteorological observations were central Africa, South America, Asia, the Arctic coastline and Antarctica. This made obtaining monthly SAT anomaly means (with respect to 1951–1970) over 5° lat. by 10° long. grids possible. Anomalies were selected to minimize observational errors caused by factors such as different data-gathering techniques, station locations, urbanization effects, representativeness dynamics (as the number of stations grows). Estimates obtained indicate that urbanization effects (urban "heat islands") on SAT trends over the 100-year period have not exceeded 0.05 °C. Since 1990, the changing number of stations have practically registered no differences in global average land SAT.

The results of shipboard SST observations from 1854 to the present, the Comprehensive Ocean-Atmosphere Data Set (COADS), constitute most of the marine observational database. Although SSTs and SATs can differ by 6–8 °C, they do have a significant correlation; therefore, SST anomalies can serve as a reliable substitute for SAT anomalies. However, it is very difficult to provide homogeneous data on SSTs due to past radical changes in sampling techniques. These changes are primarily due to past water temperature measurements being gathered by linen buckets, while more recent measurements are taken from pumped-in water used to cool ship engines.

With a general annual mean warming of about 0.5 °C in both hemispheres, beginning at the turn of the century, the dynamics of the NH SAT field has been characterized by strong spatial-temporal variability. For example, between 1940 and 1970, a hemispheric mean cooling of 0.2 °C took place in some seasons. Warming, however, was more regular in the SH. Spatial dynamics of SATs in the NH are characterized by considerable variability between warming and cooling periods since seasonal differences are much greater in the NH. Although summer and fall SATs in the 1980s were approximately the same as in the 1930s and 1940s, winters and springs of the 1990s have been the warmest on record. Summers of the 1850s and 1970s were, on average, as warm as the last two decades.

With the exception of summers, annual mean data exhibit a warming trend in all seasons. Meanwhile, cooling trends observed in the 1870s and 1880s were strongest in the summer. NH data are characterized by cooling trends in the 1780s, 1810s, and late 1830s, and by warming trends in the 1820s. A comparison of data from 1901–1945 and 1946–1990 reveals the most significant warming in April (0.28 °C) with minimal warming in August and October (0.06 °C). The data reveal a distinct annual change with the most significant warming in the spring and a minimum in the fall.

An analysis of combined land and ocean temperature anomaly data between 1950 and 1979 reveals the same, although weaker, trend over land areas. Consequently, NH land data reveal cooling periods in some seasons which are practically indistinguishable from the combined data, revealing approximately the same SAT variability in both hemispheres. Data for both the Northern and Southern Hemispheres manifest a surprisingly high correlation (0.79 from 1901–1990). The effect of the El Niño (warm years) and La Niña (cold years) events demonstrate the importance of filtering out the contribution of these occurrences when analyzing the CO_2 signal.

Table 1.1.1 shows decadal mean anomalies (1950–1979) demonstrating that the 1980s were the warmest decade in the past 40 years. The spatial and temporal dynamics of the SAT field in the NH is very inhomogeneous. Warming

Table 1.1.1. Hemispheric mean and global mean SAT (°C) trends. After Jones and Briffa (1992).

Period	NH	SH	Globe
1861–70	-0.26	-0.36	-0.31
1871–80	-0.27	-0.36	-0.32
1881–90	-0.35	-0.31	-0.33
1891–1900	-0.27	-0.25	-0.26
1901–10	-0.33	-0.37	-0.35
1911–20	-0.34	-0.24	-0.29
1921–30	-0.16	0.26	-0.21
1931–40	-0.04	-0.12	-0.08
1941–50	-0.04	-0.04	0.00
1951–60	0.05	-0.05	0.00
1961–70	0.02	0.00	0.01
1971–80	-0.05	0.06	0.00
1981–90	0.20	0.26	0.23

is experienced mainly in the winter (December, January, February – DJF) and spring (March, April, May – MAM) with maxima over northern and central Asia and northwestern North America, even though other regions of Asia demonstrate cooling trends. The principal NH regions with relatively below average, but stable, SATs comprise the central Pacific Ocean and the Northern Atlantic, including Iceland and Greenland. SH SAT anomalies are weaker, except for some Antarctic regions. As a result, climatic warming, which took place during the last century, was not homogeneous.

Kukla et al. (1992) analyzed present SAT trends between 1945 and 1986 in the context of their possible forcing by variations in orbital parameters. This period was selected because it provided the most complete information. Unfortunately, SH mid and high latitude data for both hemispheres remain insufficiently representative. An analysis of the database under discussion reveals intense winter (DJF) cooling of 3–4 °C per 100 years over the eastern part of North America, the north Atlantic and northern Europe; with the greatest cooling in the Spitzbergen region. Here, and in subsequent cases, seasons corresponding to the NH have been assumed. A significant cooling trend is observed in the 30°– 50° N latitudinal belt in the western and central Pacific. On the other hand, a strong warming trend takes place over Alaska, northwestern Canada and the northern part of Central Asia, with a maximum in northern

Siberia. The remaining regions of the world oceans are characterized by warming trends with a maximum of about 2–3 °C per 100 years in the SH middle latitudes. Spring trends resemble winters with greatest warming in the northwestern part of North America and northern Asia. There is also a weaker cooling trend in the northern Atlantic and northwestern Pacific Oceans. In the northern Atlantic, cooling is weaker in summer (June, July, August – JJA), whereas in the northwestern Pacific, it is stronger in winter. Cooling trends are manifested over most of Europe and the Mediterranean Basin. Compared to winters, areas experiencing summer cooling in the NH middle and high latitudes are larger, but the trends are weaker. Warming trends over SH oceans are more intense than in December to February.

Table 1.1.2. Linear surface air temperature trends (°C/100 yr) and NH temperature of the 300–100 and 850–300 hPa layers, from data compiled between 1950 and 1986. After Kukla et al. (1992).

Season	Level hPa	Latitude 10° S–10° N	10°–30° N	30°–60° N	60°–90° N	Meridional gradient (10° S – 10° N) (60°–90° N)
DJF	300–100	-2.76	-3.79	-0.40	-6.53	3.77
	850–300	2.05	-0.96	-2.15	0.01	2.04
	Surface	1.93	0.91	-0.40	2.71	-0.78
MAM	300–100	-1.01	-3.39	-0.93	2.09	-3.10
	850–300	2.08	-1.01	-0.63	-0.11	2.19
	Surface	2.20	2.39	0.94	1.17	1.03
JJA	300–100	-1.96	-3.62	-1.31	-2.11	0.15
	850–300	1.82	-1.07	-0.20	0.36	1.46
	Surface	1.93	0.70	-0.45	-1.47	3.40
SON	300–100	-3.08	-4.61	-1.02	-3.93	0.85
	850–300	1.65	-1.05	-1.56	-1.45	3.10
	Surface	0.90	2.60	-0.59	-2.06	2.96

On the whole, SAT trends for both hemispheres are more significant during their respective winter seasons than in summer. The warming maxima are greater over land masses than over oceans in both winters and summers. NH cooling (September, October, November – SON) predominates in most of the Atlantic, the Pacific and North America, whereas warming prevails over both the lands and oceans of the SH and in the northern part of central Asia. Areas

VOLCANIC ACTIVITY AND CLIMATE

reflecting these trends are smaller and their intensity is less significant than in other seasons. The manifestation of linear trends in each square of the spatial grid is substantially masked by interannual variability.

The principal regularities of SAT trends can be determined by analyzing relative topography (thickness) data of the 300–100 hPa layer between 1977–1986. During the 30-year period ending in 1986, a maximum increase of free atmosphere temperature in the 850–300 hPa layer took place in low latitudes with a maximum increase in the Arctic causing considerable intensification of the meridional temperature gradient.

NH high latitude SAT anomalies were followed by considerable changes in other meteorological and oceanographic parameters, especially precipitation and salinity. As a result, zonal annual mean precipitation in the NH middle and high latitudes has oscillated since 1920. Unfortunately, the instability of observational techniques prevents one from obtaining quantitative estimates. Satellite observations indicate that the extent of sea ice cover in Hudson Bay increased between 1973 and 1980. Snow accumulations also increased in the southern part of the Greenland ice sheet and surface waters markedly freshened near Iceland and the Labrador Sea between 1968 and 1972. All of these phenomena occurred during the so-called "Great Salinity Anomaly," which was followed by the formation of thick sea ice and temporal breaking of deep-water convection.

A considerable freshening of North Atlantic waters, north of 50° N, was observed at intermediate and great depths between 1962 and 1981. The salinity level has now approached the threshold at which the downward water flow and the input of warm waters to the Norwegian and Barents Seas can markedly weaken. A comparison of SAT trend data averaged over the 65°, 45°–65°, 25°–45°, 5°–25°, and 5° N – 5° S latitudinal belts, with the data on variations of extra-atmospheric insolation during the past ~ 1000 years, reveal two general qualitative features:

(i) A relative SAT decrease in the NH high latitudes and an increase in low latitudes, which intensified the meridional temperature gradients;
(ii) A relative decrease of SAT in the fall and an increase in the spring over NH land and oceans.

Since, however, the absolute variations of extra-atmospheric insolation are very small (maximum increase of insolation in the period 1945–1986 constituted 0.18 W/m^2 in April near 10° N and maximum decrease at the South Pole in November of 0.25 W/m^2), it is impossible to determine the reason for SAT trend in insolation variations. One cannot exclude, however, that such variations could have played the role of trend accelerators which are actually determined by other factors (see Kondratyev and Nikolsky, 1995a,b). Having processed data of aerological sounding in the NH between 1977–1986, Weber (1990) came to the conclusion that, compared to the period 1951–1960, the tropical and subtropical troposphere (300–1000 hPa) was much warmer, especially in the region of the Pacific Ocean. Negative temperature anomalies, however, took place in the northern Atlantic, the northern Pacific and the European-Canadian sector of the Arctic. Data revealed an increase of the meridional temperature gradient which was followed by intensified atmospheric circulation.

Having undertaken an overview of the data on the secular trend of the global mean SAT, Bloomfield (1992) came to the conclusion that the range of possible growth of SAT should be estimated at 0.2–0.8 °C per 100-year period and noted that calculations made during the last several years revealed the possibility of practically total compensation of "greenhouse" warming by "aerosol" cooling due to the increase of sulfate aerosol in the atmosphere.

Very interesting SAT data for central England for the period between 1659 and 1990 were analyzed by Murray (1992) who discovered a strong annual mean SAT variability on a 10-year time scale before the beginning of the industrial revolution. For example, the SAT amplitude between 1731 and 1740 constituted 3.7 °C. Of course, this variability had to be due to natural causes and, interestingly, has remained practically the same (3.8 °C) 332 years later.

Estimates of the SAT data reliability in conjunction with other results presented at the Fifth Conference on Climate Change, organized by the American Meteorological Society in 1991, have proved to be most interesting. As Madden (1991) and Madden et al. (1993) showed, one of the serious causes of errors when estimating global mean SAT during the entire period of instrumental meteorological observations (from 1860 until now) is related to evolving spatial representativity of the network. Land surface covered by observational data increased from 20% to 90% between 1860 and 1965. Naturally, this situation results in errors when calculating global mean SAT values from the observational data that can be estimated using numerical climate modeling results.

Calculations using the National Center for Atmospheric Research General Circulation Model (NCAR GCM) showed that the root mean-square difference (RMSD) of true SAT global means from those of real network decreased from 0.224 °C in 1860 to 0.045 °C in 1950 and later. Therefore two techniques were suggested to correct the global mean SAT values calculated from the observational data, with the nonrepresentativity of network taken into account.

Madden et al. (1993) have shown that if these are reasonable estimates of the spatial variance and the number of independent estimates in the global distribution of actual surface temperature data, or, alternatively, their spatial spectrum, then it is possible to make first-order estimates of the spatial sampling errors.

VOLCANIC ACTIVITY AND CLIMATE

A number of recent publications have been devoted to statistical analysis of observed global temperature trends. On the basis of such an analysis Richards (1992) has arrived at the following conclusions:

(i) The global temperature increase since the last century is a systematic development;

(ii) Short-term variations in temperature do not have long-lasting effects on the first realization of the series; over time, stochastic perturbations dissipate and temperature reverts the trend;

(iii) Multivariate tests for causality demonstrate that atmospheric CO_2 is a significant forcing factor. The implied change in SAT with respect to a doubling of atmospheric CO_2 lies in a range of 2.17 to 2.57 °C, with a mean value of 2.34 °C. The contributions of solar irradiance and volcanic loading are much smaller;

(iv) In a multivariate system, shocks to forcing factors generate stochastic cycles in temperature comparable to the results from unforced simulations of climatological models;

(v) Extrapolation of regression equations predicts changes in global temperature that are marginally lower than the results from climatological simulation models.

Woodward and Gray (1993) have analyzed the reliability of the standard approach to test the presence of linear trend in observational data on the basis of the model

$$Y_i = a + bt + E_t$$

where Y_i represents the data at time t, and E_t is the deviation of the data from a straight line. It has been shown that trend tests based on such models for the response of prediction or inference concerning future behavior should be used with caution. Certain autoregressive moving average (ARMA) models may be very reasonable models. Woodward and Gray (1993) have shown that, based solely on the available temperature anomaly series, it is difficult to conclude that the trend will continue over any extended length of time. This problem has been further discussed in a number of recent publications (i.e., Lund et al., 1995).

Gunst et al (1993) have considered a very important problem of a solution of global mean temperature anomaly and assessed three specific aspects of the estimation of the mean: The use of weighted averages of anomalies, the effects of gridding on weighted averages, and the appropriateness of data reuse in the calculation of averages for neighboring grids. When used to estimate global mean temperature anomalies, linear distance weighting is shown to be no better than uniform weighting.

Philips et al. (1991) note that the analysis of the effect of adequacy of selected time series on the statistical characteristics of climate (first and second moments), which, so far, have been neglected, is complicated by the existing practice of data archiving, when only the results of calculations for time intervals not shorter than 6 hours are archived. In this connection, a simulation modeling was made of the minimum needed repeatability of the data (frequency of observations) as well as relative contributions of intradiurnal and low-frequency variability when answering the statistical characteristics of climate from the numerical modeling data using the European Centre for Medium Range Weather Forecasts (ECMWF) 19-level spectral (T-42) climate model. The results of these calculations show that the choice of a 6-hour interval can be used for many climatic parameters, but it leads to a substantial loss of information in the case of such parameters as global solar radiation, sensible and latent heat fluxes near the surface, characteristics of convection processes, when a 3-hour interval is preferable.

Van Loon (1991) compared the trends of the annual mean SAT values between 1880 and 1980 calculated as deviations from mean values for a long period (1946–1960) for the NH and the Arctic (65°– 85° N) and found out that:

(i) The variability of SAT and its amplitude are higher in the Arctic than in the NH;
(ii) Between 1880 and 1940 the SAT increased, then during the subsequent three decades it decreased, and then, again, the SAT rose, which continues to the present;
(iii) Before 1950 the SAT increases had not been monotonous and the smoothed change of SAT reflected rather a change in mean values from first 40 years to next 20 years of the period under consideration.

A detailed analysis of SAT variations for the period 1949 to 1972, when a decreasing trend of SAT could be analyzed from the data of aerological sounding, led to the conclusion that the hemispherical mean decrease of SAT formed as a difference between great regional anomalies of different signs was closely connected with variations in the atmospheric general circulation caused by the sensible heat transfer by quasi-stationary waves. Hurrell and Van Loon (1993) performed an analysis of the winter-time low-frequency climate variations over the NH to study relationships between an important climatic parameter, such as SAT, and variations of general atmospheric circulation, bearing in mind the urgency of these studies from the viewpoint of assessing the possibility of identification of an anthropogenic signal in climate change.

Though the SAT variations on interannual and longer time scales are globally not uniform, they are characterized by distinctive regular large-scale patterns. The spatial coherence of the SAT field is determined, in particular, by quasi-stationary planetary waves. So, for example, the NH wintertime heat transport to high latitudes has a pronounced effect on the averaged SAT field through local changes in temperature advection. The temperature decrease in midlatitudes is accompanied by an intensification of poleward heat fluxes in the stationary waves, which is demonstrated by a negative correlation of less than 0.9 between the average poleward heat flux at 43° N and zonal mean temperatures at 40° N.

Thus, not only local temperatures but also the zonal means are greatly affected by changes in the planetary waves. In this context the interdecadal climate variations in the wintertime northern Atlantic have been discussed, connected with the North-Atlantic Oscillation (NAO) characterized by the difference in normalized sea level pressure anomalies between the Azores maximum and the Icelandic minimum. Regional changes and trends in temperature can be explained as an effect of the NAO intensity variations due to variations in both components of NAO.

With a large meridional pressure gradient in the northern Atlantic (40° – 60° N), the westerly wind component prevails and Europe receives warmer air masses from the ocean: for example, the observational data show that a SAT increase over Europe early in the century (when the hemispheric SAT increase had taken place) coincided with a period of strong westerlies.

The typical feature of the SAT field is its inhomogeneity characterized by large-scale coherent structure with anomalous regions of opposite sign associated with changes in atmospheric general circulation (AGC). Thus, mean trends for a hemisphere (or the globe) result from the mutual compensation of regional variations of opposite sign. These regional variations manifest themselves through interdecadal oscillations in the northern Atlantic taking place during the last century. The time-dependent character of the temperature and pressure observational series is particularly important. An analysis of the data on pressure using the complex demodulation technique has shown that low-frequency variations are characterized by concentrations of variability in intervals of 6 to 15 years.

Reck (1993) has analyzed the available data on the height and temperature of the tropical tropopause to further use them as indicators of global climate change, bearing in mind the well-known difficulties in identification of anthropogenic climate signal from the data on SAT. Calculations of temperature trends at the tropopause level over the years 1958–1988 near 15° N have demonstrated the temperature increase by 0.9 K over the 30-year period with the standard deviation 0.5 K. Such a stable trend testifies to the necessity of a more adequate analysis of the observational data to assess possibilities to recognize the "greenhouse" climate signal.

An analysis of the decadal climatic oscillations attracts more and more attention. Though there is much evidence of quasi-periodic 0- to 30-year fluctuations of climate, such an interdecadal variability (IDV) has attracted, so far, little attention, compared to studies dedicated, for example, to ENSO oscillations on the time scale of 3 to 5 years. Nevertheless, studies of IDV are very interesting from the viewpoint of revealing a long-term interannual variability of the climatic system; otherwise, without this knowledge, one cannot recognize the anthropogenic climate signals (including the signal determined by the enhancement of the atmospheric greenhouse effect).

There is no doubt that the IDV results from the atmosphere-ocean-cryosphere-biosphere interaction. Though the origin of IDV remains unclear, the oceans, no doubt, play an important role in its formation as gigantic (compared to the atmosphere) reservoirs of heat and water, in which both heat and salt are transported long distances.

Variations in SAT during the last 100 years are one of the indicators of the IDV existence. In particular, it is very important whether the climate warming observed in the period 1980s to early 1990s is "greenhouse" or a manifestation of IDV (see Manabe and Stouffer, 1996; Wallace et al., 1996).

Other IDV manifestations are a series of severe coolings started in 1976 covering the central and eastern USA; "Great Salinity Anomaly" (freshening of the surface waters in the northern part of the northern Atlantic) between 1960 and 1970; the long-term series of droughts in the Sahel region of Africa; the 20- to 30-year cycles of the SH ocean temperature; the decadal fluctuations of the sea ice cover extent in the Arctic; varying frequency of occurrence of hurricanes in the tropical Atlantic, etc.

Numerous manifestations of the low-frequency climate change can be of global scale and connected with variations of the environmental characteristics affecting the natural system over the land and in the ocean. A detailed analysis of IDV in the regions of the NH Pacific and Atlantic Oceans has been undertaken (Intergovernmental Oceanographic Commission, 1992) with the use of the data of observations and numerical modeling, which suggests the following conclusions:

(i) The IDV is clearly manifested over these water basins and adjacent land surfaces;

(ii) The IDV over the Pacific Ocean is certainly connected with changes taking place in the tropics, but over the northern Atlantic this relationship is absent (in this case the interrelated processes of deep-water convection, fresh-water influx and ice cover formation play a decisive role);

(iii) The ecosystems of both the oceans are very sensitive to the IDV.

Further analysis is needed of the factors determining the IDV and especially the ratio of the contributions of natural and anthropogenic factors, based on the

VOLCANIC ACTIVITY AND CLIMATE

development of both observational systems and techniques for numerical modeling of the interactive system "atmosphere-ocean-cryosphere-biosphere".

Of great interest are the NOAA annual reports which characterize the current global climate change based on the use of available observational and satellite information. Especially informative is the 1992 report, when the climatic dynamics was determined by a combination of such powerful factors as the Pinatubo eruptions and the ENSO event. During the first half of the year a positive temperature anomaly was observed in the NH due to a powerful ENSO event taking place in 1991–92; whereas in the second half of the year a cooling took place caused by the Pinatubo eruption. Therefore the year 1992 was the coolest over the post-1986 period, though it belonged to the succession of the warmest years on the centennial scale. The eruptive effect on the ozone layer was essential, especially in the Antarctic, where the springtime total ozone content (TOC) decrease was the strongest over the whole period of observations.

1.2 Cause and Effect Studies

Although the problem of explaining causes of long-term climatic changes is still unresolved, there are various ideas concerning the nature of such changes (Phillips, 1992; Randall et al., 1993; Wang et al., 1991 and many others).

Kellogg (1993) noted that the absence of global climatic warming between 1940 and 1975 was sometimes considered a "moratorium" with respect to "greenhouse" climatic warming during the last century. He suggested that this 35-year anomaly can be explained by oceanic circulation effects. Around 1940, a large-scale change in Atlantic and Pacific oceanic circulation took place. Subsequently, this was followed by an intensified transport of relatively warm water to deeper or bottom ocean layers. However, further circulation changes around 1975 have stopped the downward motion of warm water.

This supposition is confirmed, in particular, by the fact that between 1975 and 1991, a warming of the NH Atlantic and Pacific oceans took place at depths of 0.5–3.0 km. To warm an oceanic water volume of over 100 mln km^3 by 0.1 °C during 1957–1991 should have taken about 8×10^{22} joules. Since the additional energy input, due to the greenhouse effect, constituted about 0.5 W/m^2 for a tropical ocean surface area of 10 mln km^2, the 35-year energy input would have been much greater than what was needed to warm the oceanic layer. Thus, it can be supposed that the absence of a SAT increase during this "moratorium" (1940–1975) resulted from the transport of additional "greenhouse" warmth to deeper oceanic layers.

Though some meteorologists believe this century's climatic warming has been due to anthropogenic factors (due to increasing GHG concentrations), the assumption of natural climatic warming becomes more plausible. Related to this, Gray et al. (1991) produced qualitative estimates which suggest that Atlantic conveyor belt alterations may possibly contribute to global multidecadal warming.

During periods when the Atlantic portion of the global conveyor belt intensifies, atmospheric general circulation substantially differs from that observed when conveyor belt circulation is weaker than normal. According to observational data, periods of strong conveyor belt circulation are characterized by SST increases in the northern Atlantic, enhanced precipitation in the Sahel region and hurricane formation in western Atlantic low latitudes. Due to these circulatory phenomena, greater warm-water volumes are transported from the tropical oceans to North America, where these waters sink to great depths (due to their high salinity), resulting in a global-scale SST decrease (see Table 1.2.1).

Table 1.2.1. Typical Meteorological Situations for the Following Four Multidecadal Periods: 1879–1899; 1900–1942; 1943–1967; 1968–1992. After Gray et al. (1991).

	1879–1899	1900–1942	1943–1967	1968–1992
Atlantic Conveyor Belt	Strong	Weak	Strong	Weak
Sahel rainfall	Wet period	Dry period	Wet period	Dry period
Atlantic hurricane activity	Enhancement	Weakening	Enhancement	Weakening
El Niño activity	Weakening	Enhancement	Weakening	Enhancement

Apart from these processes, for cooling to take place it is important for warm water input from the western sector (heat source) to the eastern sector of the Pacific Ocean and the SH oceans to decrease. This situation can happen during the cold phase of El Niño. The formation of strong conveyor belt circulation in the Atlantic Ocean leads to a gradual decrease of global SATs. Apparently, such conditions took place during the periods 1870–1948 and 1969–1992. Table 1.2.2 characterizes the events that took place in the Atlantic conveyor belt.

VOLCANIC ACTIVITY AND CLIMATE 41

During the last two decades, an especially strong reduction in Atlantic conveyor belt activity has taken place. This has been followed by decreased precipitation in the western Sahel, greater hurricane activity and an intensified El Niño. Table 1.2.2 shows some quantitative characteristics of the Atlantic conveyor belt with respect to mean-square deviation (MSD) for a 100-year period. A strong correlation testifies to the existence of a cause-and-effect interrelationship.

Table 1.2.2. Quantitative Characteristics of the Relationship Between Atlantic Conveyor Belt Strength and Meteorological Conditions. After Gray et al. (1991).

Time period	1870–1899	1900–1942	1943–1967	1968–1991
Atlantic conveyor belt strength	Strong	Weak	Strong	Weak
MSD for western Sahel rainfall	0.35	0.21	0.37	0.70
Western Atlantic low-latitude hurricane activity	154%	70%	148%	51%
El Niño activity	58%	129%	70%	145%
Global mean SAT trend	Decrease	Increase	Decrease	Increase

Apparently, global warming, which began in the late 1960s, has been related to weakening Atlantic conveyor belt circulation and cannot be attributed to increased greenhouse gases (GHG) concentrations. On this basis, the assumption can be made that the nature of global warming was not, and will not be (within the next 50–100 years), anthropogenic. As for the opposite conclusion drawn from numerical climatic modeling, emphasis should be placed on the problem of model adequacy, especially when parameterization of the hydrological cycle and cloud radiation interaction are concerned.

As has been mentioned above, approximate estimates have led to the conclusion that greenhouse global warming can induce NH ice sheet growth, since it is followed by increased atmospheric water content and, hence, intensified snowfall over landmasses. On the other hand, if the summer temperature increase is not sufficient to melt glaciers, they will mature. Ledley and Chu (1993) tested the reliability of these assessments for the NH using a coupled energy-balance climate model and a 3-layer thermodynamic sea ice model (CCSI) over a 10° lat. by 10° long. grid. This accounts for the hydrological cycle and land surface heat balance, making it possible to perform snow accumulation and ablation numerical modeling and, thus, ice cover thickness. The CCSI model considers four climatic system components: the atmosphere over landmasses; the atmosphere over oceans; mixed ocean and soil layers. Calculations for the control case have shown that the model reliably simulates different latitudinal SAT changes, except for 65° N, where summer SATs were underestimated, when compared to those actually observed. As for precipitation, calculations show that, in present climatic conditions, land summer ice melting should take place.

A CO_2 doubling (it is taken to be equivalent to the growth of the radiation budget at the tropopause level by 4 W/m^2) leads to the increase of the summertime SAT by 3.5 K and to a prolongation of the period of positive summertime SAT values (near 65° N from 2 to about 4 months). The temperature increase induced an enhancement of precipitation, but there is no net land ice accumulation, since in the summer the share of precipitation in the form of snow decreases, and this does not intensify the glaciers. The growth of glaciers is possible only under condition that a hypothetical mechanism can appear which would provide the snow cover preservation in the summer due to a higher albedo and the SAT values below the freezing point.

Duffy (1993) tested the motivation of the assumption that variations of solar activity and respective changes in the solar constant or in the duration of the cycle of solar activity had played the principal role as factors of global climate change during the last century. With this aim in view, analyses were made of time lags between changes in the solar forcing and observed temperature changes. The data on SAT for both hemispheres, SST, and the lifetime of the summer ice cover around Greenland served as a source of information.

An analysis of these data has shown, in particular, that variations in the 11-year cycle of solar activity could not have led to temperature changes, since the latter precede. The opposite situation is with variations in the duration of the 11-year solar cycle, which precede the temperature variations, but the time shifts do not agree with the estimates based on approximate climate models. Thus the results discussed reject the hypothesis of the sun-induced climate change.

A considerable portion of the previous studies of correlations between solar activity and climate has been based on analysis of correlation between SAT and relative sun spots number (RSSN). However, these correlations are unstable and even change their sign. Therefore Stellmacher and Menda (1992) processed the data on the NH SAT anomaly over the years 1851–1987, the sums of winter temperatures in Berlin between 1766/67 and 1989/90, and monthly mean RSSN during the period 1749–1990, using the length of the 11-year cycle as an indicator of solar activity.

The initial stage of data analysis consisted in comparison of Fourier-spectra for all observational series. The RSSN spectrum reveals periods of 10.9, 21.3 and 89.9 years, and the prevailing period 11.1 years is typical of SAT anomalies. A comparison of dependencies of the "solar melody" (time variations of the frequency of the cycle of sun spots) and SAT anomalies on the duration of the 11-year cycle has demonstrated a quasi-synchronous variability with a time shift of 7 years, on which the effects of volcanic eruptions and anthropogenic variations of atmospheric transparency are imposed.

The discussion of the causes of global mean SAT changes has drawn attention to a possible effect of extra-atmospheric solar irradiance (solar constant – SC) variations. In this connection, Schlesinger and Ramankutty (1992) calculated variations in hemispherical SAT, caused by a CO_2 doubling (ΔT_{2X}), and in the temperature of the ocean at different depths, determined by variations of the radiation budget of the surface-troposphere system.

Three versions of ΔT_{2X} calculations have been examined when considering the forcing by:

(i) GHGs with and without account of tropospheric sulfate aerosol (TSA);
(ii) SC variations with and without account of TSA;
(iii) GHG + SC variations with and without account of TSA.

The results of numerical modeling have been compared with observed global mean SAT anomalies with respect to averages for the period 1951–1980.

The principal conclusion drawn from the results is that though the intercycle variations of SC (i.e., an 11-year cycle of solar activity) have markedly contributed to SAT variations, beginning from 1856, the variability of the GHG content in the atmosphere has been a dominating factor. After 1894 the GHG contribution had exceeded the effect of SC; by the year 1982 the difference between these contributions to SAT variations had reached 0.19 °C. However, these conclusions cannot be considered reliable: direct satellite SC measurements of long duration (during several 11-year cycles) are needed, which will enable one to realize direct (and not indirect) assessments of the effect of SC variations on the SAT formation (see Kondratyev and Nikolsky, 1995a, 1995b).

1.3 Satellite Observations

The space-derived meteorological data become more and more informative. These data series, that started the "count" of the fourth decade, made some climatological generalizations possible.

Analysis of the NOAA MSU data, beginning from the late 1978, revealed the long-term stability of channels 2 (53.74 GHz) and 4 (57.95 GHz), with weighting function maxima in the troposphere and lower stratosphere, respectively, which determined the possibility to use these data for climatological generalizations. Comparisons of the data from six satellites over the years 1978–1991 revealed differences averaged over 10 years, in radiobrightness temperature (T_B) within 0.01–0.03 °C. The comparison of measured T_B monthly means for channel 2 for 10 years (1979–1988) with T_B calculated from the data of aerological measurements (averaged over squares 2.5° lat. by 2.5° long.) revealed a very high correlation (the correlation coefficient 0.95–0.97). The observed T_B, with nadir viewing, correlates best with the layer 1000–200 hPa temperature. The combination of the data for several viewing angles provides the highest correlation with the layer 1000–400 hPa.

Using the NCAR climate model CCM-1, Christy et al. (1991) and Christy (1993) calculated the change of monthly mean global mean SAT and mean TT between 1978 and 1987 with prescribed variations of SST over the grid 4.4° lat. by 7.5° long. from the observational data, which are quite scarce in the SH (and completely absent south of 40° S).

Comparisons of conventional (SAT) and satellite (TT, MSU) observations revealed a low correlation: the correlation coefficient for seasonal SAT values constituted only 0.32, which, apparently, resulted from inadequate parameterization of land surface processes. In the case of TT the correlation was higher (0.53). The model atmosphere can assimilate the heat from the ocean during strong positive anomalies of SST (1979–90, 1982–1983) but does not emit this heat to space as rapidly as the real atmosphere. On the whole, the results of numerical modeling show that SST anomalies can explain about 25% of the observed TT variability.

The data of surface observations of SAT and marine observations of SST are the most widely used source of information about the current variations of global climate. In connection with an urgency of analysis of the quality of the available database, Spencer and Christy (1992) assessed the reliability of the results of conventional marine observations, comparing them with the results of satellite observations made with the NOAA MSU.

The results of observations from six satellites from 1979 have given a homogeneous global database over the grid 2.5° lat. by 2.5° long. The data in Table 1.3.1 characterize the correlation coefficient of SAT-SST (conventional observations) and the results of satellite observations for MSU channels (1000-

200 hPa) and 2R (combination of MSU data for several viewing angles, which determines the temperature of the lower troposphere).

Table 1.3.1. Coefficient of correlation between the conventional and satellite observations. After Spencer and Christy (1992).

Region	No. 2	No. 2R	σ_{MSU}	σ_{SFC}	S
Globe	0.60	0.63	0.21	0.13	98
NH	0.52	0.65	0.21	0.19	90
SH	0.55	0.45	0.23	0.11	67
North America	0.90	0.95	0.69	0.32	100
Eurasia	0.77	0.89	0.40	0.95	96
Australia	0.50	0.78	0.47	0.49	92
Northern Atlantic	0.47	0.48	0.37	0.22	100
Southern Atlantic	0.30	0.23	0.31	0.24	77
Northern Pacific	0.37	0.49	0.40	0.32	100
Indian Ocean	0.56	0.56	0.30	0.22	95
ESTPO	0.82	0.83	0.54	0.58	80
WSTPO	0.10	0.17	0.27	0.18	93

σ_{MSU} and σ_{SFC} are MSD values (°C); S is the share (%) of the respective surface covered with observations; ESTPO and WESTPO are east and west sectors of the tropical Pacific Ocean.

Analysis of the causes of low correlation, illustrated by the data shown in this table, requires, first of all, an estimation of the errors of observations. In the case of MSU the noise level (0.005–0.023 °C^2) never exceeds 10% variability (0.09–2.3 °C^2), and the standard error for the chosen grid varies from 0.07 °C (the tropics) to 1.5 °C (high latitude continents). Estimations of the errors of conventional observations are more difficult. Calculations have shown that in this case the variability of the errors reaches 1.39 °C^2, and the standard error of an individual observation constitutes 1.2 °C (naturally, it decreases by a factor of $\sqrt{N}$ for N observations). The results obtained show that the earlier conclusions about the temperature trends in the past should be treated cautiously, since they are very uncertain due to lack of acceptable representativity of the observational data.

Hurrell and Trenberth (1992) undertook a comparison of the data on monthly mean brightness temperatures (BT) anomalies obtained from the MSU data for the period 1979–1990 (144 months), with the data of the ECMWF on the air temperature monthly means for the period 1982–1989 (96 months), both for individual pressure levels and averaged over various atmospheric levels, with account of weighting function for channel 2. The comparison revealed a high correlation (with the correlation coefficient more than 0.9) between the MSU data and ECMWF data weighted over most of the globe, except for the tropics, Southern Atlantic and NH high latitudes.

The best agreement between the data of 4-D assimilation obtained at the ECMWF, and MSU data takes place in the regions for which the most reliable results of aerological sounding are available, whereas in the regions where the objective analysis was based on satellite information, the correlation was weaker. The latter can be explained, apparently, by systematic errors of the retrieval of the temperature field from satellite data on outgoing long-wave radiation (OLR). A certain contribution was also made by the changed techniques of 4-D assimilation and forecast, regularly practiced during 10 years (this could be manifested more strongly in the tropics and in the regions of deficient data of observations).

As for the results of comparisons of the data of MSU and ECMWF for individual levels, the correlation turned out to be higher at 300 hPa (for the globe), where, probably, the effect of changing the techniques of the ECMWF data processing was weaker. In the regions with more adequate observational data (the NH land, Australia) there is a high correlation between anomalies from the MSU and ECMWF data at every level in the troposphere up to 200 hPa. Thus the conclusion can be drawn that the MSU data are a valuable source of information for the characteristics of the temperature field fluctuations in the global troposphere on time scales from a month and longer (the problem of the MSU data interpretation still stays controversial, however).

Hurrell and Trenberth (1992) performed also a special analysis of the ECMWF data on the temperature in the tropics starting from 1982. The introduction of a new initialization technique in September 1982 (with account of diabatic and linear character of normal modes) has led to a considerable temperature increase in the middle troposphere of the tropics, especially at the 500 hPa level.

In May 1985 the temperature in the tropics at 700 hPa (850 hPa) levels increased (decreased) after introducing a spectral prediction model T106 with respectively changed parameterization of physical processes. The temperature near the tropopause in the tropics decreased substantially after the vertical resolution of the model increased in May 1986 from 16 to 19 levels (including three new levels in the stratosphere). The reliability of the data on the temperature field at 1000 hPa level for the whole period testifies to the importance of processing the observational data by use of the most adequate 4-D assimilation technique.

Shea et al. (1993) have substantiated a new global $2°\times2°$ monthly SST climatology, primarily derived from 1950–1979-based SST climatology from the Climate Analysis Center (CAC), which has been modified to include data from the COADS data set to improve the SST estimates in the regions of the Kuroshio and the Gulf Stream. The Shea-Trenberth-Reynolds (STR) climatology has been compared with the Alexander and Mobley (AM) SST climatology often used as a lower boundary condition in climate models. Important differences have been discovered. Generally, the STR climatology is warmer in the NH and in the subtropics of the SH during the northern winter. It is often colder south of 45° S in all months. The largest differences are more than 5 °C in the Kuroshio and Gulf Stream regions, and in the mid- to high-latitude southern oceans, the SST's are often more than 2 °C lower. It is important that the STR climatology is temporally and spatially less noisy than the AM SST climatology.

The longest global SST anomalies during the 1982–90 time period have been associated with the El Niño (1982–83 and 1986–87) and La Niña (1988) events in the tropical Pacific. Caution must be used, however, in interpreting the CAC SST anomalies for 1982 to 1990 as true anomalies, because, in certain regions, the anomalies partially compensate for the deficiencies in the CAC climatology.

On the basis of MSU data for channels 1, 2 and 3, Spencer (1993) estimated ocean precipitation on a 2.5° grid for the period 1979–1991. Satellite data were calibrated through a comparison with data from 5 to 10 years of globally distributed low-elevation island and coastal rain accumulation measurements from 132 gauges. Comparison with other oceanic rainfall climatology has revealed several important differences. The largest discrepancy between the MSU and other climatologies is in the eastern Pacific Intertropical Convergence Zone (ITCZ), where the MSU indicates up to 88 mm/day more rainfall than the Geostationary Observation Environmental Satellite (GOES) precipitation index (GPI) data. On the other hand, MSU and GPI depictions of seasonal rainfall variability associated with the ENSO warm event of 1991–92 show good agreement in the large-scale patterns.

For some reasons of special interest are satellite data on total water content – precipitable water (PW) of the atmosphere obtained during the last 10–15 years:

(i) The dynamics of water content reflects connections among evaporation, clouds and precipitation, as well as latent heat release, which is of key importance for the formation of climate variability in the tropics and for long-distance correlations between the processes in the tropical and extra-tropical latitudes;
(ii) water vapor contributes most to the atmospheric greenhouse effect and determines the feedback functioning, whose account is of critical importance for assessing the climate warming;

(iii) the Earth's radiation budget (ERB) formation is considerably determined by the water vapor content both directly and through the effect on cloud formation.

These circumstances have stimulated the appearance in 1990 of the Global Energy and Water Cycle Experiment (GEWEX) Water Vapor project (GVaP) for studies of meteorological and hydrological processes. In this connection, Bates and Stephens (1991) performed an analysis of possibilities to retrieve the PW from the data of infrared (IR) and microwave observations using the following space-borne instruments: SSMR, channel 22 GHz (data for the period 1979 to 1983); SSM/I, channel 22 GHz (1987 to the present); HIRS/2, channels 6.7, 7.3, and 8.2 µm (1978 to the present); AVHRR/2, channels 11.1, 12.3 µm (1981 to the present).

A comparison of the observational results for the period January 31 – February 5, 1993 suggested the following conclusions: the IR data are characterized by underestimated PW (compared with microwave information) in the regions of overcast and broken cloudiness in the tropics and subtropics and overestimated values in the subtropical regions of downward motions. Since the PW field from the IR data is almost zonal (in contrast to that revealed by microwave observations), this illustrates serious differences between IR and microwave information from the viewpoint of estimates of moisture transport from the tropics to midlatitudes.

From the NOAA satellite data on the OLR for 11 years (1979–1989) in the 6.5-µm water vapor band, which serve as an indicator of water content in the upper troposphere, and for channel 11 µm in the atmospheric transparency window, Chesters and Neuendorfer (1991) considered the global dynamics of water content as a function of the low-latitude convection intensity characterized by the brightness temperature (T_B) at the wavelength 11 µm. The data of the IR remote-sounding radiometer HIRS for channels 6.7 and 7.2 µm were used to calculate T_B corresponding to the center of the 6.5 µm band using the relationship $T_B(6.5) \approx 1.5 \cdot T_B(6.7) - 0.5 \cdot T_B(7.2)$. This T_B combination corresponds to a weighting function with a maximum near 350 hPa.

An analysis of global maps of T_B monthly means (for the whole 11-year period) reveals the regions of $T_B(11)$ minima in low latitudes which reflect the presence of high-top powerful convective clouds (for example, in the regions of monsoons in the northern Indian Ocean or Central America in July – September and in Indonesia in January – February). As a rule, low values of $T_B(6.5)$ are observed, except for the regions of prevailing downward motions in the upper troposphere (the Middle East, southern Indian Ocean. southeastern sector of the Pacific Ocean in June – September; Central Pacific Ocean in the NH in January – February). The dynamics of the $T_B(6.5)$ global field contains almost no sign of the effect of ENSO on the water content of the upper troposphere. The variability of $T_B(11)$ values averaged over the Eastern and Western Hemispheres is characterized by a clear-cut semi-annual change with a phase shift of 180° in different hemispheres. The semi-annual change of $T_B(6.5)$ is weakly manifested.

Han et al. (1993) developed a technique to retrieve the size distribution of cloud droplets from satellite observations of reflected solar radiation with prescribed γ-distribution of the droplet size (clouds were parameterized as horizontally homogeneous layers). This technique was used to process the NOAA-9,10 data for the period 1987–1988 covering the latitudinal belt 50° S – 50° N. The observation data considered include the results of measurements for all five channels of the advanced very high resolution radiometer (AVHRR) scanning radiometer and the needed information on the vertical profiles of temperature and humidity from the Tiros Operational Vertical Sounder (TOVS) data. Since the reflected radiation for channel 3 (the effective wavelength 3.7 µm) depends strongly on an average size of cloud droplets, the data for this channel have been used to retrieve the size of the droplets; however, the solution of this problem has been hindered by a marked contribution of thermal emission, which should be filtered out (generally speaking, it is more reliable to use 3-channel data for wavelengths 1.6, 2.2, and 3.7 µm). The solution of the considered inverse problem is based on the use of four tables containing the results of direct calculations which enable one to interactively simulate the average size of droplets, filtering out the contributions of two other factors (thermal radiation and shortwave radiation reflected from the surface).

From the January 1987 data, the global mean radii of droplets are 9.7 µm for the ocean (the smaller size of droplets for the land agrees with the data of aircraft observations). However, in the tropics the average size of the droplets of continental and marine clouds is practically the same. Apparently, this is connected with the fact that intensive precipitation around the Amazon basin and Indonesia determine the washing out of CN, which leads to the growth of droplets, and that the marine clouds near tropical Africa have similar droplets due to the effect of continental air masses.

These circumstances (and, probably, the effect of anthropogenic aerosol) determine the fact that the average size for cloud droplets in the NH is less than in the SH (this is valid for both continental and marine clouds), which brings forth some differences between the hemispheric albedos. An attempt to reveal the effect of emissions of dimethylsulfide by the ocean on the average size of droplets (through the formation of sulfate CN) has not given positive results. The data in Table 1.3.2 illustrate the estimates of total water content of droplets.

Table 1.3.2. Total Water Content of Clouds Over Land and Ocean, g/m^2. After Han et al. (1993).

Month	January	April	July	October
Ocean	76.8	80.3	85.5	74.5
Land	85.8	86.7	83.3	74.5

In this case there are no regular "land-ocean" contrasts, which testifies to the prevailing role of the regional-scale processes.

An important aspect of global climate diagnostics is the analysis of the stratospheric climate. In this context Spencer and Christy (1993) have analyzed a continuous global record of deep-layer averaged lower stratospheric temperatures during 1979–1991 on the basis of MSU channel 4 (57.95 GHz) data. A 13-year record of temperature anomalies is time averaged into pentads and months on a 2.5° grid. The monthly gridpoint anomalies are validated with 10 years of radiosonde data during 1979–1989. The calibration stability was very high: no measurable instrumental drift at the level of 0.01 °C per year. Radiosonde comparisons to the monthly gridpoint anomalies have correlations ranging from 0.90 in the tropics (where the interannual variability is smallest) to as high as 0.99 at high-latitude stations. The corresponding standard error of estimate is generally around 0.3 °C.

The largest globally averaged temperature variations during 1979–91 occurred after El Chichón (1982) and Pinatubo (1991) volcanic eruptions. These warm events are superimposed upon a net downward trend in temperature during the period. This cooling trend is more of a step function than linear character, with the step occurring during the El Chichón warm event. It is strongest in polar regions and the NH middle latitudes. Spencer and Christy (1993) have emphasized that such characteristics are qualitatively consistent with radiative adjustment expected to occur with observed ozone depletion.

From the data of Nimbus-6,7 ERB wide-angle sensors, Rutan et al. (1991) drew global maps of monthly mean albedos and absorbed solar radiation (ASR) for 140 months (1975–1987) over the grid 5° lat. by 5° long. For example, they drew a map of ASR for December averaged over 12 years, which illustrated a comparatively smooth ASR field with maxima in the tropical belts of the oceans. The map of ASR standard deviation for 12 years show maxima in the tropics of the Pacific Ocean, equatorial Atlantic Ocean and middle part of the Indian Ocean. A maximum of about 25 W/m^2 in the central Pacific is explained by the ENSO events taking place in 1982 and 1986.

Smith et al. (1991) analyzed a 12-year series of the ASR monthly means obtained from the data of Nimbus-6,7 wide-angle sensors. An expansion of time series in empirical orthogonal functions (EOF) showed that in 1983 and 1987 considerable anomalies of ASR due to the ENSO effect were observed, the events being characterized by similarly distributed ASR, though, from the viewpoint of spatial and temporal variabilities, the 1987 ASR anomaly was stronger. The contribution of the first EOF into the retrieval of the ASR variability constitutes 0.9106 (those of the second and third EOFs are 0.0283 and 0.0122, respectively), and upon filtering out the annual change - 0.102. The EOF-1 and 2 global maps calculated with the annual change filtered out, reveal a similar dipole structure of the spatial distribution, with one pole over Indonesia, the other over the equator near the date-change line. The distributions of EOF-1 and 2 are also characterized by some differences.

The data of satellite observations obtained with the ERBE, TOVS and Special Sensor Microwave/Imager (SSM/I) instruments make it possible to analyze the dependence of OLR on the total content of water vapor (WV) in the atmosphere and, thereby, to assess the role of WV as a principal GHG as well as its forcing with respect to the longwave radiation field determining the greenhouse warming of the atmosphere. Randall et al. (1993) discussed the results of analysis of the OLR values in the knots of the 2.5° x 2.5° grid for each day of July 1988 from the ERBE data and the WV values from the data of the microwave SSM/I sounder.

To assess the WV forcing, the WV monthly means were found for 12 months for the period March 1988 – February 1989, which enabled one to draw a map of the annual mean WV minima. Processing of the OLR and WV data for July revealed a negative linear correlation between these parameters: the OLR increases with decreasing WV at a rate of 1.71 W/m² per 1 mm of the WV decrease. Based on the estimates of such correlation for 5° squares, covering the globe, the global distribution of OLR values corresponding to the absence of WV in the atmosphere, was obtained using an extrapolation: The WV forcing can be calculated from

$$Wv_f = [WV_{zamin} - WV_{obs}] \cdot SN, \qquad (1.1)$$

where WV_{zamin} is the zonal average minimum WV; WV_{obs} is the July averaged water vapor; SN is the sensitivity of the OLR to WV. Drawing of the WV_f global map for July 1988 revealed a maximum in the central Pacific Ocean and in the Atlantic Ocean (45 W/m² in both cases), as well as in the ITCZ of the SH Pacific Ocean. The zonal mean profile of WV_f is characterized by maxima near 25° N and 15° S. A strong decrease of WV_f values is typical of the equatorial belt.

To give a complex characteristic of climate change from satellite observations, Campbell et al. (1993) performed a statistical analysis of the global database for the period July 1983 to June 1990 (monthly means have been considered with a spatial resolution of not more than 2.5° lat.), including the following data: the MSU channel 2 (temperature of the upper atmosphere) and channel 4 (temperature of the lower atmosphere), SST, total cloud amount, OLR.

For the analysis the time series used were obtained after filtering out the annual change and normalizing against the global mean standard deviation. The existence of such a homogeneous initial database has made it possible to accomplish a complex analysis of the whole totality of data from the viewpoint of spatial and temporal variability in terms of EOFs.

An analysis of time series revealed a strong variability, in which the contribution of MSU data for polar regions prevail. The confinement to the belt 30° S – 30° N revealed in the variability of the first EOF a distinct manifestation of El Niño in 1982, 1987 and 1991. Calculations of the global distribution of the first eigenfunction (EF) demonstrated their similarity for the meteorological parameters mentioned above. The SST variations contribute considerably to the total variability.

In the case of EOF the results for individual parameters are similar, but not identical, since in the tropical latitudes the variability is determined mainly by the contribution of El Niño. This explains a specific character of the temporal variability of the first EF for stratospheric temperature (interacting weakly with the troposphere), in contrast to correlations between all the other parameters. To a certain degree, the El Niño effect influences however, the stratosphere, bringing forth temperature fluctuations in counter-phase with those observed in the troposphere.

Though the sensitivity of climate changes to variations in the snow cover extent has been adequately analyzed from the data of numerical climate modeling, the respective data of observations have been fragmentary. In this connection, Gutzler and Rosen (1992) studied the regularity of the interannual variability of the snow cover extent (SCE) across the NH, as well as the relationship between the SCE dynamics and large-scale circulation anomalies. The analysis was based on satellite data on the SCE monthly means for December, January and February, when the SCE was at a maximum, and the relationship between the SCE dynamics and the AGC anomalies manifested most clearly.

The most important result from analysis of the SCE maps (89x89 grid knots within the NH) drawn from the results of a subjective analysis of images in the visible consists in detecting a considerable positive correlation of the interannual variability of the wintertime SCE over the continent of North America and Eurasia.

Assessments of the SCE trends for a 19-year period have not given positive results in view of negligible trends: the intercontinental correlations of SCE are determined by interannual fluctuations and not by the presence of the trends. As has been shown above, the multiyear trends of SCE are observed in other seasons. So far, it is difficult to judge which factors determine such seasonal differences: the annual change of the present global warming; the seasonal character of relationship between SCE and temperature, or simply the observational errors. This seasonal dependence is, however, very important from the viewpoint of monitoring the SCE as an indicator of climate change.

An analysis of SCE variability on subcontinental scales revealed very strong interannual SCE variations in Europe, and a high intercontinental correlation between Europe and North America, stronger in December and January than in February. The SCE anomalies averaged over the continents are characterized by a high correlation in each of the three winter months as well as anomalies averaged over the eastern sector of North America and West Europe.

The highest correlation between SCE anomalies and AGC fluctuations revealed from analysis of time series, is connected with the Pacific-North American (PNA) field of geopotential anomalies, whose centers are located over the Central Pacific, northwestern North America, and southeastern USA. The structure of this field of anomalies correlates best with the SCE anomalies in the western sector of North America. The correlation between PNA fluctuations and SCE values averaged over either the territory of Eurasia or the NH, is weak, as a rule. Therefore the intercontinental correlations of SCE anomalies should not be governed by PNA fluctuations.

On continental or hemispherical scales, there is a marked but not very stable correlation between SCE anomalies and geopotential field in the region of the NAO. Previous studies had revealed a substantial correlation between fluctuations of hemispherical mean anomalies of SCE monthly means and SAT, whereas the continental regions are characterized by a distinct interannual variability of SAT, contrasting with the oceanic regions, where the variability of geopotential is at a maximum.

Based on the use of an advanced technique for NOAA data processing in the visible to calculate the SCE monthly means over the NH land, a data set has been accumulated for the period 1972–1992 over the 89x89 grid, with 7921 cells with a size from 16,000 to 42,000 km^2 (a cell is considered totally covered with snow if more than 50% is under snow). The data obtained by Robinson (1993) and given in Table 1.3.3 characterize the annual change of SCE averaged over the whole period considered, with a maximum in January and a minimum in August (in the latter case the snow was observed mainly in Greenland).

The last two decades are characterized by almost normal distribution of SCE with MSD varying from 0.9 mln km^2 in August to 2.9 mln km^2 in October. The year 1979 was most snowy (27.4 mln km^2) and 1990 was least snowy (23.2 mln km^2). Of 58 months between August 1987 and May 1992 only 5 months are characterized by a positive SCE anomaly (January 1988, September 1989, December 1989, December 1990, and November 1991). A substantial deficit of SCE was observed in the spring in Eurasia (North America) during the last 5 (6) years. During this period a decrease of SCE also occurred in the fall and summer, but the winter level was moderate.

Table 1.3.3. The annual mean course of SCE over the NH (mln km^2) for the period 1971–1992. After Robinson (1993).

SCE Month	Maximum (year)	Minimum (year)	Average	MSD
January	49.8 (1985)	41.7 (1981)	46.5	1.8
February	51.0 (1978)	43.2 (1990,92)	46.0	2.0
March	44.1 (1985)	37.0 (1990)	41.0	1.9
April	35.3 (1979)	28.2 (1990)	31.3	1.8
May	24.1 (1974)	17.4 (1990)	20.8	1.9
June	15.6 (1978)	7.3 (1990)	11.6	2.1
July	8.0 (1978)	3.4 (1990)	5.3	1.2
August	5.7 (1978)	2.6 (1988,89,90)	3.9	0.9
September	7.9 (1972)	3.9 (1990)	5.6	1.1
October	26.1 (1976)	13.0 (1988)	17.6	2.9
November	37.9 (1985)	28.3 (1979)	38.0	2.3
December	46.0 (1985)	37.5 (1980)	42.5	2.3
Year	46.0 (1985)	23.2 (1990)	25.5	1.1

Though the duration of satellite observation series constitutes only about 30 years or less, the accumulated data open up various possibilities of its climatological interpretation. Gruber and Arkin (1992) have made an overview of the available observation series useful for such an interpretation, which requires, however, a consideration of special features of satellite information. This concerns, first of all, the unsolved problems of the in-flight calibration of instruments and provision of homogeneous observation series, often produced by installation of more advanced instruments. Only in the case of ERB observations was the problem of in-flight calibrations solved, on the whole, satisfactorily.

The SCE is one of the climatic parameters supplied with adequate observational series. From November 1966, the NOAA compiles weekly digitized maps of SCE for the NH over the grid 128x128 knots (which is equivalent to the resolution 190x190 km^2 at 60° N) with the use of polar stereographic projection. There are also fewer adequate data on the thickness and water equivalent of snow cover.

The Joint Data Center for ice cover (U.S. NAVY/NOAA) issues weekly maps characterizing the sea ice cover extent (SICE) in the Arctic (from January

1972) and in the Antarctic (from January 1973). The ice cover concentration and thickness are also assessed from the data of various instruments for the visible, IR and microwave spectral intervals.

Considerable information has been accumulated on the normalized difference vegetation index (NDVI). From May 1982, weekly global maps of NDVI are being drawn, whose reliability, however, is questioned in view of the absence of absolute calibration of AVHRR sounding radiometers, the principal source of information.

During more than two decades (beginning from 1970) the NOAA issues global maps of SST, but only from 1981 has the SST data reliability markedly increased due to the use of the multichannel retrieval algorithm for processing the two-channel data in the visible and three channels in the IR spectral regions, which permitted a reliable filtering out of the effect of clouds and atmospheric correction. A comparison with the data of ship and buoy observations gave systematic differences within 0.1–0.3 °C and MSD of about 0.6 °C.

Various formats are used to archive the SST data: maps of SST monthly means over the grid 2.5° x 2.5° in the latitudinal belt ± 70°, the data of global (regional) objective analysis over the grids 1° x 1° (0.5° x 0.5°). Such information has proved very useful to analyze the SST trends and such an important event as ENSO.

However, it is still difficult to solve the problem of atmospheric correction, especially after large-scale volcanic eruptions. In this connection, the ERS-1 ATSR along-the-track scanning radiometers data played an important role in reducing the errors in SST retrievals. this scanning radiometer ensured the viewing of the oceanic areas at two different angles.

Most successful is the long-history experience of the analysis of observational data on ERB and its components (including the solar constant), as well as the use of respective results of observations to retrieve global solar radiation (GR), surface albedo (A) and atmospheric emission (AE). An experience of using the statistical (empirical) models of GR retrieval has shown that the relative MSD of calculated daytime sums of GR varies within 5% – 12% and suggested the conclusion that for reliable estimation of the daytime sums, four GR estimates are needed for different sun elevations.

The geostationary satellite data processing has ensured regular information on GR from June 1980 for 40x40 km^2 squares in grid knots 1° x 1° for the USA territories, as well as (not totally) México and South America. As in the case of GR, a considerable experience has been gained in the surface albedo retrieval.

Still more difficult is the problem of the AE retrieval in view of the lack of a simple relationship between AE and OLR. With the available data on GR, AE and A, it is possible to calculate the surface radiation budget (SRB). An analysis of the observational data has shown also a high correlation between

ERB and SRB, which makes it possible to retrieve the SRB from satellite observations of ERB.

Possibilities of retrieving the atmospheric composition from the data of observations in the visible, IR and microwave spectral regions are rather diverse. Firstly, it refers to the retrieval of TOC and vertical profile of its concentration. By the present time, a vast database on TOC has been accumulated, which makes it possible to adequately characterize its dynamics and, in particular, to monitor the variability of the "ozone hole" in the Antarctic.

The space-derived information on total moisture and water content of the atmosphere is rather representative as well as on rain rate (IR and microwave data). An extensive data series on the amount and some other characteristics of cloud cover has been accumulated for about 30 years, in this connection of particular importance being an accomplishment of the International Satellite Cloud Climatology Project (ISCCP).

Since 1987, the NOAA regularly issues distribution maps of the aerosol optical thickness of the atmosphere retrieved from the AVHRR data (channel 0.58–0.68 µm). The long series of observations of the vertical profiles of temperature and land and ocean surface temperature have been accumulated from data of temperature soundings of the atmosphere in the IR and microwave regions. The data on the motion of cloud cover fragments have been widely used to retrieve the field-of-wind vector. The prospects for obtaining satellite information on climate parameters are broad enough, especially in connection with the planned Earth Observation System (EOS) and development of small satellites to study climate (CLIMSAT).

1.4 Recognition of Anthropogenic Signals

The problem of identification of "anthropogenic signals" in global climate change is still urgent. In this connection of interest is the dynamics of the experts' opinion concerning the "greenhouse signal."

Changnon et al. (1992) compared the results of the opinion poll among American specialists in atmospheric sciences working in the northeastern USA (delphi experiments) undertaken in 1982 (62 respondents) and in 1992 (47 respondents, including 12 nonspecialists), to analyze the dynamics of perceptions of fundamental problems concerning global climate change: Is climate changing? How will it change? Has a climate change begun due to the atmospheric greenhouse effect intensification? Will climate change cause serious social conflicts? Shall we be able to adapt ourselves to climate change or measures that are needed to modify it? One group of questions referred to climate change in Illinois and Indiana during the past 10–20 years. The expressed opinions reflected correctly the observed transition (in the period

VOLCANIC ACTIVITY AND CLIMATE

around 1980) from the SAT decreasing trend to its increasing, but considerations on the precipitation dynamics happened to be less reliable.

The second group of questions requested the addressee's opinion about the quality of available information on climate change and gave the following results (in per cent with respect to the number of respondents; a bar stands for "no poll"):

	1982	1992
1. Do you consider available publications on climate intricate?		
Yes	61	63
No	39	37
2. Do you think such publications confuse people in general?		
Yes	93	87
No	7	13
3. Do you believe (upon getting acquainted with available publications) that they are convincing and that the climate will change?		
Yes	18	47
No	82	53
4. Do you think climate change will affect you seriously?		
Yes	87	13
No	13	87
5. Do you believe the greenhouse impact on climate has intensified?		
Yes	–	51
No	–	49

The results of this opinion poll are symptomatic: on the one hand, most of the respondents consider the existing perceptions of global climate change to be intricate; and on the other hand, the number of specialists believing in the reliability of conclusions about future climate change has increased, and half the respondents in 1992 believe that the "greenhouse" climate change has begun (though, in fact, the problem of detecting the greenhouse signal in climate change is far from being solved).

The recognition of the greenhouse signal is impossible without adequate observational data and understanding of those causes that are responsible for atmospheric greenhouse trapping (AGT) variability. Karl et al. (1993) have analyzed the present-day situation with regard to weather observing and the management of weather data to formulate principles to be accomplished so that weather observing systems could serve (through their enhancement) as climate observing systems with a perspective to recognize the greenhouse signal.

Hallberg and Inamdar (1993) have investigated the cause and effect aspect of the AGT variability, the AGT being defined as the difference between infrared emissions from the Earth's surface and infrared emissions from the top of the atmosphere OLR under clear sky conditions. The surface emission is calculated from prescribed SST data and OLR is taken from satellite measurements. The calculations show that the AGT at the same SST is greater in the winter than in the summer over temperate oceans. The opposite is true for subtropical latitudes. As Hallberg and Inamdar (1993) have emphasized, at surface temperatures above approximately 298 K, atmospheric greenhouse trapping is found to increase even more rapidly from regions of lower SST to regions of higher SST than surface emissions.

Four cases of such a "super" greenhouse effect have been explored. The two significant contributing processes are water vapor continuum absorption and the growth of the atmospheric water content with rising temperature (at constant relative humidity). Another process required to explain the observed AGT enhancement is that the atmosphere (in particular the upper and middle troposphere) must be increasingly moist over the highest SST, while the atmospheric vertical temperature profile becomes increasingly unstable. Regions with such high SST are also increasingly subject to deep convection, which suggests that the convection moistens the upper and middle troposphere in regions of convective activity relative to nonconvective regions, resulting in the super greenhouse effect.

By the initiative and with financial support of the U.S. Department of Energy, an international group of experts Pennell et al. (1993) undertook development of a methodology to detect the greenhouse signal in climate change due to increasing concentrations of CO_2 and other GHGs in the atmosphere. This methodology foresees both an analysis of observational data and numerical climate modeling. Though the fact of global climate warming during the last century raises no doubts, it is still unclear (as has been mentioned above), to what extent the observed climate change is determined by natural variability (internally determined low-frequency variability of the climate system) and anthropogenic forcing. In this connection, an important problem consists in substantiating the signatures, most sensitive to the greenhouse signal, with account of the signal-to-noise (S/N) ratio.

First attempts to apply the signature approach have not led to a reliable detection of the greenhouse signal. Hopes were set, for example, on the use of the latitudinal enhancement of the greenhouse forcing. It has become clear, however, that the low-latitude warming is also substantial.

The principal reasons of the abortive attempts to recognize the greenhouse signals are the following:

(i)   Climate model inadequacy;
(ii) Climate models do not consider the mutual compensation of some important factors (especially the greenhouse warming and aerosol cooling);

(iii) Consideration of a very limited set of climate parameters (mainly SAT);
(iv) Lack of knowledge about climate noise, especially on the scales of decades and centuries, whose consideration is essential to account of the greenhouse signal being masked with natural variability;
(v) Inadequately coordinated efforts in solving this problem.

A new stage of development can be based on serious progress reached in recent climate studies:

(i) Considerable progress in development of coupled climate models, considering the interaction among the components of the system "atmosphere-ocean-land-cryosphere-biosphere";
(ii) Assessing the substantial climate-forming role of sulfate aerosol, especially in the NH, as well as the fact that the use of the equivalent CO_2 concentration conception (instead of a separate consideration of all important GHGs) is incorrect;
(iii) Considerably improved techniques to parameterize various climate-forming processes;
(iv) Successful efforts to compare and verify various climate models;
(v) Development of more powerful computers which make it possible to increase the spatial resolution of the models, to increase the time range of calculations and apply a more detailed parameterization of physical and chemical processes;
(vi) Progress in climate diagnostics based on the use of more complete observational data archives, including the paleoclimatic evidence;
(vii) New achievements in development of mathematical techniques to recognize the signals;
(viii) Broadened and deepened cooperation in climate studies.

In addition to the scientific and technical facts enumerated above, attention to climate problems must grow in view of their serious socio-economic aspects which require scientifically grounded political decisions.

The developments aimed at solution of the problem of an early recognition of anthropogenic climatic signal (ACS) are being realized in the four principal directions: numerical modeling, diagnostic analysis of the observation data, development of techniques and their application. The conceptual aspect of the problem is quite clear: the numerical models of the climate system should be used to simulate a signal determined by the prescribed forcing during a certain historical period, and then the present techniques should be used to process the signals and to recognize fingerprints to answer the question— "can the simulated signal be detected from the observation data?"

Of course, with such a conceptual approach (its system scheme has been proposed), of critical importance is the choice of a model or models (to reduce the level of systematic errors in the numerical modeling) with subsequent calculations for a period of the order of one century with prescribed various natural and anthropogenic forcing. The accomplishment of a test experiment for a period of several centuries (to reveal the natural low-frequency climatic variability, whose consideration is very important for the reliable identification of anthropogenic signal with the additional use of paleoclimatic data) is of special interest.

A very important stage is an assessment of the reliability of numerical modeling based on analysis of the spectra of errors for each model, which forms the basis for the determination of spatial and temporal scales for which the model is most reliable. In this case the choice of parameters most reliably simulated will make it possible to substantiate their combination (fingerprint) whose spatial and temporal variability will be a reliable indicator of anthropogenic changes.

A difficult problem is the absence of representative (long-period, diverse and reliable) observational data. In this context an actual problem is a simulation (from indirect data) of climate change for the last thousand years (Manabe and Stouffer, 1996). Obviously, the detection of an anthropogenic signal will be successful only with broadly diverse procedures taking into account such factors as S/N ratio, signal robustness, and estimation of observation errors.

Consideration of the role of cloud cover dynamics is especially important for an early detection of ACS. The interactive climate-forming role of the cloud cover dynamics is particularly manifested in low latitudes characterized by the round-the-year high levels of insolation and sea surface temperature, T_s. The importance of the analysis of the cloud-climate feedbacks in the tropics is also connected with the fact that here the forecast of greenhouse climate warming is most uncertain.

In this connection Molnar (1993) undertook a numerical modeling of climate with the use of a 18-layer 2-D (zonal) AGCM with three latitudinal zones: tropical (30° S – 30° N), and extratropical zones in the Northern (30° S – 90° N) and Southern (30°–90° S) Hemispheres. The numerical climate modeling was aimed at assessing the sensitivity of greenhouse climate change caused by a CO_2 doubling, to the water vapor content in the tropical atmosphere and to changes in the properties (particle size distribution) of cirrus clouds.

The atmospheric greenhouse effect (including clouds) is defined as $G = E - F$, where E is the thermal emission of the ocean surface, F is the OLR. The contribution of a cloudless atmosphere is determined as $G_a = E - F_c$, where F_c is the OLR for clear-sky conditions, depending mainly on the moisture content of the atmosphere.

The results of the earlier analysis of satellite observations have shown that over the tropical oceans (for $T_s > 299-300$ K) the runaway greenhouse effect can occur: a very strong increase of G (up to 20–30 W/m²) with rising T_s, compared to the regions of low values of T_s (these values are denoted as T_g). Such a "super-greenhouse" effect can be compensated, however, due to the formation of cirrus clouds with a high albedo, which determines the functioning in the tropics of the "thermostat" mechanism mentioned above, due to which the T_s does not exceed ~ 305 K. Since the conclusion about the presence of the thermostat has been drawn from a comparison of the data for periods of high and low levels of T_s in the tropical Pacific Ocean, and the existence of the runaway greenhouse effect has been revealed from the data on the annual and latitudinal change, it is clear that both these phenomena are connected with insolation variations, that is, with the forcing, which has nothing to do with the CO_2-induced greenhouse effect.

In this connection, the first problem of numerical modeling was an analysis of possibility for the run-away greenhouse effect in case of a 2xCO_2 by comparing the results of calculations with and without consideration of changing cirrus clouds over the tropical ocean as well as through calculations of climate change with a 2% increase of the solar constant (to reveal the possibility for analogy between this case and the 2xCO_2 case).

The notion of "feedback force" has been introduced

$$S = [\Delta T_f/\Delta T - 1] \times 100,$$

where ΔT_f and ΔT are the SAT changes due to a CO_2 doubling with and without consideration of the cloud feedback determined by cirrus clouds in different conditions. The data in Table 1.4.1 illustrate the most important results of numerical modeling. Since the observed values of $\Delta G_a/\Delta T_g$ and $\Delta G/\Delta T_g$ are equal, respectively, to 6.5 and ~ 25 W/m²·K, the calculated (for 2xCO_2) values of $\Delta G_a/\Delta T_g$ are close to those observed in cases 1 and 3 but not 5, to which $\beta = 2$ corresponds. The data in Table 1.4.2 characterize the variability of S.

The results obtained show that a consideration of possible feedbacks between cirrus clouds in the tropics and T_s from the data of satellite observations can substantially change the predicted greenhouse warming even on global scales. A deep convection enhancement (vertically developed clouds) can cause both strong positive and strong negative cloud-climatic feedbacks, depending on microphysical properties of cirrus clouds. It is of interest that both these possibilities more ($\beta = 4$) or less ($\beta = 2$) agree with the observed runaway greenhouse effect, pointing to the fact that the hypothesis concerning the cirrus clouds-induced "thermostat" needs further verification from the observational data.

Table 1.4.1. The atmospheric greenhouse effect sensitivity. After Molnar (1993).

Conditions for numerical modeling	$\Delta g_a/\Delta T_g$	$\Delta G/\Delta T_g$	ΔT_g
Case 1: Standard version, 2xCO_2 without cloud feedback.	6.226	4.402	1.044
Case 2: Same, but with a 2% increase of SC.	1.900	0.763	1.280
Case 3: 2xCO_2, interactive parameterization of cirrus clouds (with their fixed cloud top height), consideration of DCC/SST relation ship from observation data (DCC - deep convective cloudiness); the ratio of efficiency of SW radiation scattering to that of LW radiation absorption, $\beta = 4$.		7.641	0.612
Case 4: As case 3, but with a 2% increase of SC.	0.385	20.423	0.842
Case 5: As case 3, $\beta = 2$.	3.727	17.900	1.611
Case 6: As case 5, with a 2% increase of SC.	1.146	0.000	0.907

Table 1.4.2. Annual mean value of "feedback force" (%) for the tropics (TR), extratropical latitudes of the NH and SH, as well as for the globe (G). After Molnar (1993).

Cases	NH	TR	SH	G
1	0.00	0.00	0.00	0.00
2	-0.81	21.89	0.69	7.99
3	-4.19	-29.32	-1.50	12.47
4	1.96	-5.13	0.65	1.07
5	29.67	64.70	13.59	37.08
6	36.38	98.22	18.13	52.86

According to the preliminary results considered, it is not likely that changes in cirrus clouds due to the greenhouse effect will lead to the formation of a negative feedback. The situation should, however, be reversed, provided the particle size distribution of cirrus clouds is characterized by prevailing large particles, and this will manifest itself stronger with a 2% increase of SC than in the ($2xCO_2$) case.

It is doubtful that the "Lindzen effect" (the assumption that the deep convection enhancement that accompanies the predicted greenhouse warming) can lead, on the average, to a decrease of moisture content of the middle and upper troposphere in the tropics and, hence, to a reduction of the global and extratropical greenhouse warming. So, for example, even in the hypothetical absence of warming in the tropics an increase of annual mean temperature in higher latitudes decreases only within 20%–40% compared to the standard $2xCO_2$ case (this result depends on the assumed scheme of parameterization of the meridional heat transport). Thus the "Lindzen effect" cannot lead to a decrease of the global greenhouse warming. Further important studies in this context have been accomplished by Pierrehumbert, 1995, Schneider et al., 1996, Soden and Fu, 1995, Waliser, 1996, and others.

The results obtained testify to the necessity to continue such a complex observational programme as Tropical Ocean and Global Atmosphere (TOGA) as well as to obtain more reliable information on the moisture content of the upper troposphere through both direct measurements and satellite remote sounding.

Gutzler (1993) has attracted attention to the fact that the greenhouse effect is very sensitive to uncertainties in vertical humidity profile. The uncertainty in clear-sky outgoing infrared radiance due to water vapor uncertainties is comparable in magnitude to the purely radiative response of the tropical atmosphere to CO_2 doubling.

Though the global mean increase of SAT within 0.3–0.6 °C observed during the last century does not contradict the estimates of the warming due to the greenhouse effect enhancement, other causes of the warming are possible, including the internal variability of the climatic system. The problem of recognition of the greenhouse effect only by the SAT data is extremely difficult. The natural variability of climate can be filtered out more easily with the use of complex indicators in the form of most sensitive sets of parameters, which can be found by comparing the results of numerical modeling for real conditions and a double CO_2 concentration.

Such a comparison has demonstrated that one of the most characteristic and persistent features of climate variability in the ($2xCO_2$) case consists in tropospheric warming and cooling in the lower stratosphere, the warming being at a maximum in the upper troposphere in the tropics and in the high-latitude lower troposphere in winter. In this connection Karoly (1991) suggested using variations in zonal mean air temperature at different altitudes and latitudes with a doubled CO_2 as a complex indicator (CI) of climate change.

Based on the Geophysical Fluid Dynamics Laboratory (GFDL), NCAR and U.K. Meteorological Office (UKMO) climate models, calculations of the annual mean values of this indicator were made, which gave very close results, despite a substantial difference of the models. Table 1.4.3 gives the results of calculations of correlation between calculated and observed temperature values (data of radiosondes averaged over two 10-year intervals: 1970–1988 and 1963–1973), showing the statistical significance (%) averaged over one year and over two seasons (June-July-August, December-January-February).

Table 1.4.3. Correlation of calculated and observed air temperatures. After Karoly (1991).

Period of averaging	Model		
	GFDL	NCAR	UKMO
Year	0.65 (10%)	0.74 (5%)	0.72 (5%)
JJA	0.51	0.63 (10%)	0.63 (10%)
DJF	0.45 (10%)	0.56 (5%)	0.60 (1%)

As seen from the table, the correlation is positive; about half (one third) of the observed variability of the annual mean (seasonal) temperature values can be explained by the contribution of the greenhouse effect. An analysis of the observational data has not revealed, however, any enhancement of the warming both in high latitudes and in the upper troposphere of tropics, as follows from the numerical modeling results. Since an important property of the complex indicator (CI) should be its unequivocal property as an indicator of the greenhouse effect manifestations, an analysis has been made of the possible effect of such factors as ENSO, total increase of SST, decrease of TOC in the stratosphere. This analysis has led to the conclusion that the effect of each of the factors mentioned above differs from the complex manifestations of the greenhouse effect, though, for example, a simultaneous rise of SAT and decrease of TOC can cause similar variations.

From the data of observations at 5328 stations (including the Antarctic), Díaz (1991) performed an analysis of precipitation variability on regional (North America, Australia) and global scales. The North American and Australian continents were divided into regions with account of the amplitude and phase of the annual change of monthly mean precipitation. The data of observations point to different trends of precipitation in the tropical and extra-tropical regions of the NH (the same but weaker effect is observed in the SH). After World

War II the precipitation intensified in high latitudes, while an opposite trend was observed in the tropics. The precipitation amount near the equator varied weakly.

Since water vapor is the principal GHG, of extreme importance is an assessment of its trends in the context of anthropogenic global climate warming. In this connection, Wu and Newell (1993) analyzed variations of specific humidity in tropical regions at the surface level (from the data of 233 ground stations) as well as at levels 850 hPa and 700 hPa (from the data of radiosondes of 51 and 45 stations, respectively) for the period 1961–1989.

Table 1.4.4 illustrates the statistics of the specific humidity trends. As is seen, 99 surface stations (48% of the total) recorded a negative trend, and 134 stations a positive trend, which does not suggest the conclusion of the existence of a single trend in the increase of the atmospheric moisture content during the last 30 years.

At the 850 hPa level at 16 stations (31%) a negative trend was observed. If, however, the whole period under consideration is divided into several intervals, then, as a rule (see the table) the trends change their sign even during 20–30 years, which reflects the influence of still poorly studied (especially on a decadal scale) long-period fluctuations in the climate system. Also, some contribution could be made by changed observations techniques.

Table 1.4.4. The statistics of the trend of specific humidity. After Wu and Newell (1993).

Level	Total number of stations	Stations with negative trend			
		Total	Type 1	Type 2	Type 3
Surface	233	99 (42%)	44 (19%)	50 (21%)	5 (14%)
850 hPa	51	16 (31%)	5 (10%)	11 (22%)	0
700 hPa	45	17 (38%)	2 (4%)	15 (33%)	0
		Stations with positive trend			
Surface	233	134 (58%)	38 (16%)	82 (35%)	14 (6%)
850 hPa	51	35 (69%)	7 (14%)	26 (51%)	2 (4%)
700 hPa	45	28 (62%)	11 (24%)	17 (38%)	0

Note: Type 1 – stations with the sign-varying trend (in various periods); Type 2 – stations with either different or statistically insignificant trends in some periods; Type 3 – stations with a constant trend.

1.5 Numerical Climate Modeling Results

The most important objective of the climate diagnostics is an accumulation of the observational data to check the reliability of the climate numerical modeling results. The respective results turned out to be rather disappointing. Jastrow et al. (1990) noted that an agreement between the results of numerical modeling and the data of observations is confined to the closeness of calculated and observed increase of global mean SAT during the last century. On the whole, there are serious differences discussed in detail elsewhere:

(i) Contrary to the "greenhouse" logic, the largest share of cumulative warming by about 0.5 °C had fallen in the period before 1940, whereas an intensive increase of GHG concentration began later. From 1940 to 1960 there was a decreasing trend of SAT.

(ii) According to the data of surface observations, the 1980s were characterized by a temperature increase. This conclusion has not been confirmed, however, by satellite data for the free atmosphere. Contrary to the conclusion from numerical modeling about a more intensive warming in the NH than in the SH, the observations do not reveal such differences.

(iii) One of the most important conclusions of the "greenhouse" theory consists in the assumption about the warming increasing with latitude. However, it follows from the observations that after 1940 there was almost no warming trend in the NH high latitudes, but a substantial warming in the equatorial belt ± 20° lat. was observed.

(iv) From the "greenhouse" point of view, it is a puzzling fact that during the last 100 years no statistically substantial warming took place over the continental territory of the USA, judging from the most complete and reliable observational data.

These calculations have been confirmed by and supplemented with the results of the subsequent numerical climate modeling accomplished within Model Evaluation Consortium for Climate Assessment (1993) (MECCA Phase 1). Here is a summary of key preliminary findings from Phase 1:

(i) As CO_2 increases, the rate of global warming decreases, the increase in mean temperatures in midlatitudes being less than linear. Apparently, this effect of "saturation" (calculations have been made with the use of the CCM-1 model) is related to the saturation of the water vapor absorption bands more than those of CO_2. This result is very important from the viewpoint of determination of the acceptable upper limit for the CO_2 concentration increase.

(ii) During the last 30 years of observations in the Arctic no greenhouse warming trend has been detected, hence, the natural variability inherent in the Arctic climate should thoroughly be taken into account.

(iii) The IPCC concept of "global warming potentials" (GWP) should be revised, since CO_2 and other GHGs [methane, chlorofluorocarbons (CFC), nitrous oxide and tropospheric ozone] affect the climate differently. Assessments of globally averaged SAT values with prescribed effective CO_2 concentration and with a separate consideration of increasing concentrations of various GHGs agree well, but the regional climate changes differ substantially. So, for example, an increase of SAT in the region of the northwestern sector of North America was 20%–30% weaker in winter with a separate consideration of GHG (Model, 1993). It is also important to take into account manifestations of "synergism". For example, since reductions in CFC concentrations lead to increases in the upper tropospheric ozone, this should be followed by the enhancing greenhouse effect.

(iv) A consideration of mesoscale processes based on the use of nested models plays an important role in adequate simulations of regional climate changes.

The enumerated differences and contradictions suggest the conclusion that the theory of global greenhouse warming has still not been verified by the data of observations.

The situation in the field of numerical climate modeling determines an urgent need for further developments. In this connection, as MacCracken (1991, 1992) noted, with U.S. Department of Energy (DOE) support and with the participation of some leading Research Centers (National Laboratories in Livermore, Los Alamos and Oak Ridge, NCAR and GFDL), the Climate Change Monitoring and Modeling Program (CHAMMP) has been launched aimed at using the most powerful computers and the latest mathematical methodology to further develop the numerical modeling of climate based on an adequate account of the climate forming processes (Table 1.5.1).

For a number of reasons the present climate models cannot be used to obtain reliable enough climate forecasts. The objective of the CHAMMP, to be accomplished in close coordination with the DOE-supported Program of climate modeling for diagnostics and intercomparison, is to develop, verify and apply the new climate models. These models would completely consider the physical, biogeochemical and ecological processes on the basis of the use of current mathematics and computers, which eventually would require the computers' power to be raised by a factor of 10^2–10^4 and more. By using faster computers and improving the calculation procedure of algorithms, the power can be

Table 1.5.1. The present-day and perspective possibilities of numerical climate modeling. After MacCracken (1992).

Principal characteristics of present climate models	Perspective characteristics of climate models	Required growth of computers power factor
1. Horizontal resolution of hundreds of kilometers does not permit a simulation of the processes on the scales of regions and catchment areas.	1. Horizontal resolution of tens of kilometers sufficient to consider the processes over the catchment areas.	10^2-10^3
2. Inadequately verified and too simplified parameterization of climatically important processes (radiation, clouds, convection, hydrology, etc.).	2. Physically substantiated and thoroughly verified parameterization of all essential processes, including cloud dynamics, land surface processes, snow and ice cover dynamics, etc.	2–10
3. Limited consideration of processes in the ocean (usually only in the mixed layer); inadequately reliable consideration of the atmosphere-ocean interaction.	3. Completely interactive consideration of the atmosphere-ocean system including deep layers of the ocean and horizontal transport by sea currents.	2–5
4. Not interactive but prescribed chemical composition of the atmosphere, chemical reactions and aerosol effects.	4. Interactive approach to account for the role of the gas and aerosol composition of the atmosphere.	2–5
5. Too schematic consideration of biospheric-climatic feedbacks affecting the albedo, hydrological and other processes.	5. Thorough consideration of an interactive contribution of continental and marine biosphere to the formation of climate.	About 2
6. "Single-model" numerical modeling for a period of tens of years with a low spatial resolution.	6. "Multimodel" computations for hundreds of years to analyze the variability and predictability of climate and the climatic sensitivity to the variability of the most poorly known input parameters.	$10-10^2$

increased 10^2 times in the mid-1990s and 10^4 times by the end of this decade (only the use of the variable grid step can provide a 10-fold saving of computer time).

During the first stage (3 years) of the accomplishment of CHAMMP emphasis will be placed on improving the parameterization schemes for climatically important processes, substantiating an optimal detailed elaboration of the model and transition to the use of massively parallel computers. The next 3-year stage will be aimed at testing the reliability of new models and analyzing the role of various factors affecting the climatic predictability. The third (last) stage will result in development, verification and initial application of a new system of numerical climate modeling with a speed of about 10 terraflops. An important problem is to ensure its availability for the scientific community.

During the last 150 years, the most intensive systematic impact on global climate has been produced by the enhancing greenhouse effect of the atmosphere resulting from the growth of concentrations of CO_2 and other GHGs, reaching about 2.4 W/m², accompanied, apparently, by an opposite effect of sulfate aerosol (~ 0.6 W/m²).

Since the most serious uncertainty of the numerical climate modeling results is connected with assessments of its sensitivity to the forcing, one of the principal objectives of the UKMO program of numerical modeling was (Senior and Mitchell, 1993) an analysis of the factors that determine the climatic sensitivity (e.g., cloud parameterization) and improvement of the respective parameterization techniques, as well as studies of the long-term natural climate change from the data of numerical modeling.

Of great importance is the problem of verification of the calculated climate and its change. Since the ability of the model to simulate correctly the present climate does not guarantee the correct assessment of climate change, a comparison of the results of the numerical modeling of paleoclimate with the observational data plays an important role. In the context of this problem, Senior and Mitchell (1993) performed calculations of climate sensitivity to the parameterization of clouds through an accomplishment of four numerical experiments, which differed only in the parameterization of cloudiness within a global coupled climate model (with account of a simplified model of the oceanic mixed layer).

Experiment No. 1 foresaw the dependence of cloud amount on relative humidity, whereas in experiment Nos. 2 and 3 a less formal parameterization was used, considering the presence of both water and ice phases of clouds. Experiment No. 4 was characterized by taking into account the dependence of radiative properties of clouds on their calculated water content.

A comparison of the amplitude of the annual change of calculated and observed zonal mean values of the ERB components has shown that the data on SW radiation for experiments Nos. 2 and 3 agree with observations better than the data for experiment No. 1, while experiment No. 4 simulated best the annual change of the OLR amplitude. The differences between the data for these experiments turned out, however, to be comparatively small.

Quite different were the results of calculations of the SAT increase due to a CO_2 doubling, revealing the SAT variations within 5.4 °C – 1.9 °C (the latter figure refers to experiment No.4), which suggested the conclusion, mentioned above, that the model, correctly simulating the present climate, can be unreliable from the viewpoint of describing the climate changes. The sensitivity of climate to changes in the boundary condition (to a homogeneous increase of SST) turned out to be practically the same for all the four numerical experiments. The range of 2.8 °C – 2.1 °C appears to be a better estimate for equilibrium response to a doubling CO_2.

An analysis of the observational data during the last decades revealed a decreasing trend of the amplitude of the temperature diurnal change over the most of the continents. In this connection an analysis has been made of calculated changes in the amplitude of the diurnal change due to a CO_2 doubling with the use of a coupled climate model for the system "atmosphere-schematic mixed layer of the ocean." The results turned out, in the case of the present climate, to be similar to those observed, but the amplitude of temperature in conditions of a warm climate was overestimated, the amplitude of the diurnal change decreased weakly (by 0.26 °C) over land, and the mean SAT increased by 6.3 °C.

These results do not agree with observed decrease of the amplitude, caused by increasing minimum temperature (maximum SAT almost did not change). Such a situation suggests the conclusion that, apart from the CO_2 concentration increase, other factors (probably an increasing content of sulfate aerosol) have affected the decrease of the amplitude of the temperature diurnal change.

The intensified wintertime storms observed during the last years brought forth an assumption that they could have been determined by growing concentrations of GHGs that caused a more "extreme" climate. A detailed analysis of the numerical modeling results did not confirm, this conclusion however. The calculated storm-tracks shift northward in the Northern Atlantic and move farther eastward reflecting only changes in the spatial distribution of the storms, but not their substantial intensification. The $(2 \times CO_2)$-induced decrease of the meridional equator-to-pole temperature gradient in the lower troposphere favors a weakening of the NH high-latitude storms, but this range is compensated by the growing moisture content of the atmosphere.

Though the current AGCMs cannot identify the tropical storms (hurricanes) because of their low spatial resolution, they can simulate weaker tropical disturbances. In the analysis of the features of a large-scale circulation leading to real hurricanes, one can introduce an index which serves as a substitute to recognize a hurricane. The use of such an index for conditions of the control experiment has made it possible to simulate a realistic distribution of hurricanes (except for east Australia, where the calculated SST turned out to be much below that observed). In the ($2\times CO_2$) case the region of "surrogate hurricanes" broadens and the frequency of occurrence of hurricanes grows, but mainly due to the increasing duration of the season of hurricanes. The intensity of precalculated tropical disturbances increases, too.

The use of numerical experiments data to simulate the climate in mid-Holocene (6,000 years ago) did not reveal substantial differences, though in the case of experiment 1 the temperature was considerably systematically overestimated. The lack of sufficiently reliable respective data hinders a reliable testing of the model through comparisons with paleoclimatic data.

Bearing in mind as an eventual objective the forecast of the GHG- and (sulfate) aerosol-induced climate changes for the period till the end of XXI century, Murphy (1993) calculated climate changes for a period of 75 years with the use of the UKMO climate model with the prescribed gradual increase of CO_2 concentration 1%/year. The principal results of the numerical climate modeling are as follows:

(i) There is an asymmetry of climate warming in different hemispheres, characterized by a more rapid warming in the NH due to the more intensive manifestation of feedbacks over the land surface and the thermal inertia of the ocean in the SH, which slow down the process of warming;

(ii) A slight warming is observed in some regions of the World Ocean, especially north of the Antarctic coastline, in the Northern Atlantic, and south of Greenland;

(iii) Though it is clear that the presence of sulfate aerosol in the atmosphere should compensate for the greenhouse warming, the quantitative estimation of the effect of aerosol on climate is still difficult;

(iv) The results obtained agree, on the whole, with the conclusions of the IPCC Report and the Supplement to it (see Kondratyev et al., 1996).

Compared to the previous calculations in the assumption of an instantaneous CO_2 doubling, new calculations have led to the following differences:

(i) The climate warming is weaker in the polar regions, particularly in the Antarctic;

(ii) There is only a weak warming over most of Europe.

Despite considerable progress achieved in the numerical modeling of climate, a number of important problems remain unsolved:

(i) The model under discussion (like other climate models) is characterized by a slow climate drift, removed by introducing a flux correction, which is an artificial change of the heat and moisture fluxes at the atmosphere-ocean interface (naturally, the flux correction should be rejected, and the level of climate drift can serve as a criterion of the model adequacy);

(ii) Differences remain between calculations and observations, especially from the point of view of simulation of monsoons, global distribution of cloud cover and, for example, such local phenomena as the frequency of blocking the anticyclones and west-east transport over most of west Europe (such differences are very important in the context of solution of the problems of predictions of regional climate changes);

(iii) No direct account exists of the effect of aerosol on climate, it is substantiated with approximate assessments with the use of an energy-balance model calibrated against the data for a 3-D global model (a more reliable approach is being developed, as well as the parameterization scheme for radiative processes for the more adequate simulation of the atmospheric greenhouse effect.)

The use of the results of numerical climate modeling with prescribed seasonal and interannual variations of the global SST field is an important issue to simulate the climate change on the seasonal-to-interannual and interdecadal scales. Of interest in this context is the ecodynamics of the Sahel region which has suffered catastrophic droughts during the last two decades. Therefore Hurrell and Van Loon (1993) undertook a numerical modeling of climate change in Sahel for the period March – October (the season of rains in Sahel is mainly July – September) with prescribed observed variations of the global SST field for the last 10 years of the period 1949–1990.

Each of the numerical experiments was accomplished twice with the use of various initial data for the atmosphere, bearing in mind an estimation of relative contribution of the internal variability of the climatic system and the effect of SST variations. The simulated anomalies of precipitation turned out to be reliable.

From the data of 20 numerical experiments the coefficients of correlation between calculated and observed precipitation in the Sahel region and in two adjacent regions south of Sahel constituted 0.93, 0.87, and 0.87. The effect of the internal variability was comparatively weak, which testifies to the possibility to predict the seasonal precipitation with the reliably predicted global

SST field. The application of the similar approach to the Indian monsoon conditions gave less reliable results (the coefficient of correlation mentioned above reached only 0.30). Probably, the reasons were that the horizontal spatial resolution of the model was insufficiently high to obtain a reliable simulation of monsoon depression over North India, as well as difficulties in a combined consideration of the coupling between large-scale orographic effects, large-scale atmospheric circulation and sub-grid processes in the atmosphere.

An analysis of possible causes of precipitation fluctuations in North Africa has led to the conclusion that for individual locations in a number of cases (during some years) the convergence of moisture in the lower troposphere is the decisive factor. However, in other cases, more important is the convergence of moisture in the middle troposphere connected with latitudinal variations of the location of the African west-east jet stream at the 700 hPa level. At larger scales the effect of the tropical west-east jet stream shows itself in the upper troposphere (near the 200 hPa level), as well as local variations of the Hadley and Walker circulation cells.

At the scale of the tropics, a mode of precipitation variability was found out which expressed the connection between fluctuations of seasonal precipitation in the principal regions of monsoons, including North Africa. So, for example, the drought in Sahel and Sudan correlates with the droughts in Indonesia, Central America and North India, as well as with intensified precipitation over Guinea Bay and the NH tropical western Pacific. During further accomplishment of the project "Climate of the 20th Century" plans are made to perform a numerical modeling of the annual climate change for time periods from one decade to a century.

Tett (1993) analyzed possibilities to simulate the ENSO events using the UKMO climate model. From the early 1950s the Southern Oscillation has been known to manifest itself over the tropical Pacific as a strong correlation between high pressure anomalies over the western Pacific and low-pressure anomalies over its eastern sector with a periodicity of 3–5 years. Later on (in 1960) it was found out that the SST increase in the eastern Pacific during an El Niño event correlates with high-pressure anomalies in the western sector. This finding made it possible to establish a correlation between the El Niño (EN) and southern oscillation (SO) events.

An analysis of the results of this numerical modeling (with the prescribed gradual increase of CO_2 concentration by 1%/year) has shown that the calculated EN intensity (SST variations) for conditions of 1976 is half as much as that observed. Though the total pattern of the EN development is simulated relatively reliably (e.g., in both cases there is the westward motion of SST anomalies), the model cannot reflect the observed transitions of SST from rise to fall and vice versa. The numerical modeling results show that one of the reasons responsible for the westward motion of SST anomalies is, apparently, a change in the upwelling intensity, but other causes are also possible.

Murphy et al. (1993) discussed two approaches for using the results of numerical modeling of global climate with a large-scale model of low spatial resolution, to analyze the regional climate change:

(i) The use of nested high-resolution models;
(ii) A statistical analysis of climate change in individual locations or in small regions with account of empirical correlations between local variability and large-scale modeling data.

So far, a preliminary experience of realization of the first approach based on the use of the UKMO regional prognostic model with the prescribed boundary conditions in the outlying areas of the region under consideration and at the sea-surface level (SST and sea ice cover distribution) from the data of a large-scale numerical modeling has given positive results. This ensures, in particular, a more realistic simulation of wind fields and precipitation over the North Atlantic due to increased spatial resolution up to 50 km (this made it possible to more reliably consider the surface topography and the shape of the coastline as well as the soil and vegetation characteristics than in a large-scale model).

At the initial stage of numerical modeling the regional model was used to predict the weather. This experience has demonstrated for example that the regional model can simulate the orographically induced cyclones, which is impossible to achieve with a large-scale model. The reliability of the spatial distribution of precipitation including the details of the orographically induced precipitation has markedly increased. In this case the results have been examined of the numerical modeling for month 8 with a 1-year period of integration, which illustrates the stability of the regional model in the case of calculations for a long time period. The latter conclusion is confirmed by the results of analysis of time series of such parameters as water content and mass of the atmosphere.

A numerical modeling is being realized to study the dependence of calculated regional climate on the location of external boundaries of the region and the technique of prescribing the boundary conditions to reveal an optimal configuration of the regional model to calculate the climate in Europe and its changes, and at later stages in other global regions.

To exemplify the realization of the statistical approach, Murphy et al. (1993) considered an interaction between monthly mean pressure field at the sea level in the North Atlantic in December, January and February from the data of large-scale numerical modeling and monthly mean precipitation in four points of the grid on the Iberian Peninsula, which were simulated from the data of surface observations. To assess the correlations, a technique of canonic correlation was used, consisting in selection of such linear combinations of variables in the field of predictor (atmospheric pressure) and in the field of predictant (precipitation) to which the highest correlation corresponds. The results obtained testify to successful statistical forecasts of precipitation.

Bossert et al. (1992) demonstrated the usefulness of the nested approach with the use of the Colorado State University Regional Atmospheric Model for the region 27.5°–52.5° N, 97.5°–127.5° W (the northwestern USA). The calculations with a 21-layer model (with a resolution of 300 m near the surface and 1000 m at the top level of the atmosphere considered) with a horizontal resolution of 0.5 °C were made for a period of 1 month for January 1980.

A comparison of monthly mean SAT and precipitation values with the data of observations revealed a satisfactory agreement. The spatial distribution of precipitation over the area with a complicated topography was successfully simulated, and an adequate consideration of the cloud microphysics provided a realistic distribution of liquid and solid precipitation. There were, however, some differences, to remove which a more reliable parameterization of the processes of meso-β/meso-α scales is needed. In this connection plans are made to increase the spatial resolution of the model.

A consideration of an interrelationship between the dynamics of climate and atmospheric ozone is an important new step in further development of the numerical climate modeling. New assessments of the radiative forcing (RF) caused by decreasing ozone content in the lower stratosphere at middle and high latitudes suggested the conclusion about the possibility of a considerable negative RF (cooling) for the surface-troposphere system, contrary to the positive greenhouse RF due to the increasing content of CFCs. However, the calculations of anthropogenic climate changes did not take into account the recent growth of the ozone content in the troposphere, which should cause a greenhouse warming.

No doubt, when considering the effect of the observed ozone changes on RF and climate, it is necessary to take into account both the decrease of the ozone content in the stratosphere and its increase in the troposphere. Moreover, the specific features of the vertical ozone profile should be taken into account, since the SAT is very sensitive to the varying ozone content in the upper troposphere. In this connection, Wang et al. (1993) analyzed the results of observations of the vertical ozone profile at seven stations at middle and high latitudes (from Tateno 36° N, to Resolute, 75° N) to study the ozone variations taking place during the last several decades, and to assess climate changes in the context of increased concentrations of other GHGs: CO_2, CH_4, CFC–11, CFC–12, and H_2O.

The results obtained by Wang et al. (1993) show that the observed changes in the vertical ozone profile (especially in the lower stratosphere and upper troposphere) can markedly affect the climate. While the causes of the ozone decrease in the lower stratosphere are known and connected with increasing concentrations of GHGs, there are still many uncertainties with regard to the increasing trend of tropospheric ozone at the NH middle and high latitudes (up to 2%–2.5%/year in the upper troposphere in spring and fall), though, probably, it is determined by the effect of such ozone precursors as CO, NO_x and hydrocarbons. (Much effort is needed to solve this problem, especially bearing in mind the importance of the forecasts of the tropospheric ozone trends.)

Table 1.5.2 shows the results of calculations of RF due to the SW and LW radiation for the surface-atmosphere system, determined by changes in the ozone content over Hohenpeissenberg and Payerne for the period 1971–1980 to 1981–1990. Calculations were made with the use of two techniques: at a fixed air temperature (FT) and with account of adjusting to the temperature profile with a prescribed fixed dynamic heating (FD).

A comparison of the data for the troposphere and stratosphere make it possible to analyze the effect of the vertical ozone profile on the formation of RF. As is seen, the increase of the ozone content in the troposphere determines a positive RF for both SW and LW radiation, the troposphere contributing most to the formation of total RF in July.

Table 1.5.2. RF (W/m^2) due to changes in the ozone content for the period 1972–1980 to 1981–1990. After Wang et al. (1993). [© by American Geophysical Union]

Month	Stratosphere		Troposphere		Total	
	SW	LW	SW	LW	SW	LW
Hohenpeissenberg						
January:						
FT	0.207	-0.019	0.062	0.16	0.270	0.140
FD	0.207	-0.103	0.062	0.07	0.270	0.057
July:						
FT	0.084	-0.004	0.098	0.39	0.182	0.388
FD	0.084	-0.026	0.098	0.36	0.182	0.366
Payerne						
January:						
FT	0.141	-0.027	0.056	0.18	0.198	0.158
FD	0.141	-0.193	0.056	0.01	0.198	-0.008
July:						
FT	0.346	-0.045	0.027	0.07	0.373	0.024
FD	0.346	-0.306	0.027	-0.19	0.373	-0.237

Table 1.5.3 contains the estimates characterizing relative contributions of ozone and other GHGs to RF. The data in this table, again, demonstrate a marked contribution of the tropospheric ozone dynamics to the formation of RF.

Table 1.5.3. The relative role of contributions of ozone and other GHGs to the formation of RF (W/m^2). After Wang et al. (1993). [© by American Geophysical Union]

Season	Changes in ozone content				Combined effect of ozone and other GHGs	
	Total		Troposphere			
	FT	FD	FT	FD	FT	FD
Hohenpeissenberg						
January	0.410	0.327	0.222	0.136	0.734	0.710
July	0.570	0.548	0.490	0.467	1.050	1.027
Payerne						
January	0.356	0.190	0.242	0.074	0.797	0.630
July	0.397	0.136	0.097	-0.0167	0.879	0.617
Resolute						
Winter	-0.038	-0.085	0.242	0.074	0.797	0.630
Summer	0.397	0.136	0.097	-0.0167	0.879	0.617
Goose Bay						
Winter	0.025	-0.066	0.013	0.012	0.379	0.289
Summer	0.043	-0.218	-0.153	-0.415	0.497	0.235
Churchill						
Winter	0.080	0.069	-0.008	-0.158	0.275	0.127
Summer	0.110	-0.002	0.078	-0.031	0.361	0.252
Edmonton						
Winter	0.080	-0.069	-0.008	-0.158	0.275	0.127
Summer	0.110	-0.002	0.078	0.031	0.361	0.252
Tateno						
Winter	0.239	0.096	0.084	-0.060	0.491	0.348
Summer	0.304	0.287	0.295	0.279	0.585	0.569

The strong sensitivity of the ozone-induced RF to the lower stratospheric temperature points to the need of studies of climate consequences of the ozone content variations based on the use of AGCMs with an adequate account of the radiation-dynamics interactions. Also of interest in this context is the atmospheric cooling at the NH high latitudes, caused by an increase in the sulfate aerosol content.

In connection with a substantial effect of the atmospheric ozone dynamics on climate, studies of the factors determining the variability of atmospheric ozone, especially in the Antarctic, are very important. According to the present theory, the formation of the springtime TOC minimum in the Antarctic is governed by catalytic reactions with participation of chlorine oxide taking place on the surface of polar stratospheric cloud (PSC) particles. After the PSCs evaporate, ClO contained in the polar atmosphere can react with NO_2, giving a "reservoir" component $ClONO_2$, whose high concentrations can be observed in lower latitudes due to either a direct transport of polar airmasses or a $ClO-NO_2$ mixing near the edge of the circumpolar vortex. $ClONO_2$ was observed in both hemispheres near the edge of the circumpolar vortex. Though $ClONO_2$ has been known to take part in the catalytic cycle of ozone depletion, the role of this process remained unclear. In this connection, Toumi et al. (1993) undertook calculations of the ozone concentration in the March–May period within the arctic circumpolar vortex and in the NH high latitudes. Calculations have shown that during this process an accelerated decrease of TOC took place.

Apparently, the $ClONO_2$ cycle is responsible for a large magnitude of TOC decrease. An important feature of this cycle consists in that its functioning is independent of the time and place of PSC formation as happens in other cycles of chlorine compounds. This supposes that the $ClONO_2$ cycle can play a substantial role in ozone depletion in the warmer stratosphere in high latitudes. Toumi et al. (1993) demonstrated that after $ClONO_2$ forms, the ozone depletion does not stop, if one supposes that:

(i) After the PSC evaporation in March all available inorganic chlorine resides in high latitudes (65° N) in the form of $ClONO_2$;
(ii) In lower latitudes (55° N) the high-latitude airmasses get mixed with those in low latitudes.

In such conditions the ozone depletion continues after the heterogenic chemical reactions cease. Two cycles of ozone depletion are possible:

$$(1)\ ClO + BrO \rightarrow Cl + Br + O_2;\ Br + O_3 \rightarrow BrO + O_2\ ;$$
$$Cl + O_3 \rightarrow ClO + O_2$$

(the rate of this cycle is determined by the ratio of ClO - BrO reaction);

$$(2)\ ClONO_2 + h\nu \rightarrow Cl + NO_3;\ NO_3 + h\nu \rightarrow NO + O_2;$$
$$Cl + O_3 \rightarrow ClO + O_2;\ NO + O_3 \rightarrow NO_2 + O_2;$$
$$ClO + NO_2 + M \rightarrow ClONO_2 + M;\ Total:\ 2O_3 \rightarrow 3O_2$$

(where $M = N_2 + O_2$).

Though the cycles considered are representative also in the SH conditions, a rapid depletion of ozone is only possible with a high concentration of ClO. The results obtained enable one to characterize the evolution of the process of ozone depletion. During the polar winter in both hemispheres, heterogenic chemical reactions take place which ensure the formation of ClO from reservoir compounds of chlorine. There are three stages of the process of ozone depletion:

1. With a high concentration of ClO, at an early stage, a 2-D mechanism mainly prevails.
2. After the ClO concentration decrease the cycle ClO + BrO starts to dominate.
3. As the concentration of $ClONO_2$ grows, further depletion of ozone is determined by the second of the cycles mentioned above.

The transport of airmasses with a high content of $ClONO_2$ to middle latitudes causes a depletion of ozone. The process of $ClONO_2$ transformation into nonactive HCl is slow, and this prolongs the period of TOC disintegration; this can most markedly affect the biosphere in view of the growing intensity of UV radiation.

Austin and Butchart (1993) developed a 3-D photochemical model of the stratosphere which was used to assess the effect of atmospheric dynamics on the ozone holes in the Antarctic and the Arctic. Various numerical experiments have been made with the geopotential field 316 hPa prescribed as a boundary condition. In each case the integration for the Antarctic began on August 1 and continued for 100 days, according to the duration of the period of formation and disintegration of the ozone hole.

The results of TOC calculations in both polar regions turned out to be very sensitive to the prescribed dynamic forcing (in this case the amplitude of long waves in the stratosphere) and, on the whole, satisfactorily agree with the data of observations. An important factor limiting the TOC decrease in the Arctic is the long wave-induced lowering of airmasses in the polar region (even with a high concentration of chlorine in the stratosphere).

Assessments of the effect of greenhouse warming in the Arctic on the TOC led to the conclusion that the springtime TOC decrease did not exceed 20% in the current conditions, and with a CO_2 doubling the usual circulation in the stratosphere was transformed, which caused a 2-month delay of stratospheric warming at the end of winter. Combined with a low temperature, such an effect caused a prolongation of the period of ozone depletion and the formation of the ozone hole in the Arctic, with ozone almost completely depleted at a height of 20 km. The formation of the Arctic ozone hole does not take place, however, under every condition, but only during winters that are cooler than usual and not followed by the stratospheric warming. Therefore in the ($2 \times CO_2$) case the ozone holes can form in not more than 20% of the winters. We shall continue the consideration of the ozone-climate problem in connection with volcanic impacts in Chapters 3 and 5.

In conclusion, it must be emphasized that climate diagnostics data (on the basis of both observations and numerical modeling) still contain many controversies, which greatly complicates the identification of volcanic climatic signals.

REFERENCES

Austin, J., and Butchart, N., 1993: *Three-dimensional photochemical modelling.* The Hadley Centre for Climate Prediction and Research, Progress Report 1990-1992 and future programme of research, U.K. Met. Office, Bracknell, 62.

Bates, J.J., and Stephens, G.L., 1991: Global Integrated Water Vapor from Satellite Data: Preliminary Results from a GVAP Pilot Study. *Fifth Conf. on Climate Variations,* Amer. Meteorol. Soc., Boston, Massachusetts., 303-304.

Bloomfield, P., 1992: Trends in global temperature. *Climatic Change* **21** (1), 1-16.

Bossert, J.E., Kao, C.-Y.J., Winterkamp, J.L., Roads, J.C., and Chen, S., 1992: Development of a regional-scale climate model with a detailed microphysics parameterization. *Second ARM. Sci. Team Meeting,* October 26-30, 1991. U. S. Dept. of Energy Conf.- 9110366, Washington, D.C., 55-62.

Campbell, G.G., Randell, D.I., and Vonder Haar, T.H., 1993: View of the monthly mean climate with many parameters: Precipitable water, precipitation, cloudiness, radiation budget and SST. *Fourth Symp. on Global Change Studies,* January 17-23, 1993, Anaheim, California, Amer. Meteorol. Soc., Boston, Massachusetts, 306-307.

Changnon, S.A., Wendland, W.M., and Changnon, J.M., 1992: Shifts in perceptions of climate change: A delphi experiment revisited. *Bull. Amer. Meteorol. Soc.* **73** (10), 1623-1627.

Chesters, D., and Neuendorfer, A., 1991: The Climatology of Upper Tropospheric Water Vapor Observed in the 80s using the TOVS 6.5 micron bands. *Fifth Conf. on Climate Variations,* Amer. Meteorol. Soc., Boston, Massachusetts, 299-302.

Christy, J.R., McNider, R.T., and Robertson, F.R., 1991: Comparison of MSU and CCM1 tropospheric temperatures for 1979-1986. *Fifth Conf. on Climate Variation,* Amer. Meteorol. Soc., Boston, Massachusetts, 28-31.

Christy, J.R., 1993: Impacts on global temperatures due to Mt. Pinatubo and the 91-92 ENSO. *Proc. Seventeenth Ann. Climate Diagnostics Workshop*, NOAA, Washington, D.C., 13-17.

Díaz, H. F., 1991: Changes in Regional-Scale Precipitation for Global Land Areas During the Past Century. *Fifth Conf. on Climate Variations*, Amer. Meteorol. Soc., Boston, Massachusetts, 35-37.

Duffy, Ph. B., 1993: Comments on possible causes of recent global warming. *Fourth Symp. on Global Change Studies*, January 17-23, 1993. Anaheim, California, Amer. Meteorol. Soc., Boston, Massachusetts, 256-262.

Gray, W.M., Sheaffer, J.D., and Knaff, J.A., 1991: Hypothesized mechanism for stratospheric QBO influences on ENSO variability. *Fifth Conf. on Climate Variations*, Amer. Meteorol. Soc., Boston, Massachusetts, 101-104.

Gruber, A., and Arkin, P.A., 1992: Reviews of modern climate diagnostic techniques. Satellite data in climate diagnostics. *WCRP-786 (WMO/TD-N519)*, WMO, Geneva, 1992. 54 pp.

Gunst, R.F., Basu, S., and Brunell, R., 1993: Defining and estimating global mean temperature anomalies. *J. Climate* **6**, 1368-1371.

Gutzler, D.S., 1993: Uncertainties in climatological tropical humidity profiles: Some implications for estimating the greenhouse effect. *J. Climate* **6**, 978-982.

Gutzler, D.S., and Rosen, R.D., 1992: Interannual variability of wintertime snow cover across the Northern Hemisphere. *J. Meteorol.* **5** (12), 1441-1447.

Hallberg, R., and Inamdar, A., 1993: Observations of seasonal variations in atmospheric greenhouse trapping and its enhancement at high sea surface temperature. *J. Climate* **6**, 920-931.

Han, Q., Rossow, W.B., and Lacis, A.A., 1993: Near-global cloud droplet size data and analysis. *Fourth Symp. on Global Change Studies*, January 17-23, 1993. Anaheim, California, Amer. Meteorol. Soc., Boston, Massachusetts, 115-120.

Hurrell, J.W., and Trenberth, K.B., 1992: An evaluation of monthly mean MSU and ECMWF global atmospheric temperatures for monitoring climate. *J. Climate* **5** (12), 1424-1440.

Hurrell, J.W., and Van Loon, H., 1993: Analysis of low-frequency climate variations over the Northern Hemisphere using historical atmospheric data. *Fourth Symp. on Global Change Studies*. January 17-23, 1993, Anaheim, California, Amer. Meteorol. Soc., Boston, Massachusetts, 355-360.

Intergovernmental Oceanographic Commission, 1992: Oceanic Interdecadal Climate Variability. Joint SCOR-IOC Committee on Climatic Changes and the Ocean by the Ad Hoc Study Group on Oceanic Interdecadal Climate Variability, IOC Techn. Ser. No. 40, 1-40.

Jastrow, R., Nierenberger, W., and Seitz, F., 1990: *Scientific Perspectives on the Greenhouse Problem.* The Marshall Press, Jameson Books Inc., Ottawa, Illinois, 254 pp.

Jones, P.D., and Briffa, K.R., 1992: Global surface and temperature variations during the twentieth century. Part I: Spatial, temporal and seasonal details. *The Holocene* **2** (2), 165-179.

Karl, T.R., Tarpley, J.D., Quayle, R.G., Diaz, H.F., Robinson, D.A., and Bradley, R.S., 1989: The recent climate record: What it can and cannot tell us. *Revs. Geophys.* **27**, 405-430.

Karl, T.R., Quayle, R.G., and Groisman, P. Ya., 1993: Detecting climate variations and change: New challenges for observing and data management systems. *J. Climate* **6**, 1481-1494.

Karoly, D.J., 1991: On fingerprint detection of greenhouse climate change. *Fifth Conf. on Climate Variations*, Amer. Meteor. Soc., Boston, Massachusetts, 1-4.

Kellogg, W.W., 1993: An apparent moratorimum on greenhouse warming. *Fourth Symp. on Global Change Studies*, January 17-23, 1993, Anaheim, California, Amer. Meteorol. Soc., Boston, Massachusetts, 265-267.

Kondratyev, K.Ya., 1992: *Global Climate.* St. Petersburg, NAUKA, 359 pp. (in Russian).

Kondratyev, K.Ya., and Nikolsky, G.A., 1995a: Solar activity and climate., 1. Observation data: Condensation and ozone hypothesis. *Studying the Earth from Space*, No. 5, 3-17 (in Russian).

Kondratyev, K.Ya., and Nikolsky, G.A., 1995b: Solar activity and climate, 2. Direct impact of extra-atmospheric solar radiation spectral distribution. *Studying the Earth from Space,* No. 6, 3-16 (in Russian).

Kondratyev, K.Ya., Johannessen, O.M., Melentyev, V.V., 1996: High Latitude Climate and Remote Sensing. Chichester e.a. Wiley/PRAXIS, 200 pp.

Kukla, G., Knight, R., Gavin, J., and Karl, T., 1992: Recent temperature trends: Are they reinforced by insolation shifts? In *Start of a Glacia*, G. Kukla and F. Went (Eds). Proc. Mallorca NATO ARW, NATO ASI Series I. **3**, 291-305.

Ledley, T.Sh., and Chu, Sh., 1993: Can greenhouse warming induce ice sheet growth?. *Fourth Symp. on Global Change Studies*, January 13-17, 1993, Anaheim, California, Amer. Meteorol. Soc., Boston, Massachusetts, 272-275.

Lund, R., Hurd, H., Bloomfield, P., and Smith, R., 1995: Climatological time series with periodic correlation. *J. Climate* **8** (11), 2787-2809.

MacCracken, M.C., 1991: The key questions indicating the level of current uncertainty in forecasting climatic change. *Livermore, UCRL*-1D- 106243, 16 pp.

MacCracken, M.C., 1992: The challenge of identifying greenhouse gas-induced climatic change. In *Modeling the Earth System*, T. Ojima (Ed.), UCAR-OIES, Global Change Institute, Boulder, Colorado, **3**, 356-376.

Madden, R.A., 1991: The effect of changing spatial coverage on estimates of the global mean surface air temperature. Fifth Conf. on Climate Variations, Amer. Meteorol. Soc., Boston, Massachusetts, 8-11.

Madden, R.A., Shea, D.J., Branstator, G.W., Tribbia, J.J., and Weber, R.O., 1993: The effects of imperfect spatial and temporal sampling on estimates of the global mean temperature: Experiments with model data. *J. Climate* **6**, 1057-1066.

Manabe, S., and Stouffer, R.J., 1996: Low-frequency variability of surface air temperature in a 1000-year integration of a coupled ocean-atmosphere model. *J. Climate* **9**, (in press).

Model Evaluation Consortium for Climate Assessment, 1993: *MECCA Phase 1*. Key Preliminary Results, Brookfield, Wisconsin, 1 p.

Molnar, G., 1993: Greenhouse sensitivity to tropical water vapour distribution and cirrus property changes. *Fourth Symp. on Global Change Studies*, January 13-17, 1993, Anaheim, California, Amer. Meteorol. Soc., Boston, Massachusetts, 104-111.

Murphy, J.M., 1993: The first Hadley Centre transient climate change experiment. The Hadley Centre for Climate Prediction and Research., Progress Report 1990-1992 and future programme of research, U.K. Met. Office, Bracknell, 18.

Murphy, J.M., Cookmartin, G., Jones, R.G., and Noguer, M., 1993: Regional Climate. The Hadley Centre for Climate Prediction and Research, Progress Report 1990-1992 and future programme of research, U.K. Met. Office, Bracknell, 24-26.

Murray, R., 1992: Some notable features of Manley's Central England temperatures with special reference to very warm years. *Weather* **47** (3), 98-103.

Pennell, W.T., Barnett, T.P., Hasselman, K., Holland, W.R., et al., 1993: The detection of anthropogenic climate change. *Fourth Symp. on Global Change Studies*, January 17-23, 1993, Anaheim, California, Amer. Meteorol. Soc., Boston, Massachusetts, 21-28.

Phillips, T.J., 1992: An application of a simple coupled ocean atmosphere model to the study of seasonal climate prediction. *J. Climate* **5** (10), 1079-1096.

Phillips, T.J., Gates, W.L., and Arpe, K., 1991: Temporal sampling considerations in global climate modeling. *Fifth Conf. on Climate Variations*, Amer. Meteorol. Soc., Boston, Massachusetts, 20-23.

Pierrehumbert, R.T., 1995: Thermostat, radiator fins, and the local runaway greenhouse. *J. Atmos. Sci.* **52** (10), 1784-1806.

Plantico, M.S., Karl, T.R., Kukla, G., and Gavin, J., 1990: Are recent changes of temperature, cloudiness, sunshine, and precipitation across the United

States related to rising levels of anthropogenic greenhouse gases? *J. Geophys. Res.* **95**, 16617-16632.

Randall, D.A., Campbell, G.G., and Vonder Haar, T.H., 1993: The effect of atmospheric water vapor on the observed outgoing longwave radiation: An observational study of water vapor forcing. *Fourth Symp. on Global Change Studies,* January 13-17, 1993, Anaheim, California, Amer. Meteorol. Soc., Boston, Massachusetts, 93-97.

Reck, R.A., 1993: Tropical tropopause trends as a test of global change. Fourth Symp. on Global Change Studies, January 13-17, 1993, Anaheim, California, Amer. Meteorol. Soc., Boston, Massachusetts, 335-338.

Richards, G.R., 1992: Change in global temperature: A statistical analysis. *J. Climate* **5** (3), 546-559.

Robinson, D.A., 1993: Recent trends in Northern Hemisphere snow cover. *Fourth Symp. on Global Change Studies*, January 13-17, 1993, Anaheim, California, Amer. Meteorol. Soc., Boston, Massachusetts, 329-334.

Rutan, D., Smith, G.L., and Bess, T.D., 1991: A 12-Year Data Set of Albedo and Absorbed Solar Radiation from the Nimbus-6 and Nimbus-7 Satellites. Fifth Conf. on Climate Variations, Amer. Meteorol. Soc., Boston, Massachusetts, 380-383.

Schlesinger, M.E., and Ramankutty, N., 1992: Implications for global warming on intercycle solar irradiance variations. *Nature* **360**, 330-333.

Schneider, N., Barnett, T., Latif, M., and Stockdale, T., 1996: Warm pool physics in a coupled GCM. *J. Climate* **9** (1), 219-239.

Senior, C.A., and Mitchell, J.F.B., 1993: Carbon dioxide and climate: The impact of cloud parameterization. *J. Climate* **6**, 393-418.

Shea, D.J., Trenberth, K.E., and Reynolds, R.W., 1993: A global mean monthly sea surface temperature climatology. *J. Climate* **6**, 982-1001.

Smith, G.L., Rutan, D., and Bess, T.D., 1991: Variability of Absorbed Solar Radiation. *Fifth Conf. on Climate Variations*, Amer. Meteorol. Soc., Boston, Massachusetts, 384-387.

Soden, B.J., and Fu, R., 1995: A satellite analysis of deep convection, upper-tropospheric humidity, and the greenhouse effect. *J. Climate* **8** (10), 2333-2351.

Spencer, R.W., 1993: Global oceanic precipitation from MSU during 1979-91 and comparisons to other climatologies. *J. Climate* **6**, 1301-1326.

Spencer, R.W., and Christy, J.R., 1992: Precision and radiosonde validation of satellite grid point temperature anomalies. Pt. II: A troposphere retrieval and trends during 1979-90. *J. Climate* **5**, 858-866.

Spencer, R.W., and Christy, J.R., 1993: Precision lower stratospheric monitoring with the MSU: Technique, validation, and results 1979-1991. *J. Climate* **6**, 1194-1204.

Stellmacher, R., and Menda, W., 1992: Sonnenaktivität und Klimaänderungen. *Wiss. Z. Humboldt-Univ. Berlin R. Math. Naturwiss,*. 1992, Bd. 41 (2), 37-41.
Tett, S.F.B., 1993: Simulation of El Niño-Southern Oscillation events. *The Hadley Centre for Climate Prediction and Researc,*. Progress Report 1990-1992 and future programme of research, U.K. Met. Office, Bracknell, 19.
Toumi, R., Jones, R. L., and Pyle, J. A., 1993: Stratospheric ozone depletion by $ClONO_2$ photolysis. *Nature* **365**, 37-39.
Van Loon, H., 1991: Temperature trends and circulation changes in the Northern Hemisphere in winter. *Fifth Conf. on Climate Variations,* Amer. Meteorol. Soc., Boston, Massachusetts, 1991, 12-15.
Waliser, D.E., 1996: Formation and limiting mechanisms for very high sea surface temperature: linking the dynamics and thermodynamics. *J. Climate* **8** (1), 161-188.
Wallace, J.M., Zhang, Y., and Bajuk, L., 1996: Interpretation of interdecadal trends in Northern Hemisphere surface air temperature. *J. Climate* **9** (in press).
Wang Wei-Chyung, Dudek, M.P., Lian Xin-Jong., and Klehl, J.T.A., 1991: A general circulation model study of the climatic effect of atmospheric trace gases. *Fifth Conf. on Climate Variations.* Amer. Meteorol. Soc., Boston, Massachusetts, 1991, 5-7.
Wang Wei-Chyung, Zhuang, Y.-Ch., and Bojkov, R.D., 1993: Climate implications of observed changes in ozone vertical distributions at middle and high latitudes of the Northern Hemisphere. *Geophys. Res. Lett.* **20** (15), 1567-1570.
Weber, G.-P., 1990: Tropospheric temperature anomalies in the Northern Hemisphere 1977-86. *Int. J. Climatol.* **10** (1), 3-19.
Woodward, W.A., and Gray, H.L., 1993: Global warming and the problem of testing for trend in time series data. *J. Climate* **6**, 953-962.
Wu Zhong-Xiang and Newell, R.E., 1993: Is moisture increasing in the tropical regions? *Proc. Seventeenth Ann. Climate Diagnostics Workshop,* NOAA, Washington, D.C., 158-163.

CHAPTER 2

SOME PAST ERUPTIONS:

VOLCANIC AEROSOLS AND CLIMATOLOGICAL EVIDENCE

Volcanic histories are prone to error because of the rudimentary nature of the historical data available, which can only provide a crude classification of eruption size and climatic significance. Considering that the reliable meteorological data even from the earliest instrumented stations do not extend further back than about 200 years, this and the next chapters refer mainly to a few major eruptions of that period preceded by a brief summary of earlier scientific views of the relation between volcanic eruptions and climate. A number of very informative surveys of volcanic eruptions, volcanic hazards and their mitigation has been published (Tilling, 1989a, 1989b; Tilling and Dvorak, 1993; Tilling and Lipman, 1993; Tilling et al., 1987, 1990). Wright and Pierson (1992) have discussed the U.S. Geological Survey's Volcano Hazards Program; Dai et al. (1991) have pointed out that at the present time a number of chronologies of explosive eruptions exist: the dust veil index (DVI), the volcanic explosivity index and the atmospheric aerosol optical thickness estimates (AOT). Each of these chronologies has limitations and becomes increasingly deficient or inaccurate in the older part of the volcanic record. A broad spectrum of environmental records is preserved in ice cores, often with very high resolution. Acidity signatures in ice cores have provided a means for augmenting and extending the historical record of globally significant volcanic events. It has been discovered, for example, that one of the largest volcanic eruptions since the end of the last glacial stage may have occurred around A.D. 1259. In this context, a glaciological volcanic index based upon polar ice core records has been proposed and could prove more valuable for assessments of climatic impacts than previous indices.

As with any other proxy indicators, possibilities to interpret DVI data are limited, because they characterize only gaseous emissions from volcanic eruptions and take into account that the magnitude of deposition on a given ice cap or a glacier depends upon variation in large-scale atmospheric circulation, season of the eruption, and local deposition processes. (It would be proper to emphasize in this context that numerical experiments are necessary to simulate the whole succession of events from initial emission of gases by a volcano to the atmosphere, through the processes of their transport and gas-to-particle conversion in the stratosphere, to their eventual deposition on snow or ice surfaces and further ageing.

Dai et al. (1991) have noted that estimates of the magnitude of a specific volcanic eruption from measured ice core H_2SO_4 concentrations must be made cautiously. The total deposition of H_2SO_4 (kg/km^2) may not provide the best basis for comparing data among different sites as the amount deposited is dependent upon the net accumulation rate, among other factors. Dai et al. (1991) have illustrated relevant complications on the basis of the analysis of a possibility to identify volcanic eruption preserved as elevated sulfur concentrations which first appeared in the layers deposited in Antarctica and Greenland in A.D. 1810 (four ice core records with reliable dating). Two prominent SO_4^{2-} horizons were found at depths corresponding to the early nineteenth century. One of them corresponds to the well-known Tambora volcano eruption (1815), but the other (1809) is an eruption which was not identified earlier. Comparison with the Tambora ice core data has led to the conclusion that this was an explosive tropical volcanic eruption with the magnitude roughly half that of Tambora. The volcanic eruption considered was comparable to other eruptions producing large volumes of sulfur-rich gases, such as Cosiguina, Krakatau, Agung, El Chichón, and Pinatubo. It is highly probable that the climatic impact of this eruption was significant.

Ivlev et al. (1992) have discussed difficulties of the identification of contributions of various sources to the chemical dust composition for ice samples from Pamir and Tyan-Shan glaciers (Central Asia) for the time period 1972–1990. The results of neutron activation elemental analysis show that it is impossible to separate contributions reliably from dust storms, volcanic eruptions (although the El Chichón signal is visible), nuclear tests in China, and other sources.

As Sigurdsson (1990a, 1990b) has pointed out, "volcanic explosive eruptions can have several effects on the Earth's atmosphere, some with opposite surface temperature signals."

1. During fallout of tephra and condensation of volcanic gases in the troposphere, volcanic activity can cause increased condensation nuclei, cloud formation and thus increased tropospheric albedo, leading to local surface cooling.
2. Stratospheric sulfuric acid aerosol formed by SO_2 gas-to-particle

conversion leads to increased backscatter, absorption of solar radiation and surface cooling.
3. Enhanced stratospheric absorption of solar radiation by the volcanic aerosol can also contribute to surface cooling.
4. A volcanic aerosol containing large (≥ 1 μm) silicate particles, which cause infrared heating, can theoretically lead to surface warming which may balance the cooling effect of a sulfuric acid aerosol. This warming is likely to be short-lived or negligible due to very rapid fallout of such coarse silicate ash.
5. Increase in greenhouse gases such as CO_2 from a volcanic source can contribute to surface warming. This factor is likely to be negligible due to the small mass of volcanically derived CO_2.

One has to add that the significant contribution to stratospheric warming comes from enhanced absorption of upwelling thermal emission of the troposphere.

Particles of eruption aerosol clouds (droplets of sulfuric acid water solution) have strong absorption in the infrared wavelength region, as depicted in Table 2.1.

Table 2.1. Complex refractive index $m = m_r + im_i$ of sulfuric acid solutions at 300 K for different wavelengths. After Kent and Yue (1991). [Copyright by American Geophysical Union]

Wavelength, μm	Mass concentration of H_2SO_4 in the solution			
	50%		75%	
	m_r	m_i	m_r	m_i
0.694	1.394	2.07×10^{-8}	1.428	2.07×10^{-8}
1.00	1.389	2.08×10^{-8}	1.422	1.53×10^{-8}
1.06	1.387	1.313×10^{-6}	1.420	1.496×10^{-6}
9.10	1.499	0.318	1.652	0.554
9.25	1.452	0.355	1.594	0.558
9.30	1.442	0.353	1.590	0.585
9.50	1.485	0.449	1.673	0.695
9.70	1.639	0.408	1.897	0.644
9.90	1.650	0.314	1.947	0.454
10.10	1.631	0.255	1.882	0.339
10.30	1.598	0.223	1.824	0.294
10.50	1.568	0.202	1.764	0.272
10.60	1.545	0.194	1.730	0.274
10.70	1.517	0.196	1.696	0.289
10.90	1.465	0.241	1.666	0.366
11.10	1.483	0.308	1.733	0.458

Recent observational and theoretical results to be discussed in subsequent chapters show that even such a picture of climatic impact is oversimplified. Jensen and Toon (1992) have shown, for instance, that at temperatures below about -50 °C the concentration of ice crystals in the upper troposphere which nucleate may be as much as a factor of 5 larger when volcanic aerosols are present. The presence of volcanic aerosols may increase the net radiative forcing (surface warming) of certain types of cirrus near the tropopause as much as 8 W/m^2.

A very important aspect of the problem is volcanic impact on photochemical processes in the stratosphere responsible for the ozone layer dynamics and ozone-climate coupling. Deshler et al. (1992) have pointed out, for instance, that after the Cerro Hudson eruptions (12–15 August 1991, Southern Chile: 45.90° S, 76.96° W) ozone content in the atmosphere below 13 km decreased at a rate of 4–8 ppb/day over 30 days. This change began shortly after the appearance of the volcanic aerosol, providing direct measurements correlating volcanic aerosols and ozone depletion in the Antarctic. On the basis of aircraft observations Barton et al. (1992) identified the presence of the Mount Hudson cloud over continental southeastern Australia.

Hanson (1992) has discussed the uptake of HNO_3 onto ice, NAT (the nitric acid trihydrate), and frozen sulfuric acid as an important process relevant to formation of Type I polar stratospheric clouds. Kroes and Clay (1992) mentioned specific features of adsorption of KCl on ice under stratospheric conditions. Hofmann and Oltmans (1992) studied the effect of stratospheric water vapor on the heterogeneous reaction rate of $ClONO_2$ and H_2O for sulfuric acid aerosol. Di Sarra et al. (1992) have observed a significant effect of the Pinatubo aerosols on the arctic stratosphere.

These are just a few references to illustrate a complexity of the problem: volcanic impact on the stratosphere involves a great number of interacting physical, chemical and photochemical processes.

Sigurdsson (1990a, 1990b) has emphasized that the critical source parameters for evaluation of the atmosphere and global effects of an eruption include the mass eruption rate (intensity), total eruption mass (magnitude), eruption column height, and the type and mass of volcanic aerosols and other, noncondensible, volatile components release to the atmosphere (principally CO_2, HCl, and HF).

The range of column height (10–50 km) for the plinian phase of about 50 recent explosive eruptions with magma discharge rate between 10^6 and 10^9 kg/s is shown in Fig. 2.1 after Carey and Sigurdsson (1989). Eruptions with 10^{16} kg total erupted mass (~ 4000 km^3) are present in the geological record during the past 100,000 years.

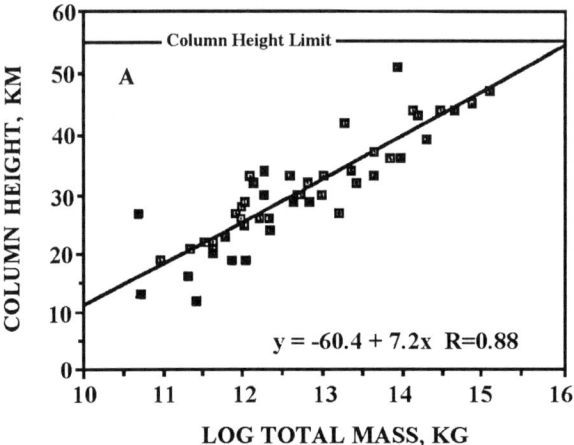

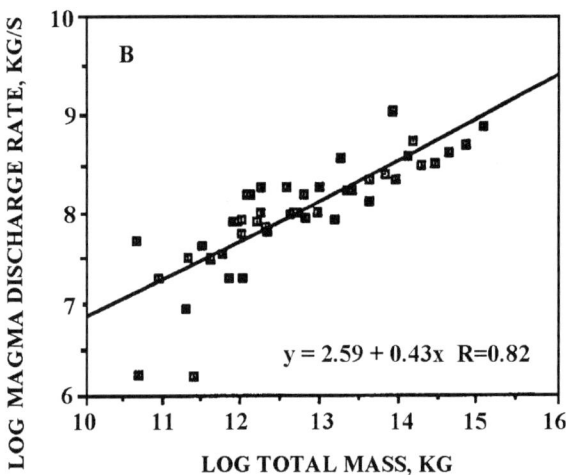

Figure 2.1. The range of column height and magma discharge rate. After Carey and Sigurdsson (1989). [© by Springer-Verlag]

Figure 2.2 demonstrates long-range atmospheric transport of tephra or volcanic ash (Sigurdsson, 1990a). The Toba eruption represents the greatest known atmospheric dispersal of tephra on the Earth, with 100-μm tephra transported over 2500-km distance from the source.

VOLCANIC ACTIVITY AND CLIMATE

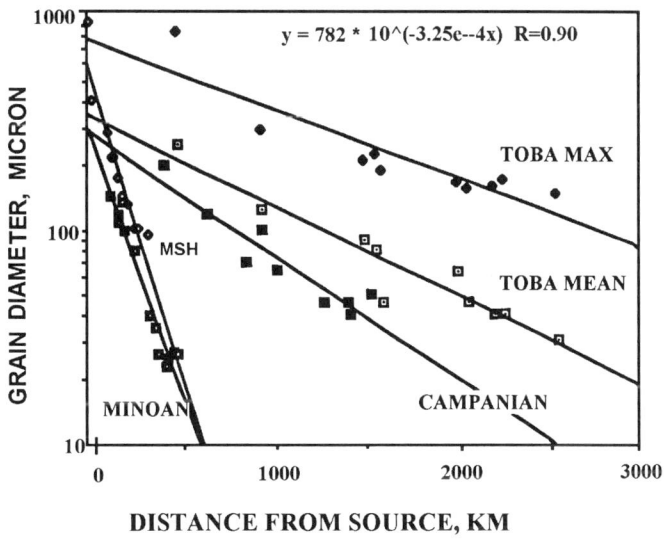

Figure 2.2. Long-range atmospheric transport of tephra or volcanic ash. After Sigurdsson (1990a).

On the basis of the analysis of volcanic rates in the geological record Sigurdsson (1990a) has come to the conclusion that hot spots (oceanic intraplate volcanism and intracontinental volcanism) are potentially an important source of volcanic aerosols to the atmosphere, as they vent volcanic gases subaerially, and because of the high sulfur content of basaltic hot-spot magmas. Various estimates of the rate of volcanic output gave values which vary from $(0.9-1.4) \cdot 10^{12}$ kg/year to $4.35 \cdot 10^{15}$ kg/year.

A problem of special interest in the context of climatic implications is the yield of sulfur to the atmosphere which strongly depends on the composition of the magma. Figure 2.3 data illustrate this conclusion in case of sulfur yield dependence on SiO_2 content in the magma (Sigurdsson, 1990a). In the case of sulfur, the yield is highly dependent on the magma composition, and eruption of basaltic and trachyandesite magmas can therefore have the greatest impact on the atmosphere.

The average volatile yield of sulfur from basaltic, intermediate and silicic eruptions is 800, 560 and 70 ppm, respectively; 65, 920 and 135 ppm for chlorine and 100, 500 and 160 ppm for fluorine (Sigurdsson, 1990b). Of course, such petrologic estimates of volatile yield give only minimal values since they take into account only the degassing of erupted magma. In the case of El Chichón and Nevado del Ruíz eruptions, degassing of the erupted magmas could account for only 10% of the total sulfur emission.

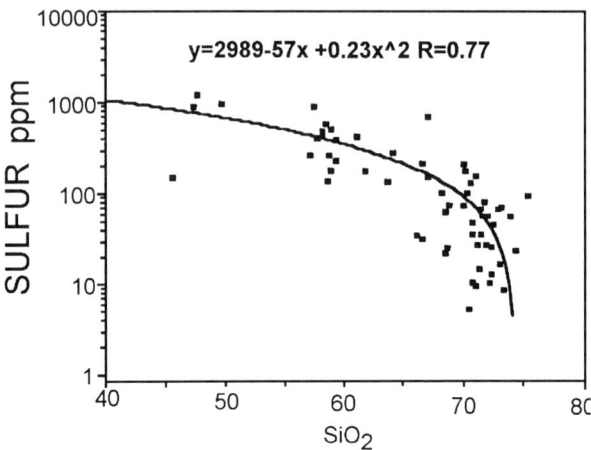

Figure 2.3. The yield of sulfur to the atmosphere in dependence of SiO_2 content in the magma. After Sigurdsson (1990b). [Reprinted with kind permission from Elsevier Science, Amsterdam, The Netherlands.]

An important contribution can be also made by the continuous sulfur global emission rate from fumarole and low-level eruptive activity. The present-day assessments of the steady-state emission rate give values within the range of 0.9 to $1.2.10^{10}$ kg/yr. Recent estimates of the ratio of eruptive sulfur release versus noneruptive (fumarole) release give values of the order of 1 in 2. However steady-state sulfur emission from volcanoes is dominantly tropospheric and scarcely reach the stratosphere. It is also small in comparison with the global emission rate of anthropogenic sulfur (3.3 to $6.5.10^{10}$ kg/yr). It should be emphasized, however that observations of continuous volcanic emissions (such as in studies of Abrams et al. (1991); Ammann et al. (1992); Mori et al. (1993)) are very important.

It is natural to expect negative correlation between sulfur yield to the atmosphere during individual volcanic eruption and average surface air temperature. Figure 2.4 data illustrate such a correlation for the NH (Sigurdsson, 1990a). As we shall see later, however, this correlation is a serious oversimplification since in the whole history of large volcanic eruptions numerous other causes influenced temperature changes. It is therefore often very difficult to identify volcanic climatic signal like it takes place in the case of anthropogenic climatic signals (Kondratyev and Grassl, 1993).

As Sigurdsson (1990b) has justly pointed out although current opinion favors the early removal of halogenes from eruption columns by rain out and adsorption on tephra particles, the fate of halogenes in the atmosphere following very large explosive eruptions remains unstudied. This problem may be very significant in the context of volcanic impact on the ozone layer dynamics, which will be considered later.

VOLCANIC ACTIVITY AND CLIMATE

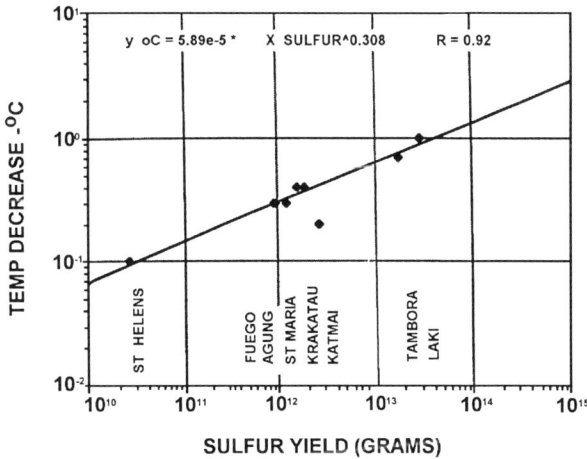

Figure 2.4. Correlation between surface air temperature and sulfur yield. After Sigurdsson (1990b). [Reprinted with kind permission from Elsevier Science, Amsterdam, The Netherlands.]

Important sources of information of climatic impacts of volcanic eruptions are paleoclimatic data from tree rings and ice core acidity analyses. Bradley and Jones (1993), Crowley and North (1991), Wigley (1991), and Crowley et al. (1993) have conducted a comparison of the Crête (Greenland) ice core acidity record with climatic variations of the last 1400 years. Some similar earlier studies suggested that changes in volcanism may have significantly influenced climate on this time scale. The problem is, however, that the ice core activity record consists of volcanically induced spikes superimposed on a background acidity of nonvolcanic origin that varied significantly in amplitude.

Because the Krakatau acidity peak in the ice core (1884) has a value of 0.8 µEH$^+$/kg, Crowley et al. (1993) adopted 0.6 as a minimum value for a volcanic spike (spikes below that value cannot easily be discriminated from noise). Spikes are defined as any event with acidity increases by 0.6 µEH$^+$/kg within one year and then decreases of at least an equivalent amount to local background levels during the next 1–2 years. Some such events are well recognized from previous volcanic studies and fall within 0–2 years of known eruptions. Tambora (1815), Krakatau (1883), and Katmai (1912). On the other hand, there is the lack of peaks for some known eruptions and other peaks of unknown origin. For instance, the 1259 peak is a global-scale perturbation which is 1.7 times the size of the Tambora peak at Crête (this was, probably, one of the largest volcanic eruptions of the Holocene). Therefore Crowley at al. (1993) have emphasized that these results indicate that our ability to correlate the ice core acidity record with the historical volcanic database is still imperfect: cross-correlations with other ice cores need to be done before this issue can be clarified. Apparently there is also a necessity to analyze the whole succession of volcanic aerosol transformations during their long-range transport, deposition and ageing.

There are 92 acidity peaks in the 1420-year record. Since at least 26% of the past-1600 eruptions (7 of 27 eruptions) are of local (Islandic) origin, Crowley et al. (1993) estimate a mean recurrence interval of 20.8 years for hemispherical-scale eruptions rich in sulfate. On decadal time scales, volcanism has varied by a factor of 3–4 over the last 1400 years. However, the climate-volcanism correlation (the tree ring record for the western USA served as the source of climatic information) is considerably less impressive (correlation coefficient $r = -0.23$) than previously concluded ($r = -0.52$). The good correspondence between volcanism and climate previously reported are due to background acidity levels that show a significant Little Ice Age increase (there are variations of almost 1.2 $\mu EH^+/kg$ in background acidity, i.e., 50% larger than the level of the Krakatau peak).

As Crowley et al. (1993) have mentioned, the peak in residual acidity during the Little Ice Age may reflect variations in productivity of the ocean, which releases DMS to the atmosphere. The subsequent gas-to-particle conversion leads to the sulfate aerosol formation. This aspect of the problem once again attracts attention to climate-biospheric interactions.

The results considered provide some support for a correspondence between the strongest volcanic eruptions and annual-scale temperature changes estimated from tree ring studies. However, in quite a number of cases acidity peaks correspond to the other ten coldest years in the tree ring record which demonstrates the importance of regional variations in the atmospheric response to volcanic eruptions.

2.1 Development of Scientific Views

Benjamin Franklin seems to have been the first scientist to suggest that "the vast quantity of smoke" and "dry fog" following a volcanic eruption might affect the weather. In his paper on Meteorological Imaginations and Conjectures read on 22 December 1784 before the Philosophical Society of Manchester he provided a vivid description and explained the low summer-autumn temperatures of 1783 and the following severe winter as possibly due to the volcanic "smoke" preventing sunlight from reaching the surface after the nonexplosive eruption of Laki crater in Iceland, which started in early June 1783 and lasted for 8 months. He pointed out that the rays of the sun were rendered so faint in passing through the volcanically introduced haze "that when collected in the focus of a burning glass, they would scarce kindle paper."

VOLCANIC ACTIVITY AND CLIMATE 95

Many other scientists have studied the relation between volcanic eruptions and weather since Franklin's observations. Most of these studies have consisted of statistical correlations between bad weather and single eruptions or between climatic anomalies and a series of volcanic eruptions. These studies do show that during some years with abnormal weather, such as 1783 and 1816, there were volcanic events; and in addition, that major climatic shifts of the past 500 years occurred many times in parallel with variations in the level of volcanic activity. On the other hand, there is no evidence that volcanic explosions preceded and initiated the ice age. Many years of droughts or severe cold occurred without evidence of any major volcanic eruptions. The volcanic eruptions are not responsible for all climate change nor all years of bad weather, but they may cause some of them.

The theme of volcanic eruptions as the cause of climatic aberrations gained strength as a hypothesis after the violent eruption of Krakatau in Indonesia in 1883 and again after the Navarupta Katmai in Alaska in 1912. It had another revival after the 1963 eruption of Mount Agung on Bali island, also in Indonesia. Its scientifically most sophisticated tests came with the May 1980 Mount St. Helens and the April 1982 El Chichón eruptions when special studies were developed to exploit instrumental aircraft and space technology as the means to understand the basic physics and chemistry of volcanic "clouds", to monitor their global distribution, and to assess their role in heating and cooling the atmosphere. Much more detailed studies have been accomplished since the Pinatubo eruption of June 1991.

Numerous theoretical studies have been conducted on the effects of volcanic aerosols on climate. Some of the earliest belong to the prominent astrophysicist Charles G. Abbott who noted shortly after the Katmai eruption changes in the solar radiation received at observatories of the Smithsonian Institution (Abbott, 1913) and speculated on the climatic significance. The well known American atmospheric physicist W.J. Humphreys (1940) supported the volcanic dust hypothesis of climate change with a physical model of optical interception of solar radiation by the particulate. In fact, he postulated as "well nigh certain" that the back scattering by volcanic dust of solar energy had lowered the earth's temperature by 5 °C after large eruptions. The pyrheliometric records for the periods after the Krakatau and Katmai eruptions showed reduced values of solar intensity at the Earth's surface. However, these were records of direct (normal incidence) radiation only. The 1963 Mount Agung eruption caused similar reductions, but the total radiation of sun and sky on the horizontal surface was not materially altered. This indicated that forward scattering by the particles compensates for much of the energy lost in the direct beam and that Humphrey's basic assumption about backscattering being large and the effect on infrared light being small were wrong, although he is frequently referenced.

The whole problem was reviewed again by Deirmendjian (1973) in the light of far more advanced models of the turbid atmosphere. He re-examined the available data for the Krakatau, Katmai, and Agung eruptions and concluded that "no significant anomalies in global radiation...clearly attributable to the volcanic dust layers have been demonstrated...". He also questioned whether the Earth's albedo had been altered, but still admits that a possibility of climatic effects of eruption exists. Finally he suggested more careful and sophisticated checks on atmospheric turbidity to arrive at an unequivocal answer.

Current research indicates that the atmospheric impact of volcanic eruptions on the Earth's climate is not simply related to the volume of erupted material or the eruption magnitude but relates also to the chemistry of the magma, specifically the concentration of the volatile constituents SO_2, H_2S, HCl and to a lesser extent Cl and F.

Pollack et al. (1976a, 1976b, 1980) discuss the sensitivity of the calculations to many of the assumed parameters such as single scattering albedo and particle size. They established that large silicate particles may warm the surface; small sulfuric acid droplets may lead to a cooling which linearly increased with optical depth. An optical depth of 0.02 to 0.03 is needed to cause a barely observable change in temperature. They also calculated the warming to be expected between the late 1800s and the middle of the 1900s due to the decline of volcanic activity. The calculation was done by using the decade average optical depths to crudely represent the response time of the Earth-atmosphere system. The calculated increase in temperature due to the decline in volcanic activity is sufficient to match the observations. These calculations show that the magnitude is also correct. Of course, further work must be done before it can be concluded (if at all), that the decrease in volcanic activity was a cause of the warming trend between 1900 and 1940. It is important to remember that the increasing level of CO_2 in the atmosphere may be considered as a major factor in the Earth's climate during these last decades.

Hansen et al. (1978) calculated the time evolution of the stratospheric and tropical tropospheric temperatures following the eruption of Mount Agung. Their calculated temperatures agree favorably with the observed ones. In their model, as well as in earlier mentioned calculations of Pollack et al. (1976a; 1976b; 1980), it was found that the absorption of infrared energy by the volcanic particles was the major factor in warming the stratosphere. The troposphere cooled because the volcanic particles reflected a small amount of sunlight back to space that would otherwise have reached the surface. Both groups assumed the volcanic particles were largely sulfuric acid and had a size distribution that is identical to nonvolcanic size distributions (Toon and Pollack, 1976). The Mount St. Helens and El Chichón eruptions provided the first opportunity to check these two important assumptions.

2.2 Prominent Eruptions During the Past 200 Years

The **Laki** crater-row eruption of 1783 is the first major flood-basalt volcanic pollution event for which there are at least a few meteorological records. The eruption broke out in the eastern volcanic zone of Iceland in June 1783 and lasted for 8 months. During this time, 12.4 km^3 of lava poured out of 115 craters on a 25-km-long fissure to produce the largest lava flow in historical time, with an area of 565 km^2 (Thorarinsson, 1969, Sigurdsson, 1990a). In addition, the eruption produced 0.3 km^3 of tephra, equivalent to the total volume of airfall ash ejected by the Mount St. Helens 1980 eruption. The vast outpouring of volcanic gases, which included an estimated 1.3 to 6.3 x 10^7 tons of SO$_2$ (Thorarinsson, 1969, Hammer, 1977), produced a bluish haze all over Iceland and led to the destruction of most summer crops. The resulting famine caused the loss of 75% of all livestock in Iceland and the deaths of 24% of the Icelanders in the following years.

The effects of the blue haze were by no means restricted to Iceland, and the easterly drift of the haze has been traced, from historical records, across Europe and as far east as the Altai Mountains in China, where it was observed 50 days later (Thorarinsson, 1979). The effects of the haze in Western Europe were first observed and documented by Benjamin Franklin, at the time the United States Ambassador to the court of King Louis XVI of France.

The seasonal temperature in the eastern part of North America was reported as below the 225-year norm by: autumn 0.8, winter 4.9 and spring 0.5 °C. Even the annual averages for 1784 and 1785 were below the normal, by 1.5 and 1.2 °C respectively, opening the way for speculation that the volcanic effect lasted for 2 years (Sigurdsson, 1982).

Yet another line of evidence of massive volcanic pollution in 1783 has been obtained from the Greenland ice sheet (Hammer, 1980, Hammer et al., 1980). Ice cores extracted by drilling reveal a continuous stratigraphic record of precipitation, resulting in the formation of an ice layer on the order of 20 to 30 cm per year. The snow precipitation was accompanied by a trace of aerosols or impurities from the atmosphere that consisted largely of sulfuric acid. High acidity layers in the ice have been correlated with volcanic eruptions, representing sudden loading of the atmosphere with SO$_2$ and H$_2$S volcanic gases, which become converted to minute aerosol droplets of H$_2$SO$_4$ (Hammer, 1977). The peak acidity values in the 1783 ice layer are higher than any others in the last thousand years (Hammer, 1980) and correspond to a global fallout of 10^8 tons of H$_2$SO$_4$ from the Laki eruption as shown in Figure 2.2.1. Various assessments gave the H$_2$SO$_4$ mass from 1 to 2.8 x 10^8 tons. Further analyses of the magma confirm that the eruption had released no less than 2.5 x 10^7 tons of sulfur, or 77 x 10^6 tons of H$_2$SO$_4$ which is in good agreement with the Hammer estimate, based on ice core analysis, of 100 x 10^6 tons H$_2$SO$_4$ (Sigurdsson,

1982). For comparison, the annual natural volcanic sulfur loading of the atmosphere is estimated (Cadle et al., 1976) to be 2.8×10^5 tons/year (90 times less than Laki); and the Mount St. Helens eruption produced only about 7.5×10^4 tons (333 times less than Laki). An indirect, and somewhat inflated estimate of the increase of the optical depth of the atmosphere by 0.5 after the Laki eruption (Sigurdsson, 1982) is three times greater than the increase of 0.14 of the optical depth measured after the Mount Agung 1963 eruption by DeLuisi and Herman (1977).

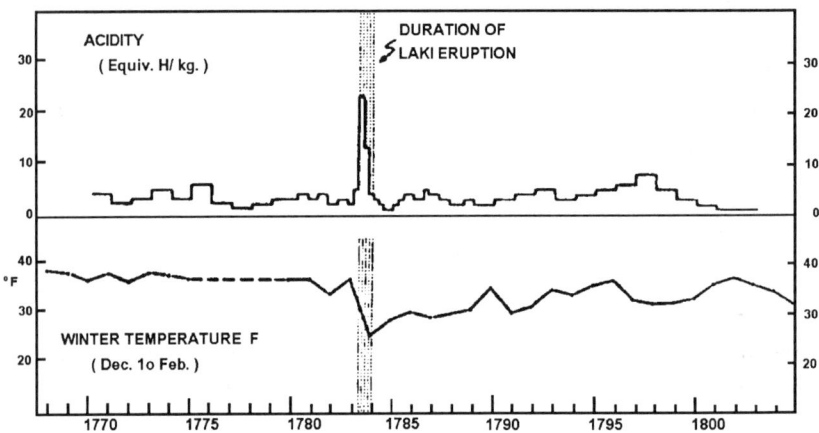

Figure 2.2.1 The effects of the Laki eruption in 1783 as reflected by acidity levels of precipitation on the Greenland ice sheet (Hammer, 1977) and the winter (Dec. to Feb.) temperature record of the eastern United States (Landsberg et al., 1968). After Sigurdsson (1982). [© by American Geophysical Union]

The largest and deadliest volcanic eruption in recorded history was that of **Mount Tambora** on the island of Sumbawa (8° S, 118° E) Indonesia, in April 1815. Around 90,000 people were killed in the immediate vicinity alone; the height of the mountain was reduced by almost 1,400 m, with an estimated volume of released magma exceeding 50 km^3; the effects of the tsunami that the eruption generated were appreciable 1,200 km away; ash was deposited some 1,300 km away; the explosion was heard 2,600 km away, and darkness was experienced for up to 2 days to a distance of around 600 km. In a comprehensive review of the evidence, Stothers (1984) estimates that about 150 km^3 of ash and 25 km^3 of ignimbrite were ejected over, at most, a 24-hour period. The force of the explosion injected material over 45 km high into the stratosphere, producing an ash loading that may have been as much as an order of magnitude greater than that resulting from the 1883 eruption of Krakatau and a sulfate aerosol loading more than twice as large–about 150 to 200 x 10^6 tons of H_2SO_4 (Rampino and Self, 1984, Stothers, 1984). The stratospheric veil produced by the eruption is believed to have had an impact on a much wider scale during ensuing years, although much of the evidence is very scattered. Stothers (1984) estimated that during the main eruption the average mass flux rate from the crater must have been devastating, 5 to 8 x 10^6 m^3/sec.

There is much evidence which, combined with theory and analogy, indicates that a mass flux rate of well over 10^6 m^3/sec had thrust material up to 50 km heights (Wilson et al., 1973, Stothers, 1984).

The clearest evidence comes from subsequent observations of the globally dispersed dust in the atmosphere. The coarser ash particles fell out within a week or two of the eruption as a result of rapid tropospheric mixing and washout. If finer ash particles, aerosols and gas molecules reached the stratosphere, where some of them might reside for months to a few years at altitudes of 12 to 30 km, they would eventually be carried by winds longitudinally around the globe and, at the same time, transported to all latitudes by meridional currents. Within several weeks of the eruption, numerous secondary aerosols would also have formed by photochemical reactions between the directly injected sulfur gases and the stratospheric ozone and water vapor.

Between June 28 and July 2, and later between September 3 and October 7, 1815, prolonged and brilliantly colored sunsets and twilights were frequently seen near London, England. These displays were explicitly differentiated by the observers from the more familiar effects of London smog. Typically, the twilight glows appeared orange or red near the horizon, purple or pink above, and were occasionally streaked with diverging dark bands resembling cirrostratus clouds. According to a New York report (of 1816) the dry fog reddened and dimmed the sun to such an extent that sunspots became visible to the naked eye. During the spring and summer of 1816, a persistent "dry fog" was seen and reported in the northeastern part of the United States (Landsberg and Albert, 1974, Old Farmer's Almanac, 1966). Since neither surface winds nor rain dispersed it, the haze must have been located above the troposphere. Its optical extinction properties, needed for climate model studies, have been estimated in several ways (see Stothers, 1984). In summary, it would appear that the stratospheric dust veil produced by Tambora spread to the latitude of England in about 3 months. The excess of visual extinction (in astronomical magnitudes) and its change as function of the time from the eruption, over 41° to 71° N, is shown in Figure 2.2.2 (Stothers, 1984). It demonstrates the typical first 3 months rapid increase followed by very slow (4–5 years) return to normal.

The rate of decline of the extinction curve can be checked with modern measurements of the record of acid fallout over the Greenland ice cap. Before

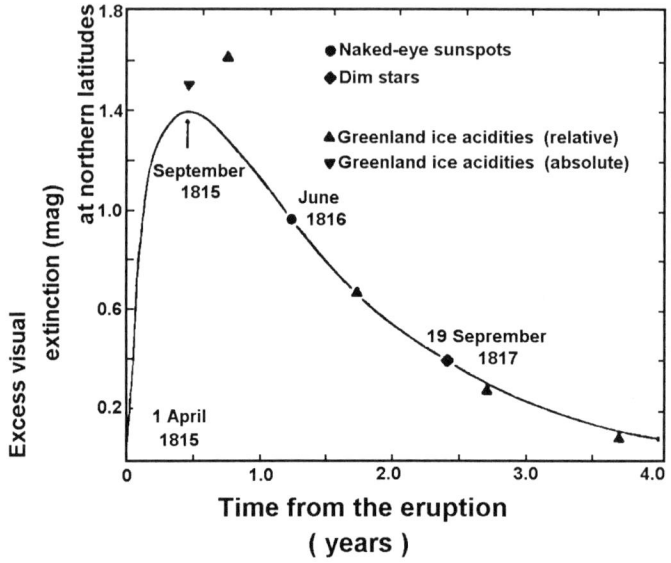

Figure 2.2.2. Excess zenithal visual extinction $(\Delta m)_{OD}$ in astronomical magnitudes (mag) at northern latitudes (41° to 71° N) as a function of time, reckoned from the date of Tambora's eruption. After Stothers (1984). [© American Association for the Advancement of Science. Reprinted with permission.]

Tambora's dust reached Greenland, most of the fine silicate ash must have dropped out leaving behind more slowly settling aerosols, principally sulfuric acid. A strong 4-year-long (1815–1818) acidity enhancement in the ice at 71° N as reported by Hammer (1980) consists of the following annual acidities: 1.8, 4.9, 2.5, and 1.6 microequivalent (μeq) of H^+ per kilogram of ice. The background acidity at this place is 0.9 μeq. After simple calculations (Stothers, 1984) these data confirm not only the shape of the extinction curve shown in Fig. 2.2.2 but also its magnitude as 1.4 ± 0.2 μeq at 41° to 71° N, which should be compared with ≈ 0.6 for Krakatau and only ≈ 0.02 μeq for Mount Agung (the latter is small because its aerosol did not penetrate over far northern latitudes).

In a later study Kelly et al. (1984) examined pressure anomaly charts covering Europe for 1815–1816 and early stations pressure records before and after the Tambora eruption. They found negative pressure anomalies centered over Europe 7–10 months after the event. However, when mean standardized pressure anomalies were calculated for the gridpoints 50°–60° N, 10° W – 10° E for all months from 1812 to December 1827, as shown in Figure 2.2.3, they concluded that the pressure drop in 1816, presumably associated with the Tambora eruption, is an unusual, but by no means a unique event. Similar (by magnitude) excursions have occurred also in "nonvolcanic" periods.

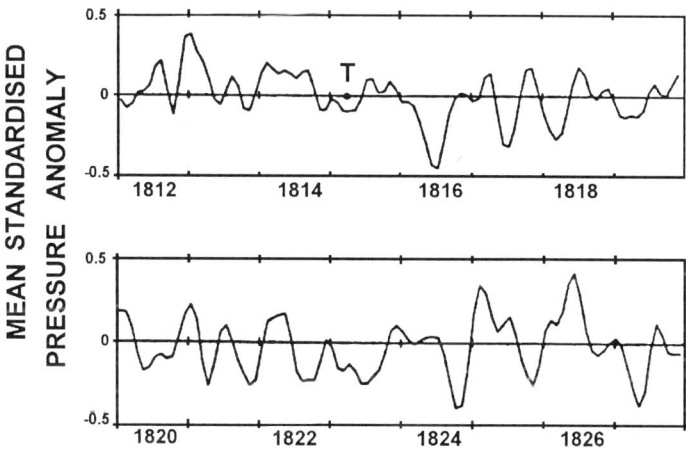

Figure 2.2.3. Standardized and smoothed monthly pressure anomalies for Europe. The date of the Tambora eruption is marked. After Kelly et al. (1984).

The Tambora dust veil is often blamed by modern studies for the cold summer of 1816, although some deny that there was any connection with the volcano (Deirmendjian, 1973). The year 1816 was called "the year without summer" in New England and eastern Canada where the daily minimum temperatures were abnormally low from late spring through early fall. It was cold with heavy rains in Europe also. The summer 1816 average temperatures on the east coast of North America were the lowest on record with 1.5–2.5 °C below the seasonal norm (Landsberg and Albert, 1974). It was a very cold summer too in Edinburgh (Δ 1.5 °C), Budapest (Δ 1.6 °C), Rome (Δ 1.4 °C), and Hohenpeissenberg (Δ 2.6 °C). However, it was milder at the Eastern European stations. In Figure 2.2.4 a compilation of annual mean temperature deviations for four NH isothermals and their average for the hemisphere as a whole is reproduced from Stothers (1984) who used Köppen's (1873) annual deviations as a basis. The average deviation for the NH in 1816 is 0.7 °C on the background of standard deviation ±0.5 °C for the period 1800 to 1840 which is not a statistically unique deviation. Although insufficient data from that period could be blamed for part of the weakness of the apparent temperature drop, it is obvious the data do not offer substantial evidence that even as gigantic an eruption as Tambora was directly responsible for global lowering of surface temperature.

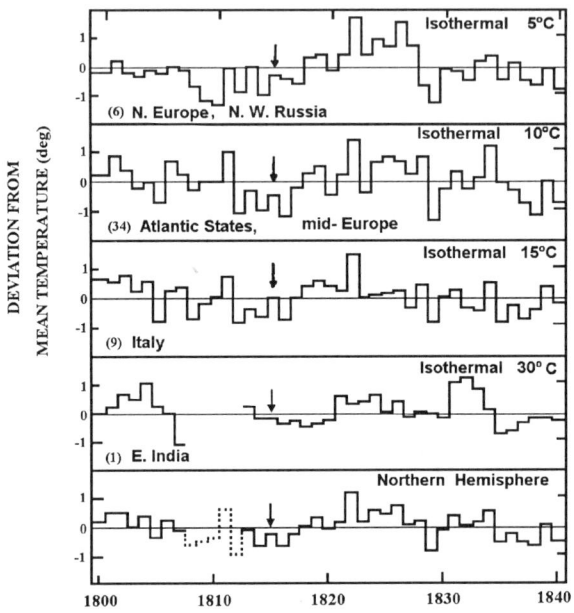

Figure 2.2.4. Annual mean temperature deviations for four NH isothermals as well as for the NH as a whole. The minimum number of stations used and the areas in which they are situated are indicated. Arrows point to 1815, the year of Tambora's eruption. From Stothers (1984). [© American Association for the Advancement of Science. Reprinted with permission]

The August 26/27, 1883 explosive eruption of **Krakatau** is probably the most famous volcanic event of the past because of the death of more than 36,000 people and the widespread devastation which became known within only a few days due to the existence of the telegraph. It was the first time an eruption was extensively documented by geodetic surveys and by geophysical instruments around the world. Every barograph in the world recorded the air-pressure wave over a period of 5 days; tide gauges reacted to the sea-wave thousands of miles away; an up to 40-m-high tsunami wave crashed onto the nearby coasts of Java and Sumatra; a blue and green sun was observed for weeks, and red sunsets persisted for up to 3 years; the volcanic dust, shielding out solar radiation, is believed to have lowered the global surface temperatures in the following few years (Simkin and Fiske, 1983).

Krakatau was located on a 5 x 9 km island Rakata in the Sunda Straits, between Java and Sumatra in the Dutch East Indies. It was a familiar landmark, both to the tens of thousands of nearby coastal residents and to the crews of thousands of ships from Europe and the Americas that passed through the Straits each year on their way to and from the Far East. The volcano had last erupted in 1681 and was not regarded as a likely site for catastrophic activity. However, in May 1883 it renewed its activity and during the explosion at the end of August most of the island of Rakata collapsed to form a caldera, about 7 km in diameter whose floor lay submerged 100 m beneath the waters of Sunda Straits. During the eruption more than 18 km^3 of volcanic debris were injected and reached the stratosphere. The amount of H_2SO_4 aerosols generated in the stratosphere was estimated (from optical depth) to be 50×10^6 tons. A similar amount (55×10^6 tons) was calculated from ice-core acidity analysis (Hammer, 1980). These amounts should be compared with the 10×10^6 tons from Mount Agung and El Chichón each and $\approx 200 \times 10^6$ tons from Tambora. In the northern latitudes (41° to 71° N) the excess visual extinction was estimated to be 0.6 astronomical magnitudes. Hunt (1977) used a three-dimensional global circulation model to follow the dispersion of the ash-cloud from the Krakatau eruption. He calculated that the cloud spread to cover approximately half of the globe in ≈ 30 days and more or less the whole globe in ≈ 150 days. In Figure 2.2.5, adopted from Kelly and Sear (1984), are plotted the superposed epoch analysis of individual sequences of standardized monthly NH temperatures relative to the zero month—the explosion data of Krakatau and Agung respectively. The values of the monthly standard deviations (°C) of the NH average surface air temperature data for 1881–1980 are as follows:

Jan	Feb	Mar	Apr	May	June	July	Aug	Sept	Oct	Nov	Dec
.64	.54	.42	.35	.31	.28	.23	.25	.27	.39	.44	.50

These are the values of one standard deviation of Figure 2.2.5 converted into degrees Celsius. Considering that Krakatau is located in the equatorial zone of the SH, a lag of a few months between the eruption and the appearance of negative deviations of temperature over the NH seems reasonable. From April until September 1884 it seems that the entire NH was about 0.5 °C colder than its 100-year norm; of course, cautions are necessary on the reliability of hemispheric temperatures a century ago.

Mount Agung represents the first major eruption of this century for which reliable meteorological data including radiosonde stratospheric temperatures are available. Mount Agung, located on Bali island (8° S), Indonesia, erupted in March 1963, killing over 1,500 people. The volume of magma was estimated at 0.4 to 0.6 km^3, much of whose products reached ≈ 18 km height, with peak rates of output of dense magma $\approx 10^4$ m^3/sec. The sulfuric acid generated as a result is estimated to be 10 to 20×10^6 tons, having a very pronounced effect on the transmission of light through the atmosphere not only over the SH but also noticeable by the Mauna Loa Observatory in Hawaii. Data on the normal incident solar irradiance apparent transmission discussed by

Mendonca et al. (1978) show that a significant decline in the annual value of the apparent transmission down to 0.915 was registered during the year following the Mount Agung eruption. This eruption is believed to be the first one since Katmai in 1912 that had a magnitude large enough to have a significant effect on the atmospheric opacity due to enhancement of the stratospheric aerosol mass. The maximum broadband optical depth attributed to the Mount Agung eruption observed at Mauna Loa was 0.02 and at Aspendale ≈ 0.15 (DeLuisi and Herman, 1977). A few months after the El Chichón eruption the optical depth observed at Mauna Loa was 0.18, exceeding by far the only other large value ever measured – that in Aspendale after the Mount Agung eruption in 1963.

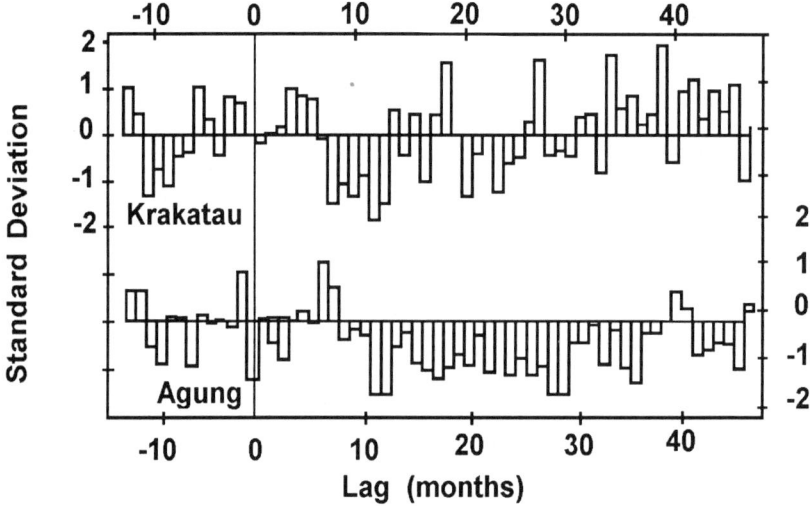

Figure 2.2.5. Individual sequences of standardized monthly NH temperatures following the eruption date of Krakatau and Mount Agung. The data are expressed as deviations from prevailing temperature levels defined by the 12-month average for the period up to but not including the eruption date (month zero). The deviations in °C are given for each month in the text. After Kelly and Sear (1984). [Reprinted with permission from *Nature*. © by Macmillan Magazines Limited].

Because volcanic particles are large enough to scatter all visible colors fairly efficiently, the sky appears milky white rather than blue after large eruptions. Dyer and Hicks (1965) presented observations of the direct, diffuse and total skylight in Australia after the Mount Agung eruption. Although the direct solar beam was reduced by nearly 25%, the diffuse skylight doubled, and thus the total sunlight reaching the surface was only slightly diminished, as shown in Figure 2.2.6.

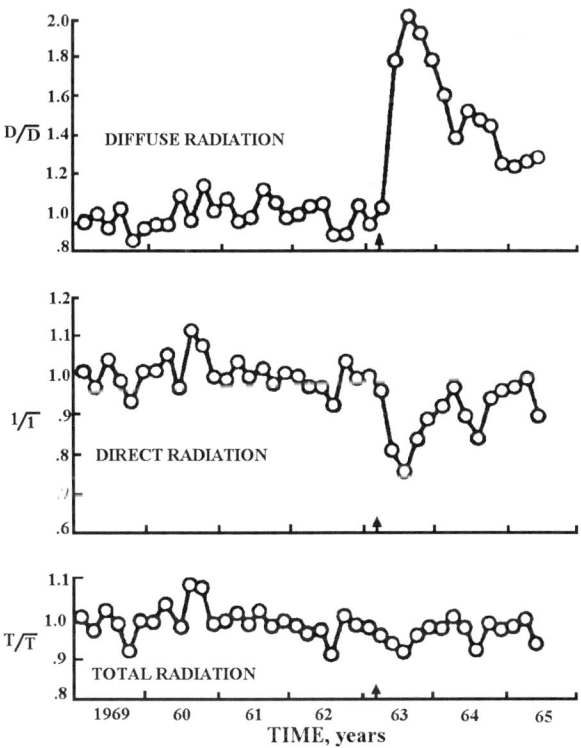

Figure 2.2.6. Solar radiation changes after the Mount Agung eruption in 1963. From Dyer and Hicks (1965). [Reprinted with permission from *Nature.* © by Macmillan Magazines Limited.]

While tropical tropospheric temperatures decreased by ≈ 0.4 °C (Angell and Korshover, 1977), the stratospheric temperatures there increased dramatically by few degrees after this eruption. The effect on the stratospheric temperature at 50 mbar (≈ 20 km) over Port Darwin (12° S) is shown in Figure 2.2.7 (adopted from McInturff et al., 1971). The quasi-biennial oscillation (QBO) with rather constant amplitude of about 1.5 °C is a usual event there. However, the maximum attained after the Mount Agung eruption with an amplitude of more than 5 °C has not been observed before and it is an extraordinary event, consistent with the observed spread of the volcanic aerosols. As already mentioned relative to the southern latitudes very little dust entered the NH. Nevertheless Figure 2.2.8 (Labitzke and Naujokat, 1983) which shows the zonal mean 30-mbar temperatures (°C) during July at 10, 20 and

30° N for the period 1962 through 1983 demonstrates an exclusive deviation well above the 3σ level only after the Mount Agung and El Chichón eruptions at these latitudes (see more in Labitzke and van Loon, 1989).

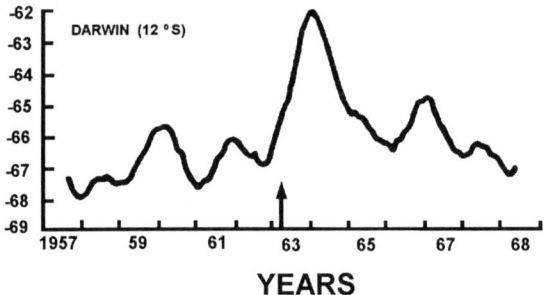

Figure 2.2.7. The course of the temperature in the stratosphere at ≈ 20 km over Darwin (12° S) during the period around the Mount Agung eruption (marked by arrow). After McInturff et al. (1971).

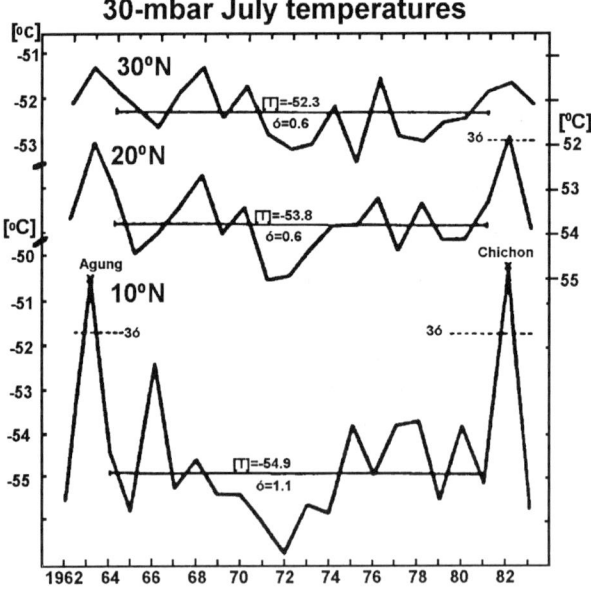

Figure 2.2.8. Zonal mean 30-mbar temperatures during July at 10, 20 and 30° N, for the period 1962 through 1983. After Labitzke and Naujokat, 1983.

Although the temperature decreases at the surfaces are larger than the standard deviations of the data, and there is good correlation among the tropical and extratropical latitude regions, the magnitudes of the temperature changes are not unusual. It is interesting to note that the largest temperature decrease at the surface of about 0.5 °C after the Mount Agung eruption which occurred in the NH (see Figure 2.2.6) could not have been caused directly by the volcanic dust. However, it should be noted that the radiation field is coupled to the atmospheric dynamics in a complicated manner and, therefore, climatic changes in regions not directly exposed to the radiative perturbation are clearly possible. The tropospheric cooling was greater (0.7 °C) in the extratropics of the SH (Angell and Korshover, 1977).

On May 18, 1980, there was an eruption of the **Mount St. Helens** volcano in the State of Washington in the USA., followed by an estimated 0.35 km^3 ejection of ashes and gases into the troposphere and lower stratosphere. Most of the material was injected laterally, laying waste to large areas out to a distance of 20–30 km. Up to 15 cm of fine ash was deposited some 200 km downstream. The wind-driven material near jet stream levels (10–14 km) circulated the globe in ≈ 16 days. The ash was fine-grained with predominant diameter of the particles <200 μm and only 1% of the mass resided with particles less than 2 μm. The material near 16 km moved eastward more slowly, crossing the U.S. eastern coast on 23 May and reaching Europe about 5 June. The material above 20 km moved westward (penetrating the altitude of wind reversal) and circulated the globe in 60–70 days (Toon, 1982). This eruption was not one of exclusive strength (volcanologists estimate there have been approximately 30 similar eruptions since 1912); however, it occurred at a location permitting extensive instrumental observations. Energetic efforts to achieve wide observational coverage in the United States and in some other countries after Mount St. Helens has made possible a broad program including six items:

1. The nature and implications of volcanic eruptions (in particular, the impact on climate, ecosystems, and water basins);
2. Direct measurements of eruption products (gaseous and aerosol components);
3. Indirect measurements using remote sounding techniques (ground-truth lidar, aircraft, and satellite measurements);
4. Transport and diffusion of eruption products;
5. The chemistry of eruption products (mainly, the processes of conversion of the gaseous components into aerosol ones);
6. Possible weather and climate implications (paleoclimatic data, the effect on the radiation budget, climatic impact).

In November 1980 a workshop was organized by National Aeronautics and Space Administration (NASA) to bring together experts in the various disciplines studying the Mount St. Helens eruption. The results of the chemical and physical measurements of the volcanic cloud from that report (Toon, 1982) are summarized in appropriate places of Chapters 3–5.

On March 28, April 3, and April 4, 1982, eruptions of the Mexican volcano **El Chichón** (17° N, 93° W) introduced gaseous and particulate matter up to ≈ 30 km into the stratosphere. Some details on the aerosols are given in Chapter 3. It was quickly recognized that this volcanic cloud ranked as one of the most massive clouds of the century; lidar returns from the cloud exceeded by large factors those obtained from the stratosphere since modern lidar soundings of the stratosphere were begun almost two decades ago (DeLuisi et al., 1983). Such values for the optical depth exceed by more than an order of magnitude those of the Mount St. Helens cloud during its first 6 months. This difference in optical depth occurred because the magma of the El Chichón volcano was much more gas-rich than the magma of Mount St. Helens and hence the El Chichón explosions contained a much larger amount of sulfur-bearing gases, the precursors of the sulfuric acid particles, and went straight into the stratosphere. Indeed, these optical depths for the El Chichón cloud are an order of magnitude larger than those recorded for any widespread volcanic cloud in the NH since the 1912 Katmai cloud and in the SH since the 1963 Mount Agung cloud. Such optical depths are large enough to cause nontrivial perturbations to the Earth's radiation budget and possibly to its climate, which is discussed in Chapter 5.

The difference in injection heights—27 to 28 km for El Chichón and 22 to 23 km for Mount St. Helens—is mostly accounted for by the 5-km difference in tropopause height at tropical and midlatitudes: a rising volcanic plume is driven by buoyancy forces, and therefore, greatly slows down once it enters the stratosphere with its large static stability. However, the static stability was about 40% larger near the base of the tropical stratosphere.

Within about 3 weeks, the main mass of the cloud circumnavigated the globe, transported by summertime easterly winds (winds that blow from east to west) that were enhanced by the quasi-biennial oscillation. By the end of May, it had spread throughout much of the northern tropics. By this time, the volcanic cloud extended from close to the tropopause (16 km) to 30-km altitude, with the peak concentration of cloud material at 27-km altitude and most of the cloud mass lying above 20 km (DeLuisi et al., 1983). During the summer of 1982, the main cloud mass above 20 km remained largely confined to the northern tropics.

For comparison of eruption characteristics of some of the most frequently mentioned volcanoes Table 2.2.1 was composed from current literature sources.

VOLCANIC ACTIVITY AND CLIMATE 109

Table 2.2.1. Comparison of Eruption Characteristics.

Volcano (lat°)/ year	Volume of magma (km³)	Height of eruption (km)	Stratospheric H₂SO₄ (9 ton x 10⁶)	Approx. ΔT° in NH
Laki (64° N), 1783	≈ 13	low	100	-1.0
Tambora (8° S), 1815	> 50	>45	150–200	-0.7
Krakatau (6° S), 1883	>10	>40	50–55	-0.4
St. Maria (15° N), 1902	≈ 9	>30	20	-0.04
Katmai (58° N), 1912	15	>27	20	-0.2
Mt. Agung (8° S), 1963	0.4 to 0.6	>18	10–20	-0.3
St. Helens (46° N), 1980	0.35	22	0.3	0 to -0.1
El Chichón (17° N), 1982	0.35	26	10–20	-0.4

A new powerful explosive volcanic eruption of **Pinatubo** (15.14° N, 120.35° E) took place on June 15, 1991 (see the survey of relevant data in Kondratyev, 1993). Table 2.2.2 contains some intercomparison data DVI.

Of all the possible causes of climate changes, volcanic eruptions are the most adequately documented and understood. For example, a stratospheric temperature increase trend was observed after the 1963 Agung eruption in Indonesia. This trend was superimposed on a growing-in-amplitude quasi-biennial stratospheric temperature variation and could be explained by the Agung eruption. This increase was clearly pronounced in the SH, where the variability of the amplitude of quasi-biennial oscillations is small, whereas in the NH (with a small input from the eruptive cloud) there was no stratospheric temperature increase. These observational data agree with theoretical estimates, from which it follows that a stratospheric warming is mainly caused by absorption of the upward thermal emission flux and, to a smaller degree, by the solar radiation absorption. In the troposphere and at the surface there was a global-scale temperature decrease (in the NH it was the largest, 0.6 °C). Hence the decrease could not have been caused totally by the eruption.

Table 2.2.2. Estimates of various volcanic eruption characteristics.

Characteristics of eruptions	Nevarupta/Katmai, 1912	El Chichón, 1982	Nevado del Ruiz, 1985
SO_2, amount, kg	$(5.2–20).10^3$	7.10^5	750
Amount of ash ejections, g	$(3.4) 10^{16}$	$3.0.10^{15}$	$4.8.10^{13}$
Gas/ash mass ratio	$(2.6).10^{-4}$	$2.3.10^{-3}$	0,015
DVI	6	4–5	3

Following the 1815 Tambora volcanic eruption in Indonesia "a year without the sun" was observed, and a summertime temperature decrease in New England and Western Europe (with respect to the climatic mean) reached 1 to 2.5 °C.

Mass and Schneider (1977) analyzed the consequences of numerous eruptions causing an increase to several tenths of the aerosol optical thickness in the stratosphere in the visible. They found that a statistically reliable temperature decrease a year after the eruption must constitute 0.3 °C; and after 2 years, 0.1 °C and more.

Although volcanoes have long been considered a cause of glaciations, this hypothesis remained unsubstantiated. Analysis of the aerosol particle content in the Greenland ice cores has not revealed any increase in the aerosol content during the early stages of the Wisconsin glaciation cycle.

The role of volcanic eruptions in the present temperature trend in the NH (Kondratyev, 1981a; 1981b; 1985; 1988; 1992a, Toon, 1982) is not clear, since the post-1940 cooling was not preceded by an intensification of volcanic activity. Theoretical estimates reveal a strong effect of the composition and size distribution of volcanic aerosol on climate changes. For example, even with a small amount of an absorbing silicate aerosol, a considerable increase of solar radiation absorption may take place, and the presence of silicate particles larger than 0.5 μm can cause a warming at the surface as a result of the thermal emission of such particles. Calculations of temperature variations at the surface and in the stratosphere made by Hansen et al. (1978) after the eruption revealed good agreement with observations.

Bryson and Goodman (1982) analyzed the "modulation" of the incoming direct solar radiation under the influence of volcanically induced changes in atmospheric transparency, using data of the longest series of actinometric observations. To estimate the temporal dynamics of the volcanic aerosol content in the stratosphere, it is assumed that the characteristic reaction time for the formation of the H_2SO_4 aerosol is 115 days, and the time for deposition of the particles about 400 days. In view of a rapid gravitational settling of ash particles, the mean annual content of the submicron H_2SO_4 aerosol is assumed to be twice as large as that of ashes directly ejected into the stratosphere. With these assumptions, one can describe the temporal dynamics of the ash and H_2SO_4 components of eruptive aerosols, typical of an explosive eruption as shown in Figure 2.2.9.

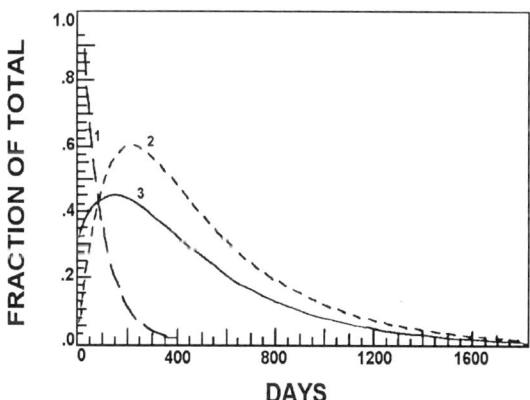

Figure 2.2.9. Temporal evolution of volcanic erupted aerosols. 1. Ash particles; 2. Sulfuric acid aerosol resulting from gas-to-particle conversions; 3. Total content of particles. After Bryson and Goodman (1982).

Until recently, observations of atmospheric transparency that were used to estimate the climatic impact of eruptions (Galindo, 1992, Galindo and Bravo, 1975, Galindo et al., 1996, Kalitin, 1935, Kondratyev, 1980a; 1980b; 1981a; 1981b; 1992a; 1992b, Loginov, 1984a; 1984b, Loginov et al., 1983a; 1983b) have been a major source of information of the atmospheric effect of volcanic eruptions. Clear-sky solar radiation data for 1880–1980, used to identify the secular trend of the atmospheric optical thickness, revealed a distinct correlation between transparency decrease and large-scale volcanic eruptions , as shown in Figure 2.2.10. These include the following eruptions among others: Pelée and Soufrière (1902), Katmai (1912), Bezymiannaya (1955), Agung (1963) and El Chichón (1982). A comparison of the secular trend of the optical thickness and a "synthesized" chronology of powerful volcanic eruptions, taking into account the moment of eruption and the subsequent decrease of the concentration of erupted aerosols for low, middle, and subpolar latitudes, yielded a correlation coefficient of 0.87. An additional consideration of moderate eruptions raised the correlation coefficient to 0.95. This permits one to believe that variations in atmospheric transparency are practically completely governed by volcanic eruptions.

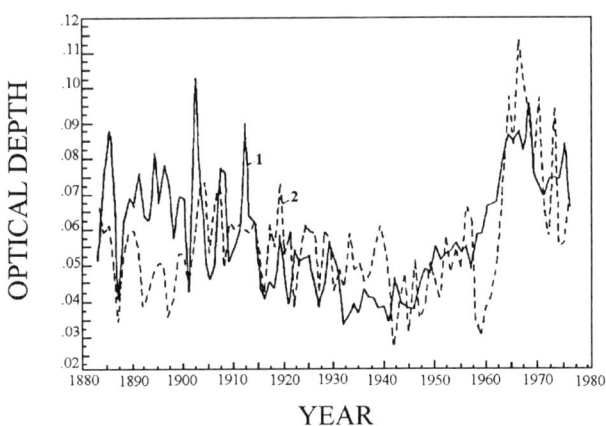

Figure 2.2.10. Secular trend of the atmospheric optical depth. Data for the year 1900 should be considered unreliable. 1. Observations; 2. Theoretical. After Capone et al. (1983). [© by American Geophysical Union]

The residual variability of the optical thickness (about 5% to 10%) reveals an increasing trend that may be ascribed to an anthropogenic impact. Attempts to find 11- or 22-year cycles in the residual variability have not been successful. This result attests to the absence of the impact of solar activity on incoming solar radiation. An analysis of data on clear-sky global radiation and mean annual surface air temperature after the year 1920 yielded a correlation coefficient of 0.91. Although this correlation should not be considered as proof of the respective cause-and-effect connection, a volcanically induced modulation of the mean hemispherical temperature secular trend can nevertheless be supposed to take place (Fig. 2.2.11).

Examination of this assumption, by calculating the temperature variations in the twentieth century with the use of an energy-balance climate model (the mean hemispheric cloud cover is assumed to be fixed) showed that with a prescribed observed trend of the optical thickness, temperature variations close to those observed are obtained. Using this as a basis Bryson and Goodman (1982) drew the conclusion that this observed variability of the mean surface air temperature in the NH is determined primarily by volcanic eruptions as shown in Figure 2.2.12. On the other hand, it follows that with the volcanically induced variability excluded, one can filter out the trend due to an increase in

VOLCANIC ACTIVITY AND CLIMATE 113

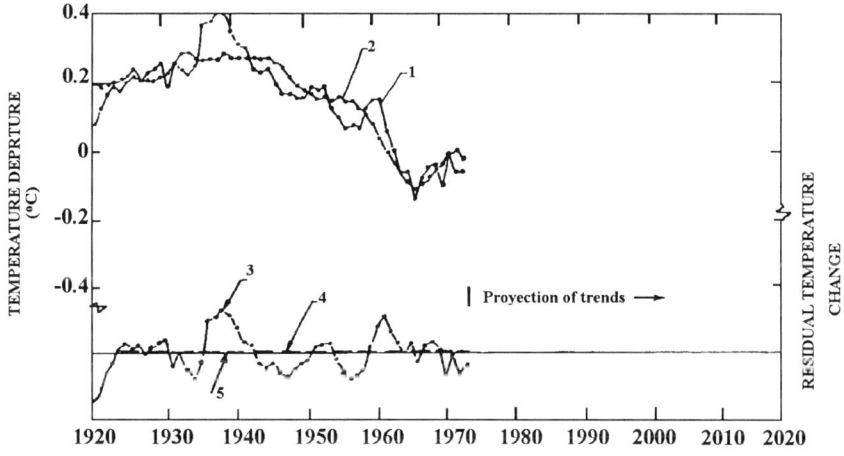

Figure 2.2.11. Variability of the NH mean surface air temperature and optical thickness of the atmosphere. 1. Observed NH surface air temperature variations from data of Borzenkova et al. (1976); 2. Secular trend of the optical thickness anomalies; 3. Residual temperature variations; 4. Linear trend (dashed line); 5. Exponential trend (solid line). After Borzenkova et al. (1976).

the atmospheric CO_2 concentration. This attempt, however, failed: the residual variability shows no trend of temperature increase. Even the forecast taking into account CO_2 concentration increases until the year 2020 (with the optical thickness of the atmosphere fixed at the level of 1975) revealed a future warming during the next 50 years less than the warming observed during the recent years. In Bryson's opinion, in forecasting possible future climate changes the problem of possible prediction of volcanic eruptions rather than the runaway CO_2-induced greenhouse effect must be of primary importance. It must be emphasized, however, that this conclusion is of questionable validity (Kondratyev, 1980b; 1981a; 1981b; 1988; 1992a).

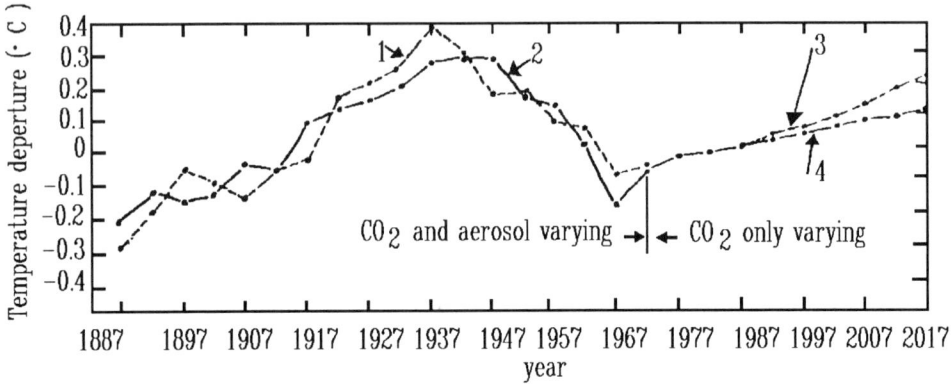

Figure 2.2.12. Secular trend of the 5-year averages of the NH mean surface temperature: 1. Observed; 2. Calculated. Calculated temperature variations at a fixed optical thickness of the atmosphere (at the level of the year 1975); 3. Due to an exponential increase of the CO_2 concentration; 4. Due to an increase calculated from data on possible dynamics of fossil fuel consumption. After Bryson and Goodman (1982).

It has to be noted in this connection that based on the use of the techniques of superimposed epochs, a correlation has been found between low-latitude volcanic eruptions and a positive SST anomaly in the eastern Pacific (90°– 190° W, 0° – 10° S), whereas the extratropical eruptions were followed by a negative SST anomaly. In particular, in three seasons after an eruption in the tropics, an SST increase had been statistically substantial at a 90% confidence level, i.e., the SST had exceeded the norm by more than 2.81 $\sigma/\sqrt{N}$ (N is the number of eruptions considered equal to 12, and σ is the SST MSD during a season).

To test the reliability of these results (with the use of the technique of superimposed epochs), Parker (1988) processed another SST database for the latitudinal band 20° S – 20° N, east of 170° W. Though the new results confirm the conclusions mentioned above, an inadequate volume of the initial database and random successions of extreme SST anomalies do not permit one to consider the new results sufficiently reliable. In this connection, the development of realistic numerical models of the situation under discussion is of great importance.

Using the technique of superimposed epochs, Nicholls (1988) analyzed the anomalies of atmospheric pressure in Darwin (12° S, 121° E) as an indicator of the ENSO event before and after the volcanic eruptions in the tropical latitudes. This analysis from the data of mean monthly pressures for the period 1898–1984 revealed clear correlations.

The results obtained reflect the possibility of a relationship between ENSO and low-latitude eruptions. However, a limited volume of samples does not permit one to draw a conclusion about the statistical reliability of correlation. There is no doubt that the assumption about ENSO being caused by volcanic eruptions cannot be considered substantiated, since the pressure anomaly and the trend of pressure mentioned above appear long before the moment of eruption. On the other hand, if a maximum positive anomaly of pressure 4 months after the eruption is the result of the pressure-increasing trend appearing before the eruption, an opposite assumption with respect to eruptions being caused by anomalies in the atmosphere and ocean, forming in the course of ENSO, can be considered more likely. The results under discussion should rather be considered as a manifestation of random fluctuations. These results demonstrate, on the other hand, that the climate system "machine" is so complicated that it is prohibitively difficult to isolate a single climatic "signal". We shall discuss this problem in more detail later on.

Assessment of posteruption variations in the Earth's radiation budget is a key aspect of the climatic impact of volcanic eruptions. Using the two-stream approximation and a three-layer model of the surface-troposphere-stratosphere system Lenoble et al. (1982) estimated the effect of a stratospheric aerosol layer constituting a 75% water solution of sulfuric acid or background aerosol particles (Table 2.2.3) on the system's radiation budget and its components, as well as on the surface temperature (the surface is considered to be an isotropic reflector). The distribution of the particles number density (n) by radius (r) is prescribed by a modified gamma function:

$$n(r) = A_r^\alpha \exp(-br^\gamma),$$

where A, α, b and γ are parameters. A comparison of the applied two-stream approximation to an accurate solution has shown that the errors of an approximate calculation of the diffuse transmission do not exceed 10%, with the single scattering albedo $\varepsilon \geq 0.6$, optical thickness $\tau \leq 0.5$, and normally incident solar rays. Here the system's albedo is calculated accurately, and the error in calculations of albedo variations, due to changed properties of the stratospheric aerosol layer, does not exceed 10%. The Table 2.2.4 data illustrate variations in the system's radiation budget (ΔB at $\tau_s = 0.03$), testifying to the fact that in both cases the radiation budget decreases since the contribution by backscattering prevails. An estimation of the mean global surface temperature decrease (at $\tau_s = 0.03$) gave -0.45 °C (volcanic aerosol) and -1.1 °C (background aerosol). An account of the albedo feedback resulted in an intensified cooling down to

-0.8 °C and -2.15 °C, respectively. A weaker effect of volcanic aerosols on climate would exist if there were a lower single-scattering albedo (e.g., stronger absorption).

Table 2.2.3. The size distribution and optical properties of stratospheric aerosols. After Lenoble et al. (1982).

Aerosol model	Parameters					
	α	b	γ	ω_s^{eff}	b_s^{eff}	$Cs=\tau_{IR}^{eff}/\tau_s^{eff}$
Background aerosol	1.0	18	1.0	0.998	0.189	0.0383
Volcanic aerosol	1.0	16	0.5	0.994	0.169	0.217

[a] ω_s^{eff}, b_s^{eff} are effective values of the single scattering albedo and backscattering: τ_{IR}^{eff}, τ_s^{eff} are effective values of the optical thicknesses for the shortwave and longwave radiation, respectively; τ_s=0.03.

Table 2.2.4. Variations in the System's Radiation Budget (RB) $\Delta B(W/m^2)$. After Lenoble et al. (1982).

Aerosol Model	RB Variations Due to:			
	Backscattering	Shortwave Radiation Absorption	Greenhouse Effect	Total
Background aerosol	-28.5	0.0923	1.56	-26.8
Volcanic aerosol	-26.4	2.28	0.887	-23.2

When discussing the results of observations of volcanic aerosols, it is important to assess possibilities of the quantitative characteristics of the eruption intensity. In this regard, Kelly and Sear (1982) analyzed the adequacy of Lamb's DVI, which was introduced to characterize quantitatively the impact of volcanic aerosols on the ERB and climate during the several years following an eruption (Johnson, 1993). An estimate of the DVI taking into account the observational data and empirical and theoretical studies of the possible climatic implications

VOLCANIC ACTIVITY AND CLIMATE 117

of the dust veils is based on the nature of such implications as understood in 1960. The DVI is calculated by averaging over the maximum possible number of DVI estimates determined with the use of various techniques.

Five available techniques were mutually calibrated (normalized) to give the DVI value 1000 for the 1883 Krakatau eruption. According to technique B, the mean global DVI is

$$DVI = 0.97 \cdot R_{max} E_{max} t_{mo},$$

where R_{max} is the posteruption maximum extinction of direct solar radiation determined as a monthly mean for the midlatitudes of the given hemisphere (maximum extinction can sometimes be reached only 2 years after the eruption); E_{max} is the geographical extent of the dust veil in conditional units as a function of the volcano's latitude (E_{max} = 1.0 for the band 20° N to 20° S and 0.3 for latitudes 40°); t_{mo} is the lifetime of the dust veil (in months). Technique C gives

$$DVI = 52.5 \cdot T_{Dmax} E_{max} t_{mo},$$

where T_{Dmax} is the mean temperature decrease (°C) in the midlatitudes of the given hemisphere during the year when the effect of an eruption is at a maximum.

Recently specified data on the chronology of eruptions and a more adequate understanding of the nature of their climatic impact (e.g., the role of the secondary H_2SO_4 aerosol, and so forth) necessitate a revision of the understanding of the DVI.

The climatic implications of volcanic eruptions have recently been of great concern. For example, an extensive complex of observations of volcanic emissions and their propagation in the atmosphere have been carried out with the use of conventional observational means (e.g., ground-based, aircraft, balloon) as well as satellites. Substantial progress has been reached in the development of the techniques for numerical modeling of climate and large-scale transport (Kondratyev, 1992a; 1992b, Kondratyev and Grassl, 1993). Of great importance was the successful numerical modeling of the processes of stratospheric aerosol formation.

Recent powerful eruptions of the Mount St. Helens (1980), El Chichón (1982) and Pinatubo (1991) volcanoes have become new "experiments" that have made it possible, for the first time, to document adequately posteruption changes in the atmosphere. Some relevant results are considered below, with the emphasis on an analysis of El Chichón and Pinatubo data.

2.3 Volcanic Aerosols

2.3.1 General Background

Volcanic explosions occasionally inject gases and particles into the stratosphere. Initially, the volcanic aerosols may be dominated by micron-sized ash particles; however, the photochemical conversion of sulfur-containing gases (e.g., SO_2) causes sulfuric acid aerosols to become the chief particulate species after a few weeks. The small size of the acidic volcanic aerosols – and hence their very low sedimentation velocities, as well as the absence of rainfall in the stratosphere, allows significant quantities of these particles to remain in the stratosphere for a year to several years. Consequently, the volcanic cloud can spread over a large fraction of the globe before it is depleted and it may alter the ERB over a long enough time interval to produce measurable perturbations to temperature and wind fields in both the stratosphere and the troposphere.

Some time ago it became clear that the long-lived stratospheric clouds produced by volcanic eruptions are composed largely of sulfuric acid aerosols (Castelman et al., 1974, Pollack et al., 1976a; 1976b). The amount of sulfur-rich volatile (for example, SO_2, H_2S) injected into the stratosphere by an explosive eruption is, therefore, a critical determinant of its atmospheric impact. The comparatively small-volume eruptions of Mount Agung in 1963 and El Chichón in 1982 both generated substantial stratospheric aerosol clouds, despite the fact that they erupted < 0.5 km³ of magma. Comparison of data from direct measurements of stratospheric optical depth, Greenland ice-core acidity, and volcanological studies show that such relatively small, but sulfur-rich eruptions can have atmospheric effects equal to or even greater than much larger sulfur-poor eruptions. These small eruptions are probably the most frequent cause of increased stratospheric aerosols.

Major injections of SO_2 into the stratosphere such as those during the 1963 Mount Agung and 1982 El Chichón eruptions probably led to a considerable decrease for many months of the hydroxyl content of stratosphere during the gas phase oxidation of sulfur compounds such as:

$$SO_2 + OH + M \rightleftharpoons HSO_3 + M \quad (2.1)$$
$$H_2S + OH \rightleftharpoons HS + H_2O \quad (2.2)$$

Shortly after an eruption, conditions in the ash-filled cloud containing much condensed water are likely to be favorable for the so-called heterogeneous oxidation of SO_2. In this process, gaseous SO_2 is adsorbed by the particles and subsequently oxidized either on the surface of solid particles or in the body of a liquid droplet. Available kinetic information indicates that H_2S is converted to SO_2 much faster than SO_2 is oxidized to sulfate by homogeneous gas phase reactions. The process by which SO_2 is chemically converted to sulfate is not precisely known, and various reaction mechanisms have been proposed. The rate-limiting step of SO_2 oxidation is believed to be reaction with the OH radical to form the $HOSO_2$ radical. Depending on the fate of the $HOSO_2$ radical, odd hydrogen may or may not be regenerated in the overall oxidation sequence (McKeen et al., 1984). Other environmental factors which are thought to influence the rate of heterogeneous oxidation of SO_2 are concentrations of H_2S_2 (or its precursor O_3), alkaline substances in the ash or ammonia in the air which enhance the absorption of SO_2 by the particles or droplets and trace metal catalysts that are present in the ash.

The heterogeneous oxidation of SO_2 in stratospheric volcanic clouds can account for early formation of sulfate, particularly in association with ash particles. The principal effect is to enhance the size of existing particles which would, in turn, enhance the gravitational settling rates of the affected particles. Once sufficient dilution of the eruption cloud occurs by mixing with previously clean stratospheric air, the conditions of lower concentrations of reactants and low temperature tend strongly to suppress the rates of heterogeneous chemical processes.

On the basis of present knowledge of chemical reactions and composition of the stratosphere, it is reasonably certain that SO_2 placed in the stratosphere is photochemically oxidized, ultimately to what is believed to be sulfuric acid, by the reaction (2.1). The mean lifetime of SO_2 for this reaction is estimated to be of the order of 25 to 100 days depending upon model estimates of OH concentrations. The supposed product of this reaction, HSO_3, has not been identified. If, indeed, HSO_3 does form from the reaction, its subsequent reactions in the atmosphere and its ultimate end products are uncertain. Friend et al. (1980) and Davis et al. (1979) have discussed possible reactions of sulfur-oxygen radicals and their hydrates to form such species as $H_2S_2O_6$ (dithionic acid) and $H_2S_2O_8$ (peroxodisulfuric acid). Turco et al. (1979a, 1979b) have suggested that HSO_3 may be converted to SO_3 by reaction with OH radicals and the SO_3 may then be converted to H_2SO_4 by reaction with H_2O. Thus, to summarize the above, though the chemical mechanisms are not known in detail, SO_2 in the stratosphere is oxidized via reaction with OH radicals to form intermediate sulfur-oxygen species which in turn form particles commonly thought to consist of H_2SO_4 and H_2O.

The question of the means by which sulfate aerosols are formed from gaseous species in the stratosphere is of particular significance in any assessment of the state of knowledge of the impact of volcanic emissions on global climate. Unfortunately, it is also an open question. The foregoing discussion illustrates the state of knowledge concerning the chemical identity of the product of SO_2 photooxidation. One concept of aerosol formation considers that H_2SO_4 is the principal product of SO_2 photooxidation and that the principal fate of the H_2SO_4 vapor is to condense on the surface of "prior existing nuclei." In this manner, sulfate aerosol is formed while essentially no new particles nucleate (because the homogeneous condensation of H_2SO_4 and H_2O vapor is far too slow). Simulations of aerosol formation by these processes have been made by Turco et al. (1979a, 1979b). Another notion of aerosol formation propounded by Friend et al. (1980) holds that, while the principal sink of gaseous sulfur oxyacid ($H_2S_2O_6$ or $H_2S_2O_8$) is scavenging by aerosol surfaces, new particles are formed by the "activationless" clustering of H_2O and HSO_x molecules.

At this juncture, there are no atmospheric data that permit unambiguous resolution to the question of particle formation. However, it does appear likely that the major chemical pathway for the conversion of most of the volcanic SO_2 in the stratosphere is the homogeneous oxidation by OH radicals. The observations of apparent new particle formation by Hofmann and Rosen (1981, 1982, 1983a, 1983b, 1984) and Hofmann and Oltmans (1992) are in accord with this concept, since the heterogeneous mechanisms perforce cannot create new particles. Further review of the processes of the stratospheric aerosol formation could be found in Kondratyev (1980b, 1985, 1988, 1992a, 1992b) and McKeen et al. (1984).

In the light of the foregoing discussion, the expected effects of volcanic emissions on the Earth's radiation balance are largely caused by ash particles and sulfur-oxyacid aerosol particles suspended in the stratosphere. Present understanding of atmospheric photochemistry implies that weeks to months are probably required for nearly complete conversion of the emitted sulfur gases to aerosol. Thus, ash particles may have been relatively more responsible for early effects on the radiation balance, but as the ash settled out, the secondary aerosol formed from H_2S and SO_2 oxidation probably played an increasingly important role.

The eruptions of Mount Agung and El Chichón provided first evidence that relatively small-volume explosive events can have noticeable impact on the content of atmospheric aerosols followed by significant effects on climate.

The measured optical-depth changes after the Mount Agung eruption suggest $\approx 10-20 \cdot 10^6$ tons of aerosol in the stratosphere (Deirmendjian, 1973, Stothers, 1984). Based on the maximum estimated bulk volume of Agung ashfall (≈ 1 km^3) the total amount of very fine (< 2 μm) ash that could have been produced by the eruption is roughly estimated as $< 8 \cdot 10^6$ tons. Direct high-altitude sampling soon after the eruption suggest $\approx 5 \cdot 10^6$ tons of ash in the stratosphere (Toon and Pollack, 1982). Such fine ash has only a brief stratospheric residence time, however, and was largely removed within a few months, leaving an aerosol cloud composed mainly of H_2SO_4 droplets.

For comparison, the El Chichón (17° N) eruption in 1982 was also small in volume, producing about 0.3–0.35 km^3 of trachyandesite magma. The

VOLCANIC ACTIVITY AND CLIMATE 121

eruption had three main subplinian explosive phases: on 28 March, and 3 and 4 April. The eruption columns penetrated the stratosphere, where the main stratum of El Chichón aerosol spread out at about 26 km altitude. Estimates based on lunar eclipse data give an average global peak optical depth of 0.12 in December 1982; the aerosol was preferentially concentrated in the NH, where the optical depth was 0.15. Conversion from optical depth to sulfuric aerosol mass loading yields about $20 \cdot 10^6$ tons of H_2SO_4 aerosol in the global stratosphere 6 months after the eruption. This agrees with estimates based on airborne lidar measurements ($12 \cdot 10^6$ tons) and on balloon-borne particle counters ($10 - 20 \cdot 10^6$ tons) (Rampino and Self, 1984).

Several studies have suggested that iron-rich basaltic and andesitic melts have a generally high capacity to carry dissolved sulfur. Melts of iron-poor dacite to rhyolite composition that commonly produce large volume explosive plinian and ignimbrite eruptions (for example, Krakatau, 1883; Santa María, 1902) are usually poorer in sulfur volatile. Where could the excess sulfur released in the 1963 and 1982 eruptions have come from? In the case of El Chichón, anhydrite ($CaSO_4$), considered rare in volcanic rocks, occurs as a common phenocryst phase in the deposits, and decomposition of anhydrite has been suggested as a source of additional sulfur during the eruption. The Agung deposits, however, show no existence of anhydrite. Another possible source of sulfur volatile is degassing of nonerupted magma (Rampino and Self, 1984).

In the following parts of this chapter we will confine ourselves to the results of some characteristics of stratospheric ions, to data from aerosol measurements after some eruptions and to a few comparative aspects with other planets.

2.3.2 Comments on the Stratospheric Ions

In very close relation to understanding the formation of the stratospheric aerosols is better understanding of the behavior of the stratospheric ions. Until recently, it was common opinion that the fate of ions in the stratosphere was determined by spontaneous neutralization as a result of recombination of charged particles of the opposite sign. Arnold (1980) has shown that such a conclusion can be incorrect with respect to complex ion clusters, since clustering promotes ion stabilization, rather than neutralization. In this case stable ion pairs should be observed. Subsequently the first simultaneous measurements of the composition of positive and negative ions in the stratosphere (at an altitude of about 36 km) revealed the presence of components more stable than those forecast theoretically. This demonstrated the need to study the probable role of ion pairs in the formation of aerosols. Arnold (1980) made an assessment of the possibilities of the formation of stable ion pairs in the light of new data on the stratospheric composition proceeding from criteria of stability formulated as:

$$E_D = E_N + E_C.$$

Here E_D is the effective energy needed to tear an electron from a negative cluster ion; E_N is the effective energy released in neutralization of a positive cluster ion by a free electron; E_C is the energy released in formation of a chemical bond which may result from interaction of ion nuclei.

An analysis of the characteristics of stratospheric ions has shown that the criterion in question is not fulfilled during the recombination of proton hydrates $H^+(H_2O_4)$, $H^+(H_2O)$, and NO_3^- $(HNO_3)_2$, i.e., in this case recombination does not lead to stable ion pairs but to a spontaneous recombination. Stable pairs of ions are formed, however, in the case of recombination of more complex cluster ions. Due to the great dipole moment of such pairs, there can be successive joining of free ions and forming of multi-ion complexes (MIC), similar in nature to ion crystals or salt particles. Mutual joining of ions is an alternative mechanism to MIC growth. The possibility of coagulation of MIC with aerosol particles should also be considered.

Thus, the formation and growth of MIC is a probable mechanism for gas-to-particle conversion, which does not require condensation and, consequently the necessary presence of oversaturated gaseous components. If, however, such components (for instance, H_2SO_4 vapor) do exist, then MIC, having reached a considerable size, can function as condensation nuclei. They also can substantially affect minor gaseous components of the stratosphere causing not only condensation but also reactions on MIC surfaces.

If the role of MIC in the formation of the stratospheric aerosol becomes substantial, it could be important in determining the mechanisms of the impact of solar activity on the atmosphere, since the solar and galactic cosmic rays are the main source of ionization in the stratosphere and troposphere. In this connection there is a need to search for MIC in the stratosphere, by means of balloon measurements using ion mass spectrometers operating in a wide range of masses.

Nitrogen oxides are known to be one of the photochemically active stratospheric components. If nitrogen oxides can react with sulfur to form constituents of stratospheric aerosols, these constituents might provide a sink for nitrogen oxides and, thus, purification of the stratosphere of pollutants. Though ammonium sulfate $(NH_4)_2SO_4$ has been reported in stratospheric aerosols, it can not be considered as the basis for a sink for nitrogen oxides, since ammonium sulfate results, apparently, from the reaction between gaseous ammonium and H_2SO_4 droplets.

Farlow et al. (1978) pointed out, however, that some studies had revealed a possibility of direct reactions between nitrogen oxides and sulfuric compounds:

$$2SO_2 + 3NO_3 + H_2O \rightarrow 2NOHSO_4 + NO, \quad (2.3)$$
$$2NO_2 + H_2SO_4 \rightarrow NOHSO_4 + HNO_3, \quad (2.4)$$
$$NOHSO_4 + SO_3 \rightarrow NOHS_2O_7 \quad (2.5)$$

In this connection, an attempt has been made to find the two above-mentioned forms of chamber crystals ($NOHSO_4$ and $NOHS_2O_7$). An X-ray analysis of 17 impactor samples of the stratospheric aerosol obtained in 1976 using a U-2 aircraft flying at 15 – 20 km altitudes, revealed the presence of both forms of chamber crystals as well as $(NH_4)_2SO_4$ and $(NH_4)_2S_5O_8$. The estimates have shown that a maximum NO content that can be absorbed by the stratospheric aerosol may range from 1/3 to 2 with respect to the content of nitrogen monoxide in the environment. Proceeding from the fact that $NOHSO_4$ constitutes 18%–19% of the precipitated aerosol, one can estimate that from $1 \cdot 10^7$ to $5 \cdot 10^7$ kg/year of nitrogen monoxide can be removed at the expense of the formation of chamber crystals. But further efforts are needed to verify the results obtained.

Another viewpoint on this matter, however, is that $(NH_4)_2SO_4$ is probably not present in the stratospheric aerosol (Hayes et al., 1980). They found that when the U-2 samples are protected by the inert gas argon from the time of collection until laboratory analysis, no solid ammonium sulfate content is found, whereas samples without argon protection changed to ammonium sulfate solids after 1 hour of exposure to laboratory air.

The problem of the origin and dynamics of the sulfate component of the stratospheric aerosol layer has evoked great interest in studies of the chemistry of sulfur compounds in the atmosphere. Earlier the carbonyl sulfide, OCS, had been supposed to be an important source of sulfur for the stratosphere (apart from the sulfur dioxide ejected during volcanic eruptions). As a result of OCS photolysis in the stratosphere the atoms of sulfur were released and then oxidized giving H_2SO_4. It had been assumed also that a reaction between OCS and OH could have been a source of the background tropospheric SO_2, and a similar reaction with participating CS_2 produced OCS and SO_2. Thus, reactions of OCS and CS_2 with hydroxyl could determine considerable sinks of these molecules in the troposphere.

In this connection, Cox and Sheppard (1980) undertook measurements of the rates of reactions of hydroxyl with a number of sulfur compounds, including carbonyl sulfide and carbonyl disultide at an atmospheric pressure of 10^5 Pa and a temperature of 279 ± 2 K. The results obtained showed that sulfur dioxide is the main product of reactions between hydroxyl and the sulfur compounds under investigation. The presence of hydroxyl in the troposphere determines the existence of the sink for all the gases reacting with it. With the $OH + CS_2$ reaction rate equal to $4.3 \cdot 10^{-13}$ cm^3 $mol^{-1}s^{-1}$, the lifetime of carbonyl disulfide in the troposphere is 0.2 years. The reaction of CS_2 with OH also leads to the formation of OCS in the troposphere. Since the measured rate of the OH + OCS reaction is small ($\leq 4 \cdot 10^{-14}$) and still estimated unreliably, there are no grounds for the conclusion that an intensive sink of carbonyl sulfide exists in the troposphere.

On the basis of the consideration of elementary ionic and molecular reactions in mechanisms of stratospheric heterogeneous analysis Burley and Johnston (1992) have suggested a general mechanism for acid catalyzed reactions, with examples that convert inactive HCl and $ClNO_3$ into photochemically active $ClNO_2$. ClNO, Cl_2, HOCl, and ClO, and showed that reactions forming all of these products are thermodynamically allowed.

2.3.3 Aerosol Measurements After Some Eruptions

Despite clear-cut evidence for the impact of volcanic eruptions on global-scale meteorological processes, studies of eruption products based on direct measurements are still not sufficient. We shall discuss below some of the relevant early results (an excellent survey was published by Hofmann in 1988). Stith et al. (1978) undertook airborne particle and gas measurements in the plumes from five volcanoes in Alaska and one volcano in the State of Washington. The detailed aerosol measurements included determination of the chemical composition (filter samples' analysis) and size distributions in a wide range of sizes from Aitken condensation nuclei (0.002 µm) and cloud condensation nuclei to 66 µm (photoelectric counters for particles of different types). Application of the flame photometry and chemiluminescence technique enabled the estimation of the mixing ratios of sulfur dioxide, ozone and nitrogen oxides. Simultaneously, measurements were made of meteorological parameters (temperature, dew point, wind turbulence, UV radiation). Most of the data were processed by the airborne computer.

The flights were carried out according to two schemes:

1. In the presence of the continuous emission plume, the aircraft crossed the plume from above, downward at distances from the volcano along the wind direction.
2. In the case of individual ejection clouds the aircraft entered the zone of ejection several times as the ejection cloud moved.

The data available on the vertical sounding and wind measurements enabled the estimation of the total fluxes of different components. The results obtained refer to the periods of before, after and during the main stage of eruption. In this connection, classification of volcanic emissions into four categories was used:

(i)   Paroxysmal, lasting several minutes;
(ii) Intra-eruptive, during the intervals between paroxysms of eruptions;
(iii) Post-eruptive, lasting from a day to a year (water vapor being the basic component of such emissions);
(iv) Extra-eruptive, nonconnected with eruptions (including fumarole emissions of active volcanoes) and lasting about 5–10 years and longer (water vapor is their basic component).

Concentrations of the sulfur dioxide, Aitken condensation nuclei, and the scattering coefficient are the most sensitive indicators of the volcanic eruption plumes. The material ejected during the main stage of an eruption is characterized by a relatively low content of particles ($r < 0.1$ μm) and sulfur dioxide as compared to emissions during intermediate periods.

So, for instance, the greatest ejections of sulfur dioxide during the January-February 1976 St. Augustine (Alaska) eruption fell in the periods during (but not at the moment of paroxysmal intensity) and after the eruption (about 10^{11} g). During the main (most intensive) stage of eruption the ejection of particles with a size ranging from 0.01 to 66 μm constituted about $6 \cdot 10^{12}$ g, including $2 \cdot 10^{11}$ g of particles ranging from 0.01 to 5 μm. In addition, after eruption about $5 \cdot 10^{10}$ g of particles in the $0.01 - 5$ μm range was also ejected.

Estimation of the emission rates for six volcanoes gave the values varying within:

$4 \cdot 10^{-1} - 2 \cdot 10^{-4}$ kg/s   (hydrogen sulfide),
$1 \cdot 10^{5} - 1 \cdot 10^{1}$ kg/s   (water vapor),
$6 \cdot 10^{5} - 2 \cdot 10^{5}$ kg/s   particles' mass),
$6 \cdot 10^{19} - 3 \cdot 10^{14}$ s^{-1} (Aitken condensation nuclei), and
$2 \cdot 10^{18} - 1 \cdot 10^{16}$ s^{-1} (cloud condensation nuclei).

The accuracy of estimating the fluxes of the above-mentioned components is about 20%.

An analysis of the aerosol sampling composition revealed in different emissions of the St. Augustine volcano (mainly, volcanic ashes), silicon, aluminum, magnesium, calcium, and iron, (sometimes with traces of potassium, titanium and sulfur). Some small particles ($0.3 - 1$ μm) contained much iron. Though the eruption of the St. Augustine volcano has not been strong as compared to many others, the estimate of the total ejection of particles ($6 \cdot 10^{12}$ g/year) is comparable with a lower limit for the range of global-scale volcanic ejections ($4 - 120 \cdot 10^{12}$ g/year). It is important, however, that only $0.25 \cdot 10^{12}$ g are long-living particles with a size less than 5 μm, and the main portion of the ejected mass are large particles which undergo rapid sedimentation.

As has been mentioned earlier, volcanic ash particles are dominant during the main stage of the eruption, while after a few months, sulfate particles, the by-product of gas-to-particle conversions, play the basic role. Estimation of the total sulfur dioxide quantity ejected during the St. Augustine 1976 volcano eruption (over about a year) has given a value which is only an order of magnitude less than that supposed for the global volcanic production of sulfur dioxide ($0.75 - 3.75 \cdot 10^{12}$ g). An analysis of the data obtained has led to the conclusion that sulfur dioxide ejected in the periods between eruptions is a substantial H_S source (about 10^{12} g/year). Since the main part of the ejected sulfur dioxide and small particles has fallen during the phases of weakly developed eruptions of the St. Augustine volcano, these emissions (but not volcanic paroxysms, which usually attract the greatest attention) may be of major interest for the effect of eruptions on tropospheric processes.

In February 1978, Cadle et al. (1979) conducted studies of eruptive clouds with an instrumented aircraft over the volcanoes Pacaya, Fuego and Santiaguito in Guatemala. In 11 flights of the "Beachcraft Queen Air" flying laboratory, the scientists of NCAR, NASA and a number of U.S. universities took part. During the observational period the Pacaya volcano was characterized by intense ejections of vapors, and Fuego erupted moderately, with subsequent ejections of ashes and gases. Several times during every day the Santiaguito erupted moderately. Though the composition of small eruptive clouds may differ from that of clouds during rare explosive eruptions, direct measurements of the composition of small clouds should be considered a more reliable source of information than the surface data on explosive eruptions.

The airborne equipment included:

- A filter sampler to determine relative contents of SO_2, HCl, HBr, HF, and SO_4, both in particles and gases.
- A 10-channel piezo-electric cascade impactor to measure the aerosol size distribution in the $0.05 - 25$ µm diameter interval (a dispersive X-ray spectrometer was used to identify particles).
- Stainless steel containers to sample the air for the subsequent analysis for the content of H_S, CSO, CO_2, CO, and SO_2, using a gas chromatograph.
- A correlation spectrometer to measure the SO_2 fluxes in the volcano's plume.
- An automatic 35-mm camera to photograph the volcanic cloud.
- Containers for air samples to be analyzed for the content of H, O, and C isotopes.
- An inertial navigation system to estimate the wind speed.
- An air sampler to analyze the minor components.

Cadle et al. (1979) considered the observational results from the point of view of estimating the impact of strong volcanic eruptions on the stratosphere. A preliminary analysis of the observational data has shown that during strong eruptions, sulfur gets into the stratosphere mainly as sulfur dioxide, which reacting with OH, can cause a decrease in the content of hydroxyl in the atmosphere and the formation of H_2SO_4 droplets.

Explosive eruptions eject into the atmosphere large amounts of HCl and SO_2, but the $[SO_2]/[HCl]$ ratio of volume mixing ratios is always greater than 1. Apparently, HCl is washed out from the atmosphere by rains following many large-scale eruptions. The amount of fine dispersed ash preserved in the stratosphere for long periods of time increases as the fraction of crystals in the magma increases. Probably, volcanoes contribute insignificantly to the carbonyl sulfate content in the stratosphere, but this problem requires further investigation.

The data of correlation spectrometry point to the fact that each volcano ejects daily 300–1,500 tons of sulfur dioxide. Comparison with the calculated results obtained using a 2-D model of the global diffusion of the volcanic cloud, which does not consider chemical reactions, shows the need to consider the conversion of sulfur dioxide into H_2SO_4 droplets, which can take place during an entire year, as well as the content of crystals in magma.

Ferry et al. (1978) performed aerosol sampling by impactors of two types from the U-2 aircraft flying in the lower stratosphere in the 7°–79° N latitudinal zone. Films of nitro-cellulose or poly-vinyl alcohol served as substrates (in the second case a soluble component of aerosol was adsorbed by the film and was not noticed in the electron-microscope analysis). The analysis of the data on aerosol size distribution has revealed neither altitude dependence nor latitudinal change of the size distribution (at a fixed level). Possible regularities of such a kind are suppressed by a strong day-to-day variability of the size distribution at a given latitude and altitude.

Nevertheless, the comparison of the data on the total concentration of particles at different latitudes and altitudes enables one to establish that despite a strong day-to-day variability, a maximum of concentration is observed at an altitude of about 18 km. In polar latitudes an aerosol layer is observed at lower altitudes. So, for instance, an abnormally high concentration of particles is observed in the samples taken at 12 km, 75° N and 15 km, 79° N. On the other hand, the samples taken at 18 km, 65° N, do not reveal any aerosol. This result was obtained for the first time by Ferry et al. (1978).

A maximum of particle concentration is observed in the tropical latitudes with an absolute maximum at 18 km, 7°–8° N. The concentration of particles at an altitude of 21 km was maximum at 8° N. These results agree with well-known features of the atmospheric general circulation, which consist in air rising to the stratosphere near the equator and subsequent meridional transport to the troposphere of the polar regions. This transport feature is clearly shown by the satellite sensor Stratospheric Aerosol and Gas Experiment (SAGE) which has determined that the peak mixing ratio lies, in general during nonvolcanic periods, approximately 10–12 km above the local tropopause height, with the peak global mixing ratio over the tropics (McCormick, 1983).

An electron-microscope analysis of samples on the nitro-cellulose film has shown that particles consist of a volatile mixture of a crystal-like substance in the liquid "matrix." A temporal change in the relative amount of liquid causes variations in particle fluidity. Though all the particles contain a considerable portion of the crystal-like substance, they are liquid at altitudes 12–30 km, and their physical properties are similar throughout the latitudinal range 7°–79° N.

Even at low stratospheric temperatures no frozen particles have been observed. Less than half of all the particles contain insoluble nuclei, which confirms the previously obtained results. Apparently, these nuclei are not active centers of nucleation for stimulating the formation of stratospheric aerosol particles. More probably, nucleation is caused by soluble crystals of sulfates, or particles are formed from the gas phase.

The results confirm once more that volcanic silicate ashes are not a typical component of stratospheric aerosol, except for short time periods after strong eruptions. A specific feature of the measurements is the more frequent observations of insoluble nuclei in particles that had been found previously. But the appearance of nuclei is about the same and single or multiple nuclei are observed. Therefore, on the whole, the results obtained are similar to the previous ones. The samples taken at high and lower latitudes do not reveal any noticeable differences in the liquid content; a decrease in the number of insoluble nuclei with height is more clearly observed at high latitudes. However, these conclusions require further verification.

Lazrus et al. (1979) and Stolarski and Butler (1979) have brought attention to the importance of the study of the products of volcanic activity in estimates of anthropogenic impacts on the ozone layer. Volcanic gases contain considerable amounts of hydrogen chloride and hydrogen fluoride, which play an important role in ozone destruction in the stratosphere. More details on this issue will be given later on.

Though hydrogen chloride is considered the basic chemical sink for chlorine atoms (in the photo-dissociation process of halocarbons of natural and anthropogenic origin), it can also be a potential source of chlorine atoms. The latter is connected with radicals in the stratosphere. This leads to a release of chlorine atoms causing catalytic destruction of ozone.

Hydrogen fluoride, chemically stable in the stratosphere, can be used as an indicator of the extent of photo-dissociation of chlorofluorocarbons in the stratosphere. Thus, a powerful volcanic eruption can cause a decrease of the ozone content and/or affect the most specific tracer of destruction of chlorofluoromethanes in the stratosphere. The eruptions of El Chichón in 1982 might potentially fit this category, not only because it placed huge amounts of materials to altitudes of at least 30 km, but in addition, it is thought that the magma chamber was beneath a salt dome providing a ready supply of chlorine. Unpublished data by Mankin and Coffey from NCAR indicate that about $40 \cdot 10^3$ tons of chlorine were injected into the stratosphere during the El Chichón eruption. The observations after the Mount. Pinatubo volcano eruption gave convincing evidence of an impact of the eruption on the ozone layer which we shall discuss later.

The October 1974 eruption of the Fuego volcano was the first powerful ejection of volcanic products into the stratosphere subsequent to major emphasis being given to the impact of halocarbons on the ozone layer. At that time, direct measurements of HCl and HF in the stratosphere were still not possible. However, Cadle et al. (1976), combining model and lidar data, estimated that it deposited no more than one fifth of the amount of material into the stratosphere as did Mount Agung, which led Stolarski and Butler (1979) to suggest that if chlorine content was ≈ 1% of the ash content, the peak Cl_x concentration of about 0.03 ppbv should have been observed with some small but calculable depletion effect on the atmospheric ozone.

Lazrus et al. (1979) performed an analysis of the aircraft and balloon measurements data, which showed that the Fuego eruption had caused a 10-fold increase in the sulfate content in the stratosphere, i.e., by 1.6×10^6 tons. Estimated residence time was 11.7 months. During the first 6 months after the eruption, the distribution of sulfates in the stratosphere resembled that of radioactive products in 6 months to a year after nuclear tests (see Kondratyev, 1988).

After the Fuego 1974 eruption, the sulfate concentration in the stratosphere continued increasing for several months, reaching a maximum in February 1975. This 4-month drift points to the fact that sulfur is mainly ejected as sulfur dioxide with subsequent oxidation resulting in the formation of sulfate aerosol particles. This conclusion is also confirmed by the absence of such typical components of volcanic ashes as silicon and others.

The sulfates observed in spring 1975 were, as a rule, H_2SO_4 droplets. The Fuego eruption has not caused measurable changes in aerosol (solid) chlorides nor in the concentration of nitric acid vapors; post-eruption observations did not reveal any ejections of hydrogen chloride into the stratosphere.

In February 1978 the first simultaneous tropospheric airborne measurements of the contents of halogens and sulfur in aerosols and gases were carried out in the plumes of the volcanoes in Guatemala (Pacaya, Fuego, Santiaguito). The observations showed that sulfur was present mainly in the form of the sulfur dioxide. The following relative concentrations (%) characterize the elemental composition of the volcanic aerosol particles: sulfur 2.5 ± 2.1; chlorine 18 ± 12; fluorine 44 ± 30. No indications were noticed of the systematic removal of HCl from the volcanic plume as compared to SO_2. The main concentration ratio $[HCL]/[SO_2]$ was 0.41 ± 0.26.

Though the content of the water vapor in the plume is rather moderate (4,000–9,000 ppm, which corresponds to the relative humidity $65 \pm 26\%$), it is extremely high as compared to the humidity of the stratosphere. The water vapor condensation in plumes can be (before their dissipation in the stratosphere) an effective mechanism for decreasing HCl content as compared to SO_2, since HCl solubility is approximately 300 times higher than that of sulfur dioxide. If in the plume of the St. Augustine volcano in Alaska, the concentration ratio $[HCl]/[HF]$ was 51 ± 25, then, in the case of the Guatemala volcanoes, this ratio was only 14 ± 12 (the latter value is close to those observed in the stratosphere). Variations in the $[HCl]/[HF]$ ratio, depending on the distances from volcanoes, were not observed. However, as has been mentioned above, a strong chemical fractioning of halogens may occur under the influence of water vapor condensation.

The volume of magma in the May 18, 1980 eruption of Mount St. Helens (the most violent of its eruptions that produced the largest atmospheric perturbation) was equivalent to about 0.3 to 0.6 km^3 of dense dacite; 0.2 to 0.5 km^3 was related to the plinian phase of the eruption which is the part with the most atmospheric relevance. They also determined that the ash was fine-grained with the vast majority of the mass made up of particles less than 2 μm in diameter, and only 1% of the mass resided with particles less than 2 μm. It was concluded that more than $2 \cdot 10^{11}$ g of sulfur was erupted on 18 May, but it was not known how this was partitioned between the atmosphere and removal by ash fall, or how much additional sulfur was contributed by intrusive (noneruptive) magma. A very important discovery (at least from a multi-discipline viewpoint) was that explosivity alone is a a poor criterion for estimating the atmospheric impact of an eruption globally. Mount St. Helens was relatively poor in sulfur, and thus its global impact on the upper atmosphere by producing a relatively lesser amount of sulfuric acid aerosols was minimized. A tentative conclusion was reached that this eruption might have had an unusual important phreatomagmatic component for a plinian eruption. This would have important chemical and physical effects on the atmosphere and needs further study.

In situ measurements showed that sulfuric acid was found in stratospheric portions of volcanic plumes very early; possibly the acid could have been part of the initial ejections. The primary gas present in the stratospheric clouds after the eruption was SO_2; OCS and CS_2 were apparently 1,000 times less concentrated. Water concentrations the day after the eruption exceeded by 10 or more times the background level. The addition of water vapor to the stratosphere has the potential to increase OH, but SO_2 will decrease OH. The mixing ratio of HCl in the 18 May plume was estimated to be about 10 ppbv to 50 ppbv which gives a ratio of (HCl/SO_2) 0.1 to 0.5 which is similar to the previously mentioned results of Lazrus et al. (1979) for Fuego and the other Guatemalan volcanoes. By August 1980 the amount of residual acid particles in the stratosphere was about three times the pre-eruption amounts. This is probably due to early scavenging and sedimentation processes, and sulfur gas to sulfate particle production.

A preliminary conclusion was that Mount St. Helens volcanic emissions have had no major consequences on the chemical composition of the atmosphere. In the next chapter, the evolution of the gas-to-particle conversion in the stratosphere from this volcano will be mentioned.

As will be shown in some of the figures in chapter 3, in the first 4 months after the El Chichón eruption the aerosol layer above 20 km was spread only over the 0–30° N latitudinal band. There were two main layers; one between the tropopause and ≈ 21 km and the upper one ≈ 5 km thick centered at ≈ 25 km. The concentration of the small particles (r < 0.15 μm) was ≈ 10 cm^{-3} during the entire May-October 1982 period; during the same time the concentration of greater particles (r ≥ 1 μm) decreased from about 1 cm^{-3} to 0.1 cm^{-3}. Hofmann and Rosen (1983a; 1983b, 1984) published data from an extensive series of balloonborne particle (0.01 to 1.8 μm) counter observations taken at Laramie (41° N) and in southern Texas (27–29° N) from which they deduced the temporal variation of stratospheric aerosol size and mass during the first 18 months after the El Chichón eruption. Their findings are summarized below. They were able to show that large excesses of H_2SO_4 vapor, which formed in the stratosphere shortly after the eruption of El Chichón, caused the nucleation of new droplets and substantial growth of the pre-eruption aerosol distribution resulting in a bi-modal size distribution.

The composition of the aerosol was determined to be approximately an 80% H_2SO_4, 20% H_2O solution at 25 km, and about 60% H_2SO_4, 40% H_2O at 18 km over Texas during October 1982. An average acid composition of 75% has been assumed for mass calculations. Using the spline size distribution fit, the total mass was calculated as a function of altitude. Integration of the mass concentration above various altitudes gives the total mass burden above those altitudes. These data are plotted versus time in Fig. 2.3.1 for all sounding since the eruption of El Chichón. It appears that the January-April 1983 period was a maximum period for the total stratospheric mass burden at 41° N, which reached approximately 0.035 gm^{-2} above 15 km.

Also indicated in Figure 2.3.1 is a total mass scale, which may be used to estimate the global mass burden assuming longitudinal homogeneity and applicable latitudinal extent. Satellite measurements (Barth et al., 1983) indicate that approximate longitudinal homogeneity of the 27-km aerosol layer was established after the first week of June 1982 and that it was concentrated between the equator and 30° N. Airborne lidar measurements (McCormick and Swissler, 1983) indicated that by October–November 1982 the upper (25 km) aerosol layer, which contained most of the mass, was concentrated between latitudes from 5°–7° S to 35°–37° N. By the end of 1982, data at Laramie indicated that a broad uniform aerosol layer extended from the tropopause to an altitude of about 30 km (Pollack et al., 1983). This suggests that at least 10 Tg (1 Tg = 10^6 metric tons) of material was present in the stratosphere in early 1983.

Decay of the total aerosol mass is evident following April 1983. Owing to the generally larger size of post-El Chichón droplets, as compared to previous eruptions, one may expect the aerosol decay to be somewhat faster than for other eruptions such as that of Fuego in 1974, which had a following decay time of about 10–11 months for the $r \geq 0.15$-µm mixing ratio (Hofmann and Rosen, 1981) and the total sulfate mixing ratio (Sedlacek et al., 1983). For the El Chichón eruption, the observed e^{-1} decay time from April through September was 8.5 months for the $r > 0.15$-µm mixing ratio and 7.6 months for both the $r > 0.25$-µm mixing ratio and the total mass burden above 15 km. The rapid decay may be partially due to the ascending tropopause during the midlatitude spring.

A possible explanation of why particles grew larger at 25 km than at 18 km is that the lifetime of H_2SO_4 vapor may have been shorter at lower altitudes. Since the concentrations of small, freshly nucleated droplets were observed to be similar in both layers over Texas 45 days after the eruption the concentration of H_2SO_4 vapor must have been similar at this time, otherwise the nucleation rate and consequent concentrations of new droplets would have been decidedly different in the two layers. It is more plausible that prompt transport northward at 18 km as compared to delayed transport at 25 km (as observed in the global aerosol distribution) caused the 18-km H_2SO_4 vapor concentration to be more rapidly dispersed. However, a tendency for growth to larger sizes at higher altitude was already noted following the eruption of Mount St. Helens in 1980 (Hofmann and Rosen, 1982) and may be a natural stratospheric phenomenon.

In conclusion, it appears that growth of droplets through H_2SO_4 vapor condensation with a decaying vapor concentration describes the average droplet radius versus time characteristics observed following the eruption of El Chichón at least in a qualitative sense. An H_2SO_4 particle concentration of about 10 cm^{-3}, 40 days after the eruption, is estimated. The net lifetime of the vapor appears to have been about 25 and 50 days at 18 and 25 km, respectively. These lifetimes are thought to reflect the net lifetime of chemical production and dispersion. Since the upper layer was relatively confined in latitudinal extent during the first 200 or so days, when all the aerosol growth in this layer is probably most closely

VOLCANIC ACTIVITY AND CLIMATE

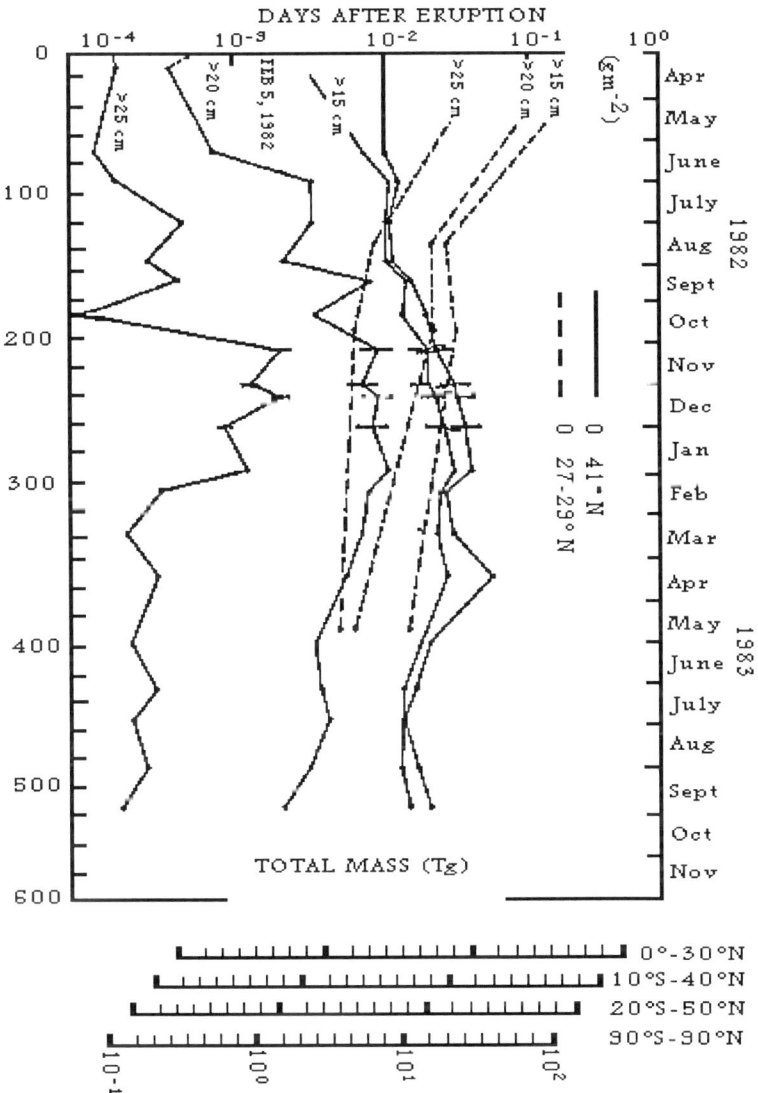

Figure 2.3.1. Stratospheric aerosol column mass loading above several altitudes (in g m^{-2}, left scale) calculated from a computer spline fit to the measured data at Laramie 41° N and in southern Texas 27°– 29° N. Error bars are for flights which did not incorporate a large particle counter and flights just before and after this period were used for these data. The total mass scale on the right assumes longitudinal homogeneity and latitudinal extent as indicated (0–30° N, 10° S – 50° N and 90° S – 90° N). From Hofmann and Rosen (1984). [© by American Geophysical Union]

associated with the chemical conversion rate of SO_2. While chemical modeling indicates a SO_2 lifetime of about 38 days at 25 km for an OH conserving reaction and much longer for chains of reactions that consume OH (McKeen et al., 1984), it is predicted to be similar at 18 and 25 km, suggesting that the apparent shorter H_2SO_4 lifetime observed at the lower altitude may be affected by other phenomena such as dispersion. In accordance with observations at Mauna Loa, two days after the April 4, 1982 eruption of El Chichón the SO_2 had reached 60 *matm* cm. The quicker growth of the aerosols on 25 km compared with 18 km, could be explained with more rapid oxidation of SO_2 due to the higher ambient temperature at 25 km and/or due to the higher concentration of H_2SO_4 there. The appearance of two aerosol layers during the first months after the El Chichón eruption could be explained, assuming that the lower one had been generated during the first two relatively weak injections (March 28 and April 3) and the higher one ($\approx$ 25 km) during the strongest injection on April 4, 1982.

2.3.4. Comparison Data with Other Planets

Of great interest are the comparative planetary aspects of this problem. A considerable amount of sulfur-containing micron-sized particles (mainly as SO_4) has been observed in the atmospheres of the Earth and Venus. On the Earth this aerosol is concentrated in a layer at 18–20-km altitude (in the lower stratosphere), with a typical number density of 1–2 cm^{-3} for particles with diameters D > 0.3 µm. The stratospheric sulfate layer consists of droplets of concentrated H_2SO_4 water solution and solid granules of sulfate salts (for instance, ammonium sulfate). In this case volcanic eruptions, industrial wastes and biological processes serve as sources of sulfur.

The global cloud cover on Venus located at 45–75-km altitude, contains particles with D > 0.5 µm and a number density of about 100–300 cm^{-3}, which are mainly droplets of concentrated H_2SO_4 water solution. Beneath the main cloud cover, a haze layer (30–45 km) is located, with a particle number density of about 1–2 cm^{-3}.

As Settle (1979) has noted, the atmosphere of Mars lacks both the persistent cloudiness and a global aerosol layer of long lifetime. There is convincing photogeological evidence, however, for volcanic activity on the Martian surface. Both the analysis of the cosmo-chemical models and the data on volcanic processes on the Earth indicate the possibility of ejecting sulfur-containing gases into the Martian atmosphere by volcanic activity occurring on the surface during earlier geological periods (the ejected products should have been primarily sulfur dioxide and hydrogen sulfide).

Due to a stable wind convergence in the lower atmospheric layers observed in the Tharsis region, the sulfur-containing gases could have moved into the upper layers of the Martian atmosphere (up to 20–30 km). An intensive circulation in the upper atmosphere would have provided for the global distribution of gases during a period of about 25 Earth days.

Surprisingly similar photochemical processes of the sulfur dioxide conversion into the sulfate aerosol are observed in the lower stratosphere of the Earth and lower layers of the Martian atmosphere (0–30 km). During the sulfur dioxide ejections into the present-day Martian atmosphere, the chemical and microphysical processes should be the same as in the lower stratosphere of the Earth. The lower layers of the Martian atmosphere in the equatorial and midlatitudes are somewhat cooler and exhibit a comparable or higher humidity and higher concentration of condensation nuclei, as compared to the Earth's lower stratosphere. The latter two factors promote faster processes of SO_3 (gas) conversion into SO_4^{2-} (particles). However, a lower temperature slows down this process.

Mutual compensation of the two effects under consideration determines the comparability of the time scales of formation of sulfate aerosol in the lower Martian atmosphere and lower stratosphere of the Earth. Therefore, by an analogy with the Earth, the principal factor limiting the formation of aerosols on Mars is, apparently, the oxidation of the sulfur dioxide resulting from chemical reactions in which O, OH, and H_2O take part.

In conditions of the present-day Martian atmosphere the formation of aerosol should take several thousand days. However, during persistent volcanic activities on the surface, the concentration of the odd hydrogen (OH, H_2O) should increase considerably and, as a result, the time of aerosol formation would shorten to several hundred days. The submicron aerosol particles formed in such a way, can remain in the atmosphere for a long time, moving dozen of times around the planet before they are removed from the atmosphere by gravitational sedimentation—the main mechanism for removing the particles.

Sedimentation of aerosols onto the surface over a wide range of equatorial and midlatitudes becomes possible due to their global distribution. The Martian volcanic sulfate aerosol should consist of the H_2SO_4 solution droplets, as well as the aggregates of the droplets with inclusions of solid particles. The chemical activity of such aerosols sedimented on the planetary surface should have stimulated the processes of leaching the material of the planetary surface. The sulfate aerosol sedimentation can be considered as a possible mechanism for the transport of sulfur to the regions of the Viking landing sites, where sulfur compounds had been found in the soil.

Numerical modeling of the general circulation of the Martian atmosphere has shown that the Chryse Planitia and Utopia Planitia regions are characterized by the presence of averaged downwelling vertical motions during most of the year, from which it follows that in these regions a prevailing sedimentation of the sulfate aerosol has taken place. Observation of a comparatively high (by the Earth's standards) concentration of sulfates in the surface layer of Viking landing sites can be explained by the global or hemispherical dispersion of the particles.

Let us consider comparative climatic impacts of volcanic eruptions on the Earth and Mars. Strong changes of climate on the Earth in its early stage of evolution, the climate change of the Earth in our days, observations testifying to a warmer climate in Mars in past epochs, and an observed impact of dust storms on the radiative regime and climates on the Earth and Mars—all these changes testify to the existence of internal factors affecting the radiative regime and climate of the planets. The analysis of these factors was made by Kondratyev and Moskalenko (1979, 1980a, 1980b, 1981, 1983), Kondratyev and Hunt (1982), Kondratyev et al. (1977, 1981, 1983a), and others. The factors mentioned above include, in particular, changes in the gas and aerosol composition of the atmospheres. The gas composition of the atmospheres varies smoothly, according to the laws of nature, due to outgassing of the solid shell of the planet and biochemical processes taking place in the atmosphere and on the planetary surface, while changes in the aerosol composition and aerosol optical properties are often sporadic, which causes strong changes in climate. The volcanic activity, whose climatic impact on the Earth and Mars has exhibited substantial temporal fluctuations, is one of the powerful factors which could cause planetary-scale climate changes.

Powerful volcanic eruptions not only change the aerosol composition of the atmosphere but also its gas composition. Note should be taken of the fact that with enhancement of the volcanic activity the amount of SO_2 injected to the stratosphere and troposphere grows SO_2, known to have strong absorption bands in the UV and IR spectral regions, as shown in Figures 2.3.2 and 2.3.3. Consequently, an increase in the SO_2 concentration decreases the planetary albedo, enlarges the shortwave radiation absorbed by the planet, and raises the effective temperature of the planet. Strong IR SO_2 bands lead to the intensification of the radiative cooling of the troposphere and to the enhancement of the greenhouse effect near the surface.

Being injected to the stratosphere, the volcanic products further evolve due to photochemical and gas-to-particle conversion of SO_2. The H_2SO_4 solution aerosol resulting from this conversion brings forth substantial changes in the content and optical properties of the stratospheric aerosol: a decreased absorption of the shortwave radiation by aerosols, and an increased activity in the longwave spectral region. The process of formation of sulfate aerosols is followed by a decrease in the SO_2 concentration. This results in an increase in the planetary albedo and a decrease in the surface temperature.

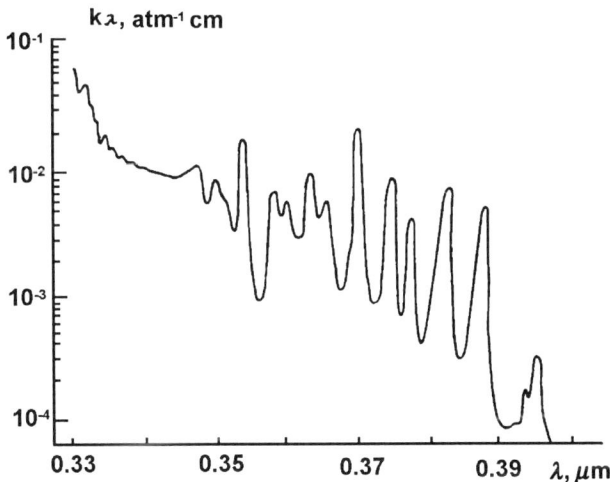

Figure 2.3.2. The spectral dependence of the SO_2 absorption coefficient in the spectral intervals 0.3–0.4 μm.

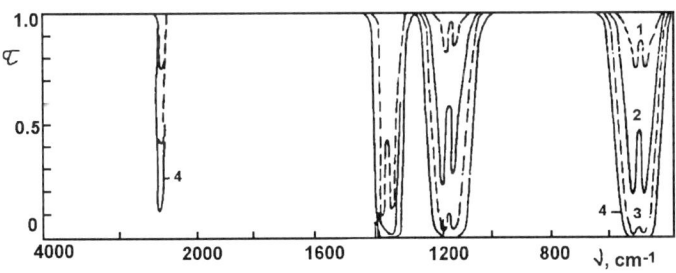

Figure 2.3.3. SO_2 absorption spectra for the contents, ω, atm.cm; 1–0.15; 2–1.2; 3–5; 4–10.

During the Earth's evolution the optical depth (thickness) of the stratospheric aerosol layer has changed considerably. According to calculations by Kondratyev and Moskalenko (1980a; 1980b), the optical thickness of the stratospheric aerosol during maximum volcanic activity reached $\tau = 4.4$. At present a temporal change of volcanic activity takes place, which causes variations in the radiative regime of the atmosphere and climate.

Modeling The Optical Properties of An Evolving Atmosphere

The major impact of the atmosphere on the planetary radiative regime is connected with the mechanism of the greenhouse effect, which causes warming of the surface and troposphere. The impact of the atmospheric greenhouse effect was calculated using a 1-D radiative-convective climate model based on radiation parameterization suggested by Kondratyev and Moskalenko, (1980a, 1980b), with account of radiation absorption by all important atmospheric gases, and the absorbing and scattering properties of cloudiness, as well as the tropospheric and stratospheric aerosols. The spectral transmission functions (STF) of gas components were calculated with the use of software and STF parameterizations after Moskalenko (1969), Moskalenko and Zakirova, 1972, and Moskalenko et al. (1978), based on data of complex field experiments and STF line-by-line calculations performed by Moskalenko and Yakupova (1978). The absorption in the 0.1 – 100 µm spectral interval by H_2O, CO_2, O_3, N_2O, CH_4, NO, CO, O_2, N_2, NO_2, NH_3, OH, and HCl was taken into account. Also used in calculations was *a priori* information about the chemical composition of the atmospheres discussed by Kondratyev and Moskalenko (1979), Moroz (1978) and Marov (1981). When necessary, the information was supplemented by respective calculation data, based on consideration of the sources and sinks of gas and aerosol components, described by Kondratyev and Moskalenko (1983).

The aerosol component of the atmospheres was simulated by superimposing the gamma distribution. In accordance with studies by Kondratyev et al. (1983b), the optical characteristics of aerosols (the coefficients of absorption and scattering, the phase functions) have been obtained with due regard to the actual chemical composition of aerosols. The representation of atmospheric aerosols and clouds as a superposition of different fractions, each having its vertical profile, size distribution and chemical composition, makes it possible to realistically consider the vertical inhomogeneity of aerosols and clouds. A data bank of optical characteristics, including about 50 types of aerosols and clouds, enables one to simulate a spatial variability of the optical characteristics of the aerosol component, with the atmospheric aerosol and clouds considered as a coupled scattering and absorbing medium. The use of the above-mentioned data bank permits the simulation of the aerosol components of the atmospheres in the process of their evolution. Figure 2.3.4 illustrates the spectral dependencies of the attenuation coefficient, $\sigma_{a,\lambda}$, as well as absorption coefficient, $\sigma_{a,\lambda}^a$ for different fractions of the volcanic aerosol. The $\sigma_{a,\lambda}^a$ and $\sigma_{a,\lambda}$ values were normalized against $\sigma_{a,\lambda}$ at $\lambda = 0.5$ µm. Figure 2.3.4 also illustrates spectral variations of the absorbing and scattering properties of volcanic aerosols, when their size distribution changes.

VOLCANIC ACTIVITY AND CLIMATE 139

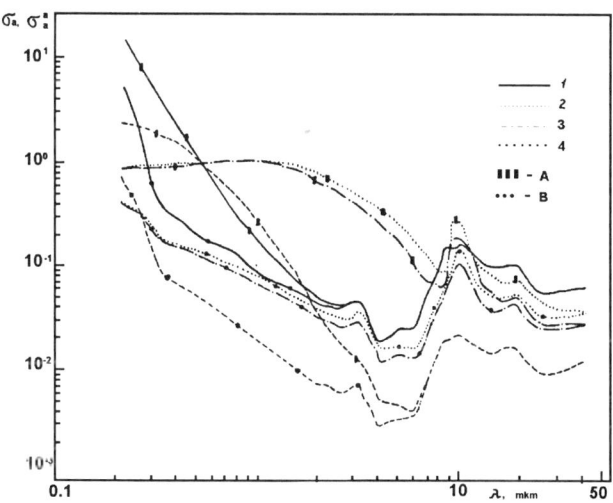

Figure 2.3.4. The spectral dependence of the coefficients of attenuation, σa, (A) and absorption, σ^a_a, (B) for different fractions of the volcanic aerosol. The size distribution (2.6) of the ensemble of spherical particles is described with the gamma distribution for the following parameters.
1 - a=1; b=50; c=0.5   2 - a=1; b=25; c=0.5
3 - a=1; b=12.5; c=0.5   4 - a=1; b= 9; c=0.5

The chemical composition and structural characteristics of an evolving atmosphere substantially changes in time, and, therefore, it is important to take into account the effect of these changes on the optical properties of the gas and aerosol components of the atmosphere. Techniques for calculation of the atmospheric absorption with due regard to selective absorption in the rotational-vibrational bands, to the continuum absorption by far wings of lines, and to the pressure-induced absorption, are described by Kondratyev and Moskalenko (1979; 1983). The authors used both the empirical spectral transmission functions and the STF calculated with the help of the line-by-line technique based on parameters after MacClatchey (1973), Kayumova et al. (1979), Moskalenko and Zotov (1977), and Moskalenko and Yakupova (1978). The application of various techniques for STF calculation to the thermal emission transfer in absorbing and scattering media has been discussed by Moskalenko and Zakirova (1972), Kondratyev and Moskalenko (1979). In STF calculations their temperature dependence has been considered both in rotational and vibrational bands and in pressure-induced absorption bands.

The importance of consideration of the temperature effect on the STF is exemplified in Figures 2.3.5 and 2.3.6 by the absorption bands for carbon dioxide and ozone.

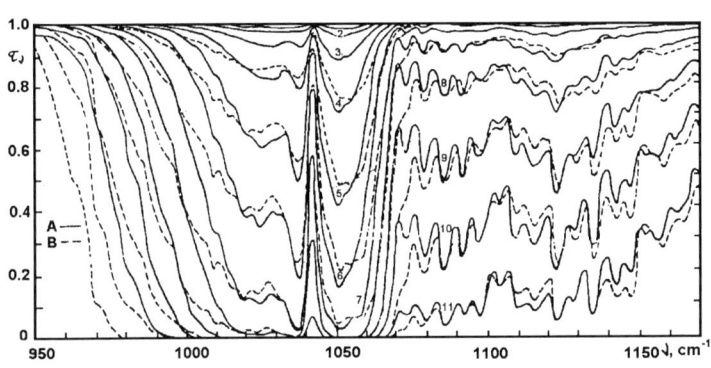

Figure 2.3.5. Transmission spectra in the 9.6 μm O_3 band at T=200 K (A) and 300 K (B). The O_3 content, ω, atm.cm: 1–0.001; 2–0.003; 3–0.01; 4–0.03; 5–0; 6–0.3; 7–1; 8–3; 9–10; 11–100. Nitrogen-broadening at a pressure of $P(N_2)$=0.1 atm.

The choice of the models of clouds and atmospheric haze is based on the following assumptions: aerosols and clouds have a multimodal size distribution and are represented by an equivalent ensemble of homogeneous spherical particles. The size distribution for $N_i(r)$ for each fraction of a given chemical composition was described in the studies by the gamma function. Then the number density

$$N(r) = \sum N_i(r) = \sum A_i(z) r^{a_i} \exp(-b_i r^{c_i}) \qquad (2.6)$$

where $A_i(z)$ depends on height (z); a_i, b_i, c_i are the parameters of the i-th fraction. Fractions $N_i(r)$ can be represented by the particles of clouds or aerosols (located below and above clouds).

VOLCANIC ACTIVITY AND CLIMATE

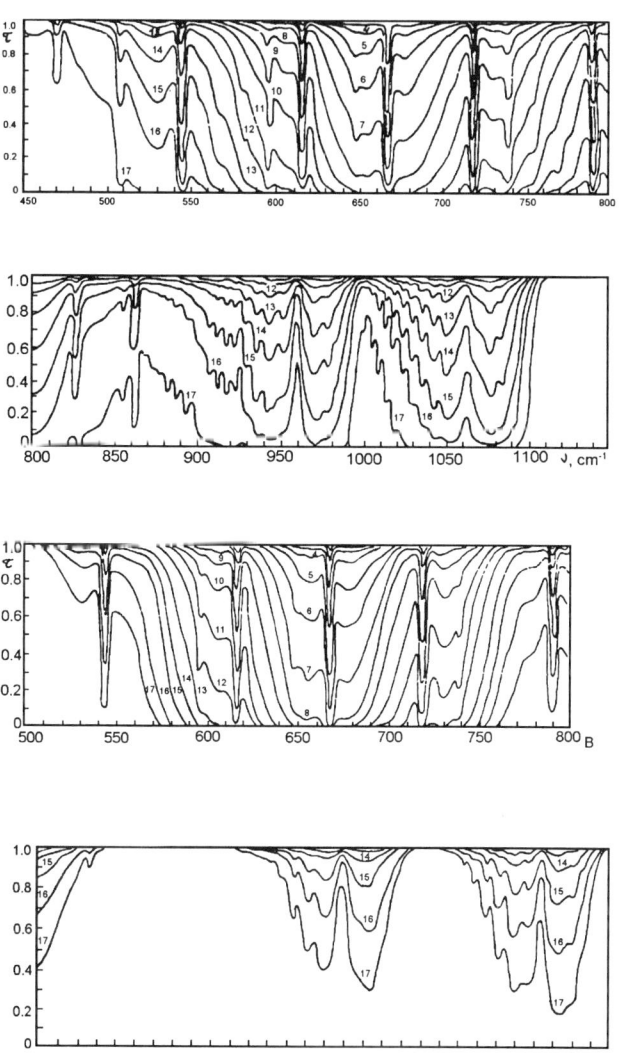

Figure 2.3.6. Spectral transmission functions for nitrogen-broadened CO_2 at T=300 K (A) and 200 K (B) and $P(N_2)$= 1 atm.
ωCO_2, atm.cm: 1–0.001; 2–0.003; 3–0.01, 4–0.03; 5–0.1; 6–0.3; 7–1; 8–3; 9–10; 10–30; 11–100; 12–300; 13–1000; 14–3000; 15–10 000; 16–30 000; 17–300 000.

In connection with the fact that most reliable information on the vertical structure of aerosols and clouds has been obtained from the optical density data (at $\lambda = 0.55$ µm), the values of the scattering coefficients $\sigma_{a,\lambda}^s$, and absorption coefficients $\sigma_{a,\lambda}^a$ are normalized against the optical density.

The models of aerosols and clouds were computed using the program of an aerosol unit. The optical characteristics of the aerosol components for different fractions were obtained from calculations based on Mie theory described by Deirmendjian (1971) for different types of aerosol formation in the atmospheres of the Earth and Mars.

The suggested models of the aerosol components of the planetary atmospheres make it possible to construct global zonal optical models of atmospheric aerosols. The same conclusion is true for the chemical composition and size distribution of cloud particles. If necessary, the consideration of the optical properties of the aerosol components of planets enables one to model a 3-D field of distribution of the optical characteristics of the planetary aerosol components.

In calculations of the STF of the gas phase of the atmosphere it is necessary to separate the contributions to the absorption from the wings of distant spectral lines for atmospheric gases, $\tau_{\Delta\lambda}^c$, from the pressure-induced absorption, $\tau_{\Delta\lambda}^i$, and from the selective absorption, $\tau_{\Delta\lambda}^s$, of spectral absorption lines, within a given spectral interval $\Delta\lambda$. This division makes it possible to substantially raise the accuracy of calculation of $\tau_{\Delta\lambda}$. Then, for the given component

$$\tau_{\Delta\lambda} = \tau_{\Delta\lambda}^c \cdot \tau_{\Delta\lambda}^i \cdot \tau_{\Delta\lambda}^s \tag{2.7}$$

the function

$$\tau_{\Delta\lambda}^c \cdot \tau_{\Delta\lambda}^i = \exp[-\beta_{\lambda c}(T) + \beta_{\lambda i}(T)\omega P] \tag{2.8}$$

The calculation of $\tau_{\Delta\lambda}^s$ using an empirical technique proposed by Moskalenko and Zakirova (1972) is based on the following equation

$$1/\left(ln\tau_{\Delta\lambda}^s\right)^2 = \left(1/(ln\tau_{\Delta\lambda}'^s)\right)^2 + \left(1/(ln\tau_{\Delta\lambda}''^s)\right)^2 + M/(ln\tau_{\Delta\lambda}'^s)(ln\tau_{\Delta\lambda}''^s) \tag{2.9}$$

where

$$\tau_{\Delta\lambda}'^s = \exp[-K_\lambda(T)\omega], \tag{2.10}$$
$$\tau_{\Delta\lambda}''^s = \exp[-\beta_{\lambda s}(T)\omega^m{}_\lambda P_e{}^n{}_\lambda] \tag{2.11}$$

$\tau_{\Delta\lambda}'^s$ determines the STF at high pressures in conditions of a smeared rotational

VOLCANIC ACTIVITY AND CLIMATE

structure; $\tau_{\Delta\lambda}''^s$ is a satisfactory STF approximation for the region of strong absorption. The parameter M in (2.9) characterizes a change in the rate of the STF growth in the transition from the region of weak absorption to that of strong absorption. An approximation of strong absorption can be expressed, for instance, as the product of functions

$$\tau_{\Delta\lambda}''^s = \exp[-\beta_{\Delta s}'(T)\omega^m_\lambda P^n_{e\lambda}][1 - (\beta_{\lambda s}''\omega^m_\lambda P^n_{e\lambda}] \qquad (2.12)$$

of the logarithm or any other dependence on $\omega^m_\lambda P^n_{e\lambda}$.

$\tau_{\Delta\lambda}^s$ along the path, l, in conditions of an inhomogeneous atmosphere as to temperature, T(l), and pressure, is calculated using the formulae

$$-ln(\tau_{\Delta\lambda}^s) = K_{\lambda s}(T_o)W_1 - ln\tau_{\Delta\lambda}''^s = \beta_\lambda(T)W_2 \qquad (2.13)$$

where

$$W_1 = \int_{(l)} \rho(l) F_{1s}[l(T)]dl, \qquad (2.14)$$

$$W_2 = \int_{(l)} \Delta\varrho(l)(P(l)/p_o)^{n_\lambda/m_\lambda} F_{2s}[l(T)]dl, \qquad (2.15)$$

$$F_{1s}(T) = K_\lambda(T)/K_\lambda(T_o); \quad F_{2s}(T) = \beta_{\lambda s}(T)/\beta_{\lambda s}(T_o) \qquad (2.16)$$

Similarly, for the pressure-induced and continuum absorptions

$$\beta_{\lambda i} = \beta_{\lambda i}(T_o) \cdot F_i(T); \quad \beta_{\lambda c}(T) = \beta_{\lambda c}(T_o) \cdot F_c(T) \qquad (2.17)$$

The best adaptation of parameters in the above given formulae leads to errors $\Delta\tau = 3\%$–4% which do not exceed the errors in STF measurements. With the use of accurate STF data obtained through line-by-line calculations, the adaptation to approximations (2.7) – (2.12) gave errors $\Delta\tau_{\Delta\lambda}^s = 2\%$–$3\%$.

Calculations of STF for the rectangular and arbitrary instrument function $\delta(\nu - \nu')$ in an inhomogeneous (by temperature and pressure) atmosphere are made using the formulae

$$\tau_{\Delta\lambda} = (1/\Delta\lambda) \int \Delta\tau_\lambda(\varrho_j(L')T(L'),L')d\lambda \qquad (2.18)$$

$$\tau_{\Delta\lambda} = \int \tau_\lambda \delta(\nu - \nu'), \; d\nu' \qquad (2.19)$$

where

$$\tau_\lambda[\rho_j(L'), T(L'), L'] = \exp[-\{\rho_{L'} (S_{ij}\{T(L'),b_{ij}[P_j(L'),T(L'), L', \nu_{ij},\nu]\}\rho_j(L')$$

$$+ \rho_{L'}\alpha L'\Sigma K_{\nu,i}^{i,c}[P_j(L'),T(L')]\}] \qquad (2.20)$$

where ρ_j is the concentration; $K_{\nu,j}^{i,c}$ are the spectral absorption coefficients determined by molecular interaction; S_{ij}, b_{ij}, ν_{ij} are the intensity, the shape and the center of the *i*-th line of the j-component. The length of the optical path for a spherical atmosphere is given as

$$dL' = dZ/\{1 - [[(R + Z_1)n(Z_i)]/[(R +Z)n(Z)]\cdot\sin\theta(Z_1)]^2\}^{\frac{1}{2}} \qquad (2.21)$$

where Z is the height; n(Z) is the refractive index; $\theta(Z_1)$ is the zenith viewing angle at a height Z_1; R is the radius of the planet. In a general case dL' is given in a numerical form. The temperature dependence of S_{ij} and d_{ij} is determined from the formulae known in spectroscopy. Moskalenko and Yakupova (1978) developed an economical algorithm to compute $\tau_{\Delta\lambda}$ with due regard to the empirical shape of the spectral lines, Foygt's shape and Lorenzian shape. They also calculated the STFs using line-by-line techniques for various absorption bands for water vapor, CO_2, N_2O, NO, CH_4, CO, and HCl for a broad range of conditions existing in an atmosphere.

Climatic Implications of the Volcanic Activity in the Process of the Earth's Evolution

Let us consider some results from studies of the impact of volcanic eruptions on the atmospheric greenhouse effect and on climate, bearing in mind the earlier results on the evolution of the Earth's greenhouse effect obtained by Kondratyev and Moskalenko (1979, 1980a, 1980b, 1981, 1983). It is of interest to consider the impact of volcanic activity on the greenhouse effect of the Earth's primordial atmosphere formed about 4.5 billion years ago. Let us assume that the sun irradiance in that period constituted 75% of that today. We assume the chemical composition of the initial atmosphere according to Hart's recommendations (1978), i.e., the atmosphere of volcanic origin includes 84.2%

water, 14.29% CO_2, 1.06% methane, and 0.23% nitrogen. The latter means that the Earth's initial atmosphere was composed mainly of carbon dioxide. The water vapor concentration in the atmosphere was limited by ambient temperature and grew as the atmospheric pressure and tropospheric temperature increased.

Upon an intensive capture of the dust cloud by the planet, its surface rapidly cooled due to low heat conductivity of soil and low density of the gas enveloping the planet. An intensive formation of the Earth's primordial atmosphere first of all had been caused by tectonics and volcanic activities. Most likely the ejected gas contained SO_2 and COS with the total concentration not exceeding 0.1%. This estimate of the concentration of the sulfur-containing components was obtained from data on the chemical composition of the present day volcanic gases. The mean planetary surface albedo was assumed to be 0.2, bearing in mind that the soil surface in that period had been weakly eroded.

Despite their relatively low concentration, the greenhouse effect of SO_2 and COS may turn out to be rather strong in connection with the shift of the absorption bands with respect to the atmospheric CO_2 bands. Of particular importance is the consideration of the greenhouse effect of SO_2 and COS at low atmospheric temperatures, when the concentration of the atmospheric water vapor is low, and the greenhouse contribution of such volatile atmospheric components as CO_2, CH_4, SO_2, and COS becomes more important. At low pressures and temperatures the lifetime of SO_2 is rather long (it may be a year or longer), and an equilibrium concentration of SO_2 becomes sufficient to produce a considerable greenhouse effect.

A model of the radiative-convective equilibrium was used. The ratio between the adiabatic and the radiative-convective temperature gradients was assumed to be constant during the formation of the primary atmosphere when the atmospheric pressure, P_s, had changed from 0.005 to 0.4 atm. The vertical profiles of pressure $P(z)$ and H_2O concentration $C_{H2O}(z)$ were calculated. The relative atmospheric humidity $r = 56\%$ was assumed to be constant from the surface level to the height of the tropopause, above which the volume concentration of water vapor was assumed to be constant and corresponding to that in the zone of the stratospheric water vapor trap.

The presence of SO_2 in the Earth's initial atmosphere had caused an increase in the stratospheric temperature in the altitude range 30–40 km due to absorption of the solar UV radiation. A heating of the troposphere had been caused by the absorption bands of methane, carbon dioxide and water vapor located in the visible and near IR spectral regions. The presence of even small amounts of SO_2 in the atmosphere had protected the planetary surface from the harmful effect of the SW UV radiation.

CS_2 and H_2S are also sulfur-containing components to be mentioned. The greenhouse effect of CS_2 is manifested at the expense of the absorption band 1535 cm^{-1} overlapped by water vapor absorption bands. Unfortunately, there is no information about the CS$_2$ concentration in the products of volcanic eruptions. A great amount of erupted H$_2$S does not directly affect the optical properties of the atmosphere, the greenhouse effect, and the planetary surface temperature, because of a short lifetime of H$_2$S in the atmosphere.

Carbon monoxide, CO, released in the process of the planetary outgassing had a weak greenhouse effect due to absorption in the 4.7-μm band. Hydrochloric acid, HCl, and hydrofluoric acid, HF, do not have absorption bands in the far IR spectral region. The HCl band is centered at 2866.04 cm^{-1} and the band of the first overtone is located near 5868.6 cm^{-1}. The intensities of the HCl bands 0–1 and 0.2 are, respectively, 132 and 3.73 atm^{-1}cm^{-2}. An HF molecule has the basic band 0–1 centered at 3961.3 cm^{-1} and the first overtone at 7750.83 cm^{-1}. In connection with the absence of absorption bands in the far IR spectral region, both HCl and HF do not practically affect the greenhouse effect of the atmosphere, but warm it at the expense of solar radiation absorption.

During the volcanic activity the transport of the products of nitrogen compounds into the stratosphere is also observed. NO is the most stable compound, but the main greenhouse effect is manifested by other nitrogen oxides: N$_2$O, N$_2$O$_5$, and N$_2$O$_4$. N$_2$O has numerous bands covering a spectral region from 0.1 to 20 μm. Detailed studies of N$_2$O absorption were discussed by Kondratyev and Moskalenko (1979, 1983), with the N$_2$O content from fractions of a millimeter to 2•10^4 atm.cm. Intensities of most of the N$_2$O absorption bands were measured and the STF was parameterized. At present concentrations of N$_2$O, the greenhouse effect is mainly observed at the expense of absorption bands 7.8, 8.6, 4.54, 4.0 and 3.86 μm.

In laboratory conditions both N$_2$O and N$_2$O$_4$ have been observed simultaneously. A number of NO$_2$ and N$_2$O$_4$ bands are overlapped and their analysis becomes possible due to a decrease in the N$_2$O$_4$ concentration taking place when the temperature grows and the (NO$_2$ + N$_2$O$_4$) pressure is reduced in the cell. These measurements allowed determination of independent contributions to the absorption from NO$_2$ and N$_2$O$_4$, and to parameterize their spectral transmission functions. Note that the NO$_2$ greenhouse effect is mainly determined by absorption in the 6-μm band. However, at the expense of solar shortwave radiation in the interval 0.3–0.6 μm, the climatic impact of NO$_2$ is quite substantial. Figure 2.3.7 exemplifies the NO$_2$ STF measured in the shortwave spectral region. An increased NO$_2$ concentration causes a decrease in the planetary albedo and the warming of the upper atmosphere. Due to a specific vertical temperature profile an increase in the NO$_2$ concentration causes a weak anti-greenhouse effect at the expense of growing intensity of the outgoing thermal emission in the 6-μm NO$_2$ band. A detailed analysis of the effect of nitrogen oxides on the Earth's climate in the process of the volcanic cloud evolution in the stratosphere requires now unavailable reliable measurements of NO and NO_2 concentrations in the gas products of volcanic eruptions.

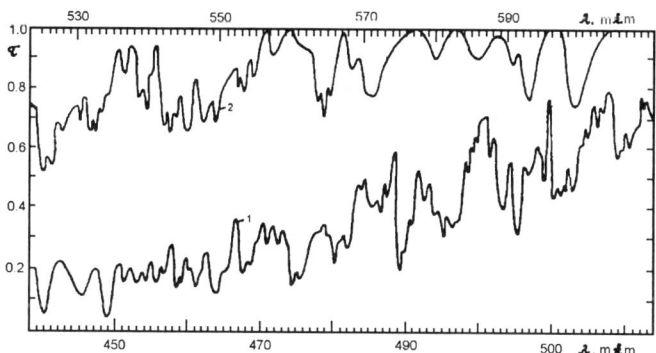

Figure 2.3.7. The NO_2 transmission function in the interval 0.3–0.6 μm at $\bar{\omega}$ = 0.1 and 1.0 atm. cm.

Note the importance of considering N_2O_5 and HNO_3 in the radiative heat exchange. The most probable model of the N_2O_5 structure is O_2NONO_2. Strong bands for N_2O_5 are centered at 1720, 1246, 743, and 557 cm^{-1}, and the respective intensities are 1650, 654, 770, and 290 $atm^{-1}cm^{-2}$. Nitric acid vapors have strong absorption bands near 20, 11.2, 7.2, 5.6, and 2.9 μm. Even if their concentration in the atmosphere is not high, they may cause a marked greenhouse effect.

To solve the problem of radiative heat exchange in the Earth's primordial atmosphere, it is also important to know the chemical composition of hydrocarbons. An analysis of available measurements data shows that methane is a prevailing component and constitutes 97%–99% of the total content of hydrocarbons. The part of other hydrocarbons (C_2H_2, C_2H_4, C_2H_6, C_3H_8) is 1%–3%. However, at high concentration of hydrocarbons in the atmosphere, constituting more than 1% by volume, minor hydrocarbons C_2H_2, C_2H_4, C_2H_6, C_3H_8 with strong absorption bands produce a strong greenhouse effect. Detailed studies of hydrocarbon absorption spectra were discussed by Kondratyev and Moskalenko (1979, 1983). The greenhouse effect of hydrocarbons was calculated on the assumption of their constant volume concentration in the primordial atmosphere.

In high layers of the Earth's initial atmosphere hydrogen cyanide (HCN) could have been formed as a result of photolysis which had produced a strong impact on the atmospheric greenhouse effect due to LW radiation absorption in strong HCN bands 712 and 1412 cm^{-1}.

During volcanic eruptions into a rarefied atmosphere the volcanic plume could reach high altitudes up to 100 km, and the resulting atmospheric aerosol encircled the planet with a thick layer, whose geometric thickness diminished with growing density of the atmosphere. With the pressure of the initial atmosphere $P_s = 0.4$ atm the aerosol layer was located within the altitude range 20–30 km.

An effective temperature of the planet is determined by its total albedo, which depends on the chemical composition of the atmosphere, its density, the vertical distribution of the aerosol and cloud components, and on the surface albedo. Figure 2.3.8 shows the albedo as a function of the optical thickness for the submicron aerosol fraction, for different surface albedo values, as well as the quantum survival probability for single scattering (single scattering albedo).

According to the calculations, the mean-planetary surface albedo was assumed to be 0.2 and to be constant in the initial atmospheric pressure range $P_s \approx 0.4$ atm. As follows from Figure 2.3.9 and from data on the optical characteristics of the stratospheric aerosol (see Figure 2.3.4 and data of Kondratyev et al., 1983b), the presence of the stratospheric aerosol layer always leads to an increase in the albedo of the surface-atmosphere system (γ) for large optical thickness of aerosol $\tau \approx 0.38$, if the single scattering albedo is 0.9. With the absorbing gas phase present, the effect of aerosols on the planetary albedo will be weaker.

In conditions of intensive volcanic activity the water vapor was ejected at high altitudes and had photodissociated giving OH, which caused the absorption of solar radiation by a system of strong NH_3 bands $A^2\Sigma\text{-}X^2\Pi(0,0)$ near $\lambda = 0.31$ μm. Vertical NH_3 profiles were calculated according to Morss and Kuhn (1978). The impact of NH_3 in the greenhouse effect is very strong and even a small NH_3 concentration in the primordial atmosphere had considerably affected the planetary albedo, thereby causing a warming of the troposphere, an enhancement of the greenhouse effect, and a rise in the planetary surface temperature.

Values of the mean global planetary surface temperature (T_s) as a function of pressure (P_{CO_2}) of the Earth's primordial atmosphere are listed in Table 2.3.1 An internal heat source is assumed to be absent, and the vertical optical thickness of the gas-dust cloud in its initial state to be about 0.01.

Table 2.3.1. The greenhouse effect of the Earth's primordial atmosphere in the absence of an internal heat source.

$P_s(CO_2)$	0.001	0.005	0.01	0.03	0.1	0.2	0.4
δ	0.2	0.203	0.207	0.22	0.26	0.32	0.45
T_e	231	231	230.6	230	227	224	219
T_s	232	237	241	255	270	284	302
ΔT	1	6	10.4	25	43	60	83

VOLCANIC ACTIVITY AND CLIMATE

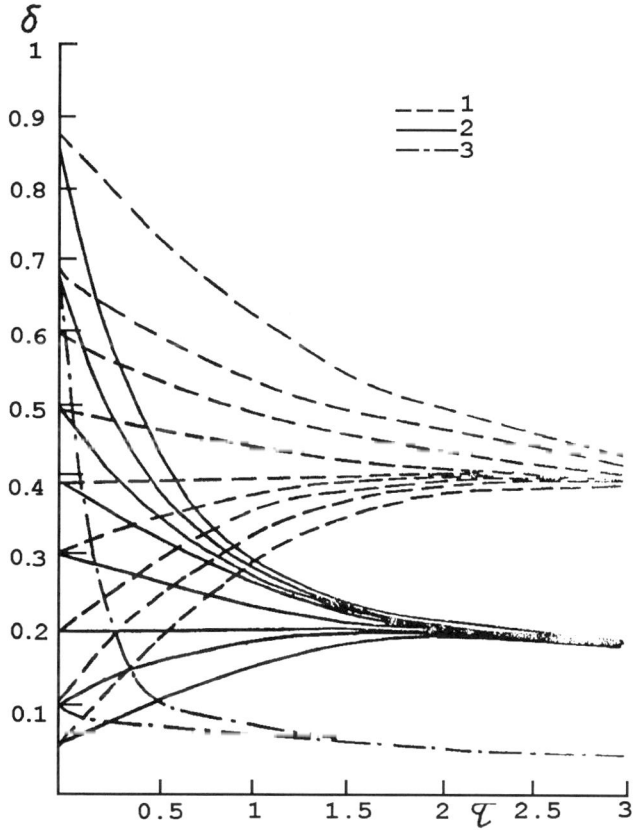

Figure 2.3.8. The dependence of the surface-atmospheric aerosol layer system albedo on the optical thickness, τ, of an aerosol cloud for stratospheric aerosol models (1), for industrial aerosol (2), and for strongly absorbing aerosol (3) at different surface albedo values, $\delta_s = \delta(\tau=0)$.

The resulting stratospheric aerosol in the altitude range 20–100 km represents a superposition of the small-sized fraction of the volcanic aerosol and the submicron fraction of the sulfate aerosol. The large-sized fraction falls out due to sedimentation and is characterized by a short lifetime. For the same reason, the large-sized aerosol fraction does not substantially influence the greenhouse effect.

An increase in the planetary albedo (γ) taking place with increasing atmospheric density, is caused by molecular scattering and by growing optical thickness of the stratospheric aerosol layer encircling the planet. Comparison of results with previous calculations made by Kondratyev and Moskalenko (1979, 1980a, 1980b) without account of absorption by minor components, suggest the conclusion that an account of the effect of minor components in the Earth's initial atmosphere leads to a mean global surface temperature increase by 6 K. However, the global greenhouse effect determined by a temperature difference $\Delta T = T_s - T_e$ (T_e is the effective mean global radiative temperature of the planet) is increased by 20 K, half the value of the greenhouse effect being caused by the feedback of an increase in the atmospheric water vapor content when the tropospheric temperature grows due to the greenhouse effect of the atmospheric minor components.

During the formation of the primordial atmosphere, volcanic activity still had not reached its maximum. An intensive formation of craters on the Earth took place about 4.3 billion years ago. Calculations show that during maximum volcanic activity about 4 billion years ago the optical thickness of the stratospheric cloud could have reached 4.4.

In the Archean period the effect of sulfur-containing components on the Earth's climate had been less manifested due to strong opacity of the atmosphere, created by CH_4, NH_3, CO_2, and by water vapor. Now, we shall consider specific features of the radiative heat exchange in the volcanically disturbed atmosphere in the anthropogenic period, when the temporal variation of the volcanic activity could have become a major factor of climatic variations.

In the present day atmosphere an intensification of volcanic activity leads to a decrease of the tropospheric temperature and an increase of the stratospheric temperature. The latter is partially caused by the SW radiation absorption by SO_2 and by stratospheric aerosols, and mainly by the absorption of the thermal emission from the surface and troposphere by the stratospheric sulfate aerosol.

In the initial stage of an explosive volcanic eruption, rather a sharp stratospheric warming is observed, which, as the volcanic cloud ejects to stratospheric levels, is followed by a slow temporal trend of cooling. The stratospheric aerosol layer with an optical thickness $\tau_a = 0.3$ leads to an increase of the Earth albedo by 7%, an increase of the stratospheric temperature of about 6 K, a decrease of the global solar radiation by 10%, of the short-wave absorbed radiation by 7%, and a decrease of the global surface temperature of 3–7 K (depending on the manifestation of the accompanying feedbacks).

The consideration of the volcanic activity impact on the radiative heat exchange in the Earth's atmosphere suggests the cooling of the surface and of the troposphere. Present day volcanic activity increases moisture exchange in

VOLCANIC ACTIVITY AND CLIMATE

the oceans-continents-atmosphere system, favoring the accumulation of ice and the broadening of the ice-covered areas. The consideration of different feedbacks in modeling the climatic impact of the volcanic activity is a problem of vital importance.

The Climatic Impact of Volcanic Activity in the Evolution of Mars

It is interesting to compare the climatic impacts of volcanic activities on the Earth and Mars. In contrast to the Earth, the atmosphere of Mars is more rarefied and mainly consists of CO_2. However, during the volcanic activity the chemical composition of the Martian atmosphere substantially differed from that of the present day, which could have led to substantial changes of climate on Mars.

In the Martian CO_2 atmosphere with low density and humidity, one could expect a lower value of the greenhouse effect $\Delta T = T_s - T_e$, averaged over the entire planet. At the same time, the diurnal temperature variations due to radiative cooling of the atmosphere and diurnal change of the vertical sensible heat exchange, can be rather great. Specific features of the Martian atmosphere manifested in low atmospheric pressure, in its temperature and humidity, could have been a cause of the strong greenhouse effect of specific components, whose temporal trends of content could have included the Martian dust storms and volcanic activity in the atmosphere's evolution.

Latitudinal variations of the vertical temperature profiles in the Martian atmosphere cause an intensive atmospheric circulation, followed by phase transformations of carbon dioxide, water vapor and by dust storms. During heavy dust loadings of the atmosphere the radiative regime of the planet changes. Strong heating of the atmosphere at the expense of solar radiation absorption by aerosols, and decrease of the surface temperature, are the main climatic consequences of dust storms, which are verified not only by Mariner–9 and Viking–1,2 measurements but also by results of numerical modeling (Kondratyev, 1990).

Note that the SW radiation absorption by aerosols favor the formation of vertical eddy fluxes in the upper and middle layers of the dust cloud which prevent dust particles from sedimentation. At a planetary surface albedo A = 0.21 the dust aerosol with a single scattering albedo $\omega \geq 0.9$ always raises the planetary albedo, and consequently, lowers the effective temperature (the thicker the dust cloud, the lower the temperature). The solution of the problem or radiative heat exchange in the 1-D approximation, has led to the conclusion about a decrease of the surface temperature with a dust load of the atmosphere increased to an optical thickness of $\tau_a \approx 1.5$.

As the dust load of the atmosphere grows further, the surface temperature slowly increases at the expense of prevailing greenhouse effect of aerosol, which screens the thermal emission of middle layers of the aerosol cloud.

Figure 2.3.9 shows the vertical temperature profiles calculated for the Martian atmosphere, whose chemical composition and density correspond to those of the present day. Curve 1 is obtained for the equatorial belt and determines the maximum possible daily mean temperature of the planet for its albedo $\gamma = 0.2$, the optical thickness of dust aerosols $\tau_a = 0.1$ (at $\lambda = 0.55$ μm), relative atmospheric humidity $f = 100\%$ and pressure $P_s = 10$ hPa.

Curve 2 corresponds to the same conditions, but for a mean global atmosphere. Curve 3 represents the most probable mean global temperature profile at the atmospheric humidity $f = 50\%$, aerosol optical thickness τ_a, pressure $P_s = 6.5$ hPa, and it corresponds to a value of the critical temperature lapse rate $\partial T/\partial Z = -2.6$ km^{-1}. This temperature profile leads to a mean global radiative temperature $T_r = 220$ K.

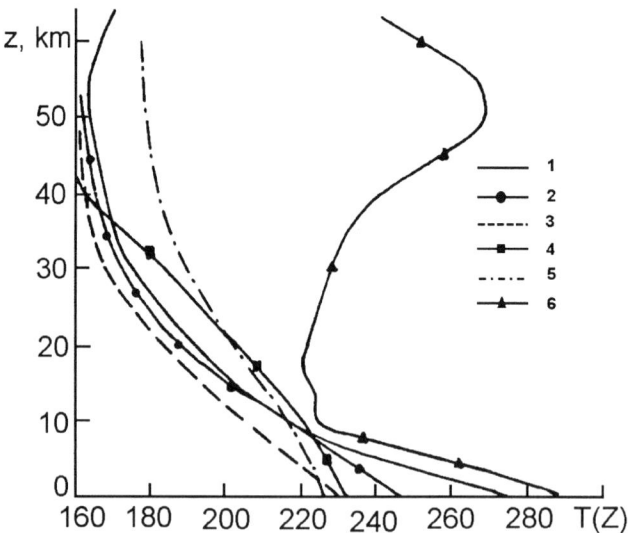

Figure 2.3.9. Vertical temperature profiles T(z) for different meteorological conditions on Mars (curves 1–5); 6 - the mean global vertical temperature profile calculated for Earth.

Curves 4 and 5 represent the vertical temperature profiles for the near-equatorial region and for the mean global Martian atmosphere, in conditions of a dust storm, with an optical thickness of the dust cloud $\tau_a = 1.0$. Even the warmest area of Mars is characterized by a mean diurnal atmospheric temperature lower than the mean global temperature of the Earth shown by Curve 6, Figure 2.3.9. The greenhouse effect of the Martian atmosphere averages about 10 K for the most probable model of the Martian atmosphere.

It is of interest to consider the impact of the Martian atmosphere on the greenhouse effect and temperature of its surface, when the density of the atmosphere and its chemical composition had changed in the process of the evolution of the planet. Of particular interest is the impact of volcanic activity on the climate of Mars.

Clark and Baird (1979) noted that the Martian lava resembles the basalt lava of the Earth, whose outgassing had been followed by release of a great amount of SO_2 and H_2S (the mixing ratio of sulfur compounds constitutes about 700 ppm). The concentration of sulfur compounds during the outgassing of the Martian lava should be higher due to the more intensive outgassing in conditions of low pressures near the Martian surface.

We assume the mixing ratio of sulfur compounds to be 1000 ppm. The lifetime of SO_2 for the present day Martian atmosphere constitutes about $3.9 \cdot 10^3$ days (in the terrestrial atmosphere the lifetime of SO_2 is about 110 days). In still earlier periods the pressure and temperature near the Mars surface could have been higher. However, the lifetime of SO_2 had always been long, which appeared to be a cause of the accumulation of a considerable amount of SO_2 in the Martian atmosphere during volcanic eruptions.

During volcanic eruptions a great amount of volcanic ash and water vapor had been injected to the atmosphere. The latter had favored the formation of clouds of H_2SO_4- water solutions. The velocity of particles' sedimentation depends on their size, and the time of sedimentation on the surface rapidly decreases with growing size of particles. For instance, particles with a size of 5 µm fall out during 30 days, and 1-µm particles during 4 years. Since the time of formation of the stratospheric sulfate aerosol on Mars constitutes about two Martian years, the long lifetime in conditions of the Martian atmosphere is only possible for submicron aerosols. The consideration of the Martian atmosphere circulation suggests the conclusion that during the formation of submicron particles, the total mixing of aerosols takes place both within the Northern and Southern Hemispheres on Mars. Thus, the Martian sulfate aerosol during the volcanic activity is a global-scale phenomenon, substantially affecting the climate on Mars.

The climate impact of the greenhouse effect on Mars was calculated by Kondratyev et al. (1981, 1983a), Kondratyev and Moskalenko (1981, 1983) in approximation of the radiative-convective equilibrium. The mean-global surface albedo was assumed to be constant and was estimated from the spectral albedo of Mars considered by Moroz (1978).

Major characteristic features of the radiative heat exchange in the atmosphere of Mars for the period of volcanic activity are determined by the following processes:

(i) The strong absorption of the shortwave radiation by the atmospheric SO_2 in the wavelength range 0.2–0.4 µm leads to a decrease of the planetary albedo and to the formation of a temperature inversion in the upper atmospheric layers.

(ii) The formation of clouds from particles of H_2SO_4 water solutions promotes the growth of the planetary albedo and the intensification of the greenhouse effect at the expense of the longwave radiation absorption in the bands of H_2SO_4 water solution.

(iii) The strong IR SO_2 bands intensify the greenhouse effect in the atmosphere and favor an increase of the temperature of the atmosphere and of the planetary surface.

Mean global Martian surface temperatures as a function of surface pressure, P_s, are listed in Table 2.3.2. The atmosphere of Mars is assumed to consist of CO_2, SO_2, and water vapor ($P_{SO_2} = P_{CO_2} \cdot 10^{-3}$).

Table 2.3.2. Mean temperature, T_s near the Martian surface as a function of $P_s(CO_2)$.

$P(CO_2)$	δ	$W_\perp(H_2)$, cm	T_s, K	T_s^*, K
0.0065	0.205	1.10–2	238	230
0.01	0.22	5.10–5	245	233
0.25	0.235	0.15	254	238
0.05	0.25	0.25	263	244
0.1	0.27	0.4	272	252
0.25	0.29	1.0	285	268

Note: $W_\perp(H_2O)$ is the water vapor content in the vertical column on Mars; T_s^* is the surface temperature in the absence of SO_2 in the Martian atmosphere

The vertical pressure profile and the albedo were calculated. The surface albedo was $A_s = 0.19$ and the relative humidity did not exceed $r = 0.5$. When the surface temperature of the planet grows due to the SO_2 greenhouse effect, the atmospheric moisture content also increases. As a result, the absorption of thermal emissions from the planetary surface and lower

VOLCANIC ACTIVITY AND CLIMATE 155

troposphere by its upper layers increases as well as the greenhouse effect. Rain and snowfall become possible even at relatively low pressures near the planetary surface $P_s \approx 0.05$–0.1 atm. The accumulation of SO_2 in great amounts shifts this level toward still lower pressures. SO_2, produced by volcanic activity, played a decisive role in the formation of the Martian climate, governing its atmospheric radiative regime and regulating its climate.

Periodic variations of volcanic activity could have been one of the factors causing temporal changes of the Martian climate. If Martian dust clouds led to an anti-greenhouse effect and lowered planetary surface temperatures, volcanic atmospheric effects during earlier Martian evolutionary periods would have created a greenhouse effect, which could have caused considerable warming. A hypothesis of a denser Martian atmosphere in the presence of powerful volcanic gas ejections permits one to explain a warmer and more humid Martian climate during its earlier stages of evolution. It should be noted here that volcanic high-altitude water vapor molecule ejections caused an increased OH-radical concentration, which subsequently led to a more intensive absorption of shortwave radiation near the UV spectral region and decreased the planetary albedo. The latter could have been an additional factor contributing to atmospheric warming during periods of increased volcanic activity.

A considerable reduction of volcanic activity resulted in strong planetary cooling. Upon evaporation, Martian water partially dissociated and some of it accumulated in the form of ice sheets near the poles. As a result of various interactions with surface minerals, acid waters released their sulfur concentrations. High sulfur concentrations in the elemental composition of the Martian surface, measured by the different Viking probes, confirmed the existence of such a mechanism being involved in Martian climate changes.

It is more likely that the warm Martian climate of the past was determined by the presence of SO_2 in the Martian atmosphere—not by a strong burst of solar activity. A broad variety of Martian landscapes was probably caused by "river" and "sea" erosive processes which are directly related to atmospheric greenhouse effects caused by an atmosphere composed of considerably different chemical components than that presently found on the planet. Some minor components which could have significantly affected past Martian climates include NH_3, CH_4, and NO_2. NH_3 produces a very strong greenhouse effect at the expense of longwave and shortwave radiation absorption. The greenhouse effect of NO_2 is determined by strong absorption of shortwave solar radiation and by narrow but intensive vibrational-rotational bands.

As a result of tectonic and volcanic activities, the Martian surface could have captured a great amount of water. Therefore, the possibility exists that a considerable amount of subterranean water is present under the Martian sands, which as of yet, remains undetectable using conventional observational means.

REFERENCES

Abbott, C.G., 1913: Do volcanic explosions affect our climate? *Nat'l. Geog. Mag.* **13**, 181-193.

Abrams, M., Glaze, L., and Sheridan M., 1991: Monitoring Colima volcano, Mexico, using satellite data. *Bull. Volcanol.* **53**, 571-574.

Ammann, M., Scherrer L., Mueller, W., Burtscher, H., and Siegmann, H.-Ch., 1992: Continuous monitoring of ultrafine aerosol emissions at Mt. Etna. *Geophys. Res. Lett.* **19** (13), 1387-1390.

Angell, J.K., and Korshover, J.K., 1977: Estimate of the global change in temperature, surface to 100 mbar between 1968 and 1973. *Mon. Weather Rev.* **105**, 375-388.

Arnold, F., 1980: Multi-ion complexes in the stratosphere—Implications for trace gases and aerosol. *Nature* **284**, 610-611.

Barth, C.A., Sanders, R.W., Thomas, R.J., Jakosky, B.M., and West, R.A., 1983: Formation of the El Chichón aerosol cloud. *Geophys. Res. Lett.* **10** (1), 993-996.

Barton, I.J., Prata, A.J., Watterson, I.G., and Young, S.A., 1992: Identification of the Mount Hudson volcanic cloud over SE Australia. *Geophys. Res. Lett.* **19** (12), 1211-1214.

Borzenkova, J.I., et al., 1976: Change in the air temperature of the northern hemisphere for the period 1881-1975. *Meteorologiya y Gidrologiya* **7**, 27-35.

Bradley, R.S., and Jones, P.D., 1993: Records of explosive volcanic eruptions over the last 500 years. In *Climate since A.D. 1500,* Bradley, R.S. and Jones, P.D. (Eds.), Rutledge, London, 606-622.

Bryson, R.A., and Goodman, B.M., 1982: The climatic effect of explosive volcanic activity: analysis of the historical data. In *Proc. Symp. on Atmospheric Effects and Potential Climate Impact of the 1980 Eruptions of Mt. St. Helens*, NASA Conference Publication **2240**, 191-202.

Burley, J.D. and Johnston, H.S., 1992: Ionic mechanisms for heterogeneous stratospheric reactions and ultraviolet photoabsorption cross sections for NO_2^+, HNO_3, and NO_3^- in sulfuric acid. *Geophys. Res. Lett.* **19** (13), 1359-1362.

Cadle, R.D., Kiang, C.S., and Louis, J.-F., 1976: The global scale dispersion of the eruption clouds from major volcanic eruptions. *J. Geophys. Res.* **81**, 3125-3132.

Cadle, R.D., Lazrus, A.L., Huebert, R.J., Heidt, L.E., Rose Jr., W.I., Woods, D.C., Chuan R.L., Stoiber, R.E., Smith, D.B., and Zielinski, R.A., 1979: Atmospheric implications of studies of Central American volcanic eruption clouds. *J. Geophys. Res.* **C84**, 6961-6968.

Capone, L.A., Toon, O.B., Whitten, R.C., Turco, R.P., Riegel, Ch.A., and Santhanam, K., 1983: A two-dimensional model simulation of the El Chichón volcanic eruption cloud. *Geophys. Res. Lett.* **10** (11), 1053-1056.
Carey, S.N., and Sigurdson, H., 1989: The intensity of plinian eruptions. *Bull. Volcanol.* **51**, 28-40.
Castelman, A.W. Jr., Munkelowitz, H.R., and Manowitz, B., 1974: Isotopic studies of the sulfur component of the stratospheric aerosol layer. *Tellus* **26**, 222-234.
Clark, B.C., and Baird, A.K., 1979: Is the Martian lithosphere sulphur rich? *J. Geophys. Res.* **84** (B14), 8395-8403.
Cox, R.A., and Sheppard D., 1980: Reactions of OH radicals with gaseous sulphur compounds. *Nature* **284**, 330-333.
Crowley, T.J., and North, G.R., 1991: *Paleoclimatology*. Oxford University Press, New York, 339 pp.
Crowley, T.J., Criste, T.A., and Smith, N.R., 1993: Reassessment of Crête (Greenland) ice core acidity/volcanism link to climate change. *Geophys. Res. Lett.* **20** (3), 205-212.
Dai, J., Masley-Thompson, E., Thompson, L.G., 1991: Ice core evidence for an explosive tropical volcanic eruption 6 years preceding Tambora. *J. Geophys. Res.* **96** (D9), 17361-17366.
Davis, D.D., Ravishankara, A.R., and Fischer S., 1979: SO_2 oxidation via the hydroxyl radical: Atmospheric fate of HSO_x radicals. *Geophys. Res. Lett.* **6**, 113-116.
Deirmendjian, D., 1971: *Scattering electromagnetic emission by spherical particles.* "Mir", Moscow, Publ. 165 pp. (in Russian).
Deirmendjian, D., 1973: On volcanic and other particulate turbidity anomalies. *Advances in Geophysics* **16**, 267-297.
DeLuisi, J.J., and Herman, B.M., 1977: Estimation of solar radiation by volcanic stratospheric aerosols from Agung using surface-based observations. *J. Geophys. Res.* **82**, 3477-3480.
DeLuisi, J.J., Dutton, E.G., Coulson, K.L., Defoor, T.E., and Mendonca, B.G., 1983: On some radiative features of the El Chichón volcanic stratospheric dust cloud and a cloud of unknown origin at Mauna Loa. *J. Geophys. Res.* **88**, 6769-6772.
Deshler, T., Adriani, A., Gobbi, G.P., Hofmann, D.J., Di Derfrancesco, G., and Johnson R.J., 1992: Volcanic aerosol and ozone depletion within the Antarctic polar vortex during the austral spring of 1991. *Geophys. Res. Lett.* **69** (N18), 1819-1822.
Dyer, A.J., and Hicks, B.B., 1965: Stratospheric transport of volcanic dust inferred from solar radiation measurements. *Nature* **208**, 131-133.
Farlow, N.H., Snetsinger, K.G., Hayes, D.M., Lem H.Y., and Tooper, B.M.,

1978: Nitrogen-sulphur compounds in stratospheric aerosols. *J. Geophys. Res.* **C83**, 6207-6211.
Ferry, G.V., Farlow, N.H., Lem, H.Y., and Dennis, M.H., 1978: Collection and analysis of stratospheric aerosols. *Atmos. and Oceanog. Res. Review* **9**, NASA. Tech. Mono. No. 80253, 119-127.
Friend, J.P., Barnes, R.A., and Vasta, R.M., 1980: Nucleation by free radicals from the photooxidation of sulfur dioxide in air. *J. Phys. Chem.* **84**, 2423-2436.
Galindo, I., 1992: Extinction of shorth-wave solar radiation due to El Chichón stratospheric aerosol. *Atmósfera* **5**, 159-168.
Galindo, I., and Bravo, J.L., 1975: On the presence of a volcanic stratospheric dust stratum over a polluted atmosphere. *Geofis. Int.* **15**, 157-167.
Galindo, I. Kondratyev, K.Ya., and Zenteno, G., 1996: Determination of the Atmospheric Optical Depth Due to El Chichón Stratospheric Aeosol Cloud in the Polluted Atmosphere of México City. *Atmósfera* **9**, 23-32.
Hammer, C.U., 1977: Past volcanism revealed by Greenland ice sheet impurities. *Nature* **270**, 482-486.
Hammer, C.U., 1980: Acidity of polar ice cores in relation to absolute dating, past volcanism and radio-echoes. *J. Glaciol.* **25**, 359-372.
Hammer, C.U., Clausen, H.B., and Dansguard, W., 1980: Greenland ice sheet evidence of post-glacial volcanism and its climatic impact. *Nature* **288**, 230-235.
Hansen, J.E., Wang, W.C., and Lacis, A.A., 1978: Mount Agung eruption provides test of a global climatic perturbation. *Science* **199**, 1065-1068.
Hanson, D.R., 1992: The uptake of HNO_3 onto ice, NAT, and frozen sulphuric acid. *Geophys. Res. Lett.* **19** (N20), 2063-2066.
Hart, M.A., 1978: The evolution of the atmosphere of the Earth. *Icarus* **33**, 1, 23-39.
Hayes, D., Snetsinger, K., Ferry, G., Oberbeck, V., and Farlow N., 1980: Reactivity of stratospheric aerosols to small amounts of ammonia in the laboratory environment. *Geophys. Res. Lett.* **7**, 974-976.
Hofmann, D.J., 1988: Aerosols from past and present volcanic eruptions. In *Aerosols and Climate*, P.V. Hobbs and M.P. McCormick (Eds.), A. Deepak Publishing, Hampton, Virginia, 195-214.
Hofmann, D.J., and Rosen, J.M., 1981: On the background stratospheric aerosol. *J. Atmos. Sci.* **38**, 168-181.
Hofmann, D.J., and Rosen, J.M., 1982: Balloon-borne observations of stratospheric aerosol and condensation nuclei during the year following the Mt. St. Helens eruption. *J. Geophys. Res.* **87**, 11,039-11,061.
Hofmann, D.J., and Rosen, J.M., 1983a: Stratospheric sulfuric acid fraction and mass estimate for the 1982 volcanic eruption of El Chichón. *Geophys. Res. Lett.* **10**, 313-316.

Hofmann, D.J., and Rosen, J.M., 1983b: Sulfuric acid droplet formation and growth in the stratosphere following the eruption of El Chichón. *Science* **222**, 325-327.

Hofmann, D.J., and Rosen, R.M., 1984: On the temporal variation of stratospheric aerosol size and mass during the first 18 months following the 1982 eruption of El Chichón. *J. Geophys. Res.* **89** (D3), 4883-4890.

Hofmann, D.J., and Oltmans, S.J., 1992: The effect of stratospheric water vapour on the heterogeneous reaction rate of $ClONO_2$ and H_2O for sulfuric acid aerosol. *Geophs. Res. Lett.* **19** (N22), 2211-2214.

Humphreys, W.J., 1940: *Physics of the Air*, McGraw Hill, 587-618.

Hunt, B.G., 1977: A simulation of the possible consequences of a volcanic eruption on the general circulation of the atmosphere. *Mon. Weather Rev.* **105** (3), 247-260.

Ivlev, L.S., Kondratyev, K.Ya., Golovin, A.V., and Kudryashov, V.I., 1992: On the dustiness of the Pamir and Ty Yan Shan glaciers. *Doklady of the Russian Acad. Sci.* **327** (6), 471-474 (in Russian).

Jensen, E.J., and Toon, O.B., 1992: The potential effects of volcanic aerosols on cirrus cloud microphysics. *Geophys. Res. Lett.* **19** (N17), 1759-1762.

Johnson, R.W., 1993: *Volcanic Eruptions and Atmospheric Change*. AGSO Issues Paper No. 1, Australian Geological Survey Organization, Canberra, 36 pp.

Kalitin, N.A., 1935: Atmospheric transparency and volcanic eruptions. *Klimat i Pogoda* **N4** (61), 8-11 (in Russian).

Kayumova, G.V., Moskalenko, N.I., and Parzhin, S.N., 1979: An atlas of parameters of the spectral absorption lines for atmospheric CO, NO, HCl. In *Abstracts of papers of the 5th. All-Union Symp. on the Propagation of the Laser Emission in the Atmosphere* (in Russian).

Kelly, P.M., and Sear, C.B., 1982. The formulation of Lamb's dust veil index. *Proc. Symp. on Atmospheric Effects and Potential Climatic Impact on the 1980 Eruptions of Mount St. Helens.* NASA Conference Publication **2240**, NASA, 293-298.

Kelly, P.M., and Sear, C.B., 1984: Climatic impact of explosive volcanic eruptions. *Nature* **311**, 740-743.

Kelly, P.M., Wigley, T.M.L., and Jones, P.D., 1984: European pressure maps for 1815-16, the time of the eruption of Tambora. *Climate Monitor. Univ. of Norwich*, **13**, 76-91.

Kent, G.S., and Yue, G.K., 1991: The modeling of CO_2 lidar backscatter from stratospheric aerosol. *J. Geophys. Res.* **96** (D3), 5279-5292.

Kondratyev, K.Ya., 1980a: Monitoring climatic parameters from satellites. *Trudy GGo,* Issue **438**, 75-102 (in Russian).

Kondratyev, K.Ya., 1980b: *Radiative Factors of the Present-Day Global Climate Changes*. Gidrometeoizdat, Leningrad, 279 pp. (in Russian).

Kondratyev, K.Ya., 1981a: Volcanoes and Climate. *Hidrometeorology, Series Meteorology, Overview,* issue **5**, Obninsk; VNIIGMI-MCD. (in Russian).
Kondratyev, K.Ya., 1981b: Stratosphere and Climate. Uspekhi Nauki i Techniki, *Meteorologia i Klimatologia,* **6**, Moscow: VINITI. (in Russian).
Kondratyev, K.Ya., 1985: *Volcanoes and Climate.* Moscow: VINITI, Itogi Nauki i Techniki, *Meteorologia i Klimatologia* **14**, 204 pp. (in Russian).
Kondratyev, K.Ya., 1988: *Climate Shocks: Natural and Anthropogenic,* Wiley, New York e.a., 296 pp.
Kondratyev, K.Ya., 1990: *Planet Mars,* Gidrometeoizdat, Leningrad, 369 pp. (in Russian).
Kondratyev, K.Ya., 1992a: *Global Climate,* Nauka, St. Petersburg, 359 pp. (in Russian).
Kondratyev, K.Ya., 1992b: Numerical Modeling of Global Change. *Izv. Russian Geographical Soc.* **124** (3), 232-240 (in Russian).
Kondratyev, K.Ya., 1993: Complex monitoring of the Pinatubo volcanic eruptions. *Studying the Earth from Space* **1**, 111-122 (in Russian).
Kondratyev, K.Ya. and Grassl, H., 1993: *Global Climate Change in the Context of Global Change.* Acad. Sci.. Publ., St. Petersburg, 195 pp. (in Russian).
Kondratyev, K.Ya., and Moskalenko, N.I., 1979: The Greenhouse Effect of the Planetary Atmospheres. *Astromicheski Vestnik* **8** (3), 129-143 (in Russian).
Kondratyev, K. Ya., and Moskalenko, N.I., 1980a: The evolution of the atmosphere and the greenhouse effect. *Izvestia AN SSSR., FAO* **16** (11), 1141-1162 (in Russian).
Kondratyev, K.Ya., and Moskalenko, N.I., 1980b: The radiative heat exchange in the atmosphere of Mars and the greenhouse effect. *Doklady AN SSSR* **255** (1), 64-66 (in Russian).
Kondratyev, K.Ya., and Moskalenko, N.I., 1981: The greenhouse effect of planetary atmosphres. *Nuovo Cimento* **4C** (6), 698-735.
Kondratyev, K.Ya. and Hunt, G.S., 1982: *Weather and Climate of Planets* Pergamon Press, Oxford, 755 pp.
Kondratyev, K.Ya., and Moskalenko, N.I., 1983: *Key problems in studies of terrestrial planets (The greenhouse effect of planetary atmospheres).* Adv. in Sci. & Technol., Ser. Studies in Space **19**, 157 pp. (in Russian).
Kondratyev, K.Ya., Moskalenko, N.I., and Terzi, V.F., 1977. Radiative cooling in the atmospheres of Mars, Venus and Jupiter. *Doklady AN SSSR,* **263, 6,** 1334-1337 (in Russian).
Kondratyev, K.Ya., Moskalenko, N.I., Terzi, V.P., and Skvortsova, S. Ya.,1981: Modelling the optical characteristics of atmospheric aerosols. In *GARP, Aerosol and Climate.* Gidrometeozdat, Leningrad, 130-153 (in Russian).
Kondratyev, K.Ya., Moskalenko, N.I., and Parzhin, S.N., 1983a: Optical properties of the Martian atmosphere and radiative heat exchange. *Adv. Space Res.* **2** (10), 39-42.

Kondratyev, K.Ya., Moskalenko, N.I., and Pozdnyakov, D.V., 1983b: Atmospheric Aerosols. Leningrad, Gidrometeoizdat, 224 pp. (in Russian).
Köppen, W., 1873: Über mehrjährige Perioden der Witterung insbesondere über die 11-jährige Periode der Temperature. *Zts. Osterr. Ges. f. Meteorol.* **8**, 257-267.
Kroes, G.-J., and Clay, D.C., 1992: Adsorption of HCl on ice under stratospheric conditions: A computational study. *Geophys. Res. Lett.* **19** (N13), 1355-1358.
Labitzke, K., and Naujokat, B., 1983: On the variability and on trends in the middle stratosphere. *Contr. Atmos. Phys.* **56**, 495-507.
Labitzke, K., and van Loon, H., 1989: The Southern Oscillation. Part IX. The influence of volcanic eruptions on the Southern Oscillation in the stratosphere. *J. Climatol.* **10**, 1223-1226.
Landsberg, H.E., and Albert, J.M., 1974: The summer of 1816 and volcanism. *Weatherwise* **27**, 63-66.
Landsberg, H.E., Yu, C.S., and Huang, L., 1968: Preliminary reconstruction of a long time series of climatic data for the eastern United States. Tech. Note BN-571. Inst. Fluid. Dynamics., Univ. of Maryland.
Lazrus, A.L., Cadle, R.D., Gansdrud, B.W., Greenberg, J.P., Hueber, B.J., and Rose, W.I. Jr., 1979: Sulphur and halogen chemistry of the stratosphere and of volcanic eruption plumes. *J. Geophys. Res.* **C84**, 7869-7875.
Lenoble, J., Tanre, D., Deschamps, P.Y., and Herman, M., 1982: A single method to compute the change in Earth-atmosphere radiative balance due to a stratospheric aerosol layer. *J. Atmos. Sci.* **39** (11), 2565-2576.
Loginov, V.F., 1984a: Possible causes and effects of volcanic eruptions. *Tr. Gl. Geofiz. Obs.* **471**, 103-107. (in Russian).
Loginov, V.F., 1984b: *Volcanic Eruptions and Climate*. Gidrometeoizdat, Leningrad (in Russian).
Loginov, V.F., Pivovarova, Z.I., and Kravchuk, E.G., 1983a: An assessment of the contribution of natural and anthropogenic factors to solar radiation variability on the Earth's surface. *Meteorol. Gidrol.* **8**, 55-60 (in Russian).
Loginov, V.F., Pivovarova, Z.I., and Kravchuk, E.G., 1983b: Variability of direct solar radiation and temperature in the Northern Hemisphere due to volcanic eruptions. *Izv. Ves. Geogr. Obschestsva* **115** (5), 401-411 (in Russian).
Marov, M.Ya., 1981: *Planets of the Solar System*. Nauka Publ. Moscow, 256 pp. (in Russian).
Mass, C., and Schneider, S.H., 1977: Statistical evidence on the influence of sunspots and volcanic dust on long-term temperature records. *J. Atmos .Sci.* **34** (12), 1995-2004.
McCormick, M.P., 1983: Aerosol measurements from earth orbiting spacecraft. *Adv. Space Res.* **2**, 73-86.

McCormick, M.P., and Swissler, T.J., 1983: Stratospheric aerosol mass and latitudinal distribution of the El Chichón eruption cloud for October 1982. *Geophys. Res. Lett.* **10**, 877-880.

McClatchey, R.R. (Ed.), 1973: APCRL Atmospheric absoption line paremeters compilation. *Environ. Res. Paper*, **434**, 78pp.

McInturff, R.M., Miller, A.J., Angell, J.K., and Korshover, J., 1971: Possible effects on the stratosphere of the 1963 Mt. Agung volcanic eruption. *J. Atmos. Sci.* **28**, 1304-1307.

McKeen, S.A, Liu, S.C., and Kiang, C.S., 1984: On the chemistry of stratospheric SO_2 from volcanic eruptions. *J. Geophys. Res.* **89** (D3), 4873-4881.

Mendonca, B.G., Hanson, K.J., and DeLuisi, J.J., 1978: Secular trends in clear sky transmissions at Mauna Loa Observatory perturbations in stratospheric aerosols. In *Third Conf. Atmos. Radiation* of the Amer. Met. Soc., Davis, CA, 28-30 June, 1978, preprints, 330-332.

Mori, T., Notsu, K., Tokjima, Y., and Wakita, H., 1993: Remote detection of HCl and SO_2 in volcanic gas from Unzen volcano, Japan. *Geophys. Res. Lett.* **20** (13), 1355-1358.

Moroz, V.I., 1978: *Physics of the Planet Mars*, Nauka Publ., Moscow, 342 pp. (in Russian).

Morss, D.A., and Kuhn, W.R., 1978: Paleoatmospheric temperature structure. *Icarus* **33** (1), 40-49.

Moskalenko, N.I.,. 1969: Experimental integral absorption functions for the bands of H_2O, CO_2, N_2O, CH_4, CO, and NO vapours. Izvestia AN SSSR, FAO, 5, 9, 962-966 (in Russian).

Moskalenko, N.I., and Zakirova, R., 1972: The calculation of the spectral, angular, and vertical distributions of the thermal emission field of the atmosphere and surface of the Earth. *Izv. AN SSSR, FAO,* **8** (8), 828-842. (In Russian).

Moskalenko, N.I., and Zotov, O.V., 1977: Latest experimental studies and specifications of the CO_2 spectral trasmission fuctions: The line parameters. *Izvestia AN SSSR, FAO,* **13**, 5, 488-498. (in Russian).

Moskalenko, N.I., and Yakupova, F.S., 1978: A solution for the problems of radiation transfer in the atmosphere by computer modelling. In *IV. All-Union Meeting on Molecular Spectroscopy of High and Super-High Resolution,* Novosibirsk, 178-182 (in Russian).

Moskalenko, N.I., Ilyin, Yu.A., and Parzhin, S.N., 1978: Continuum and pressure-induced absorptions in the spectra of CO_2 and water vapor. In: *5th All-Union Meeting on Molecular Spectroscopy of High and Super-High Resolution,* Novosibirsk, 187-191 (in Russian).

Nicholls, N., 1988: Low latitude volcanic eruptions and the El Niño-Southern Oscillation. *J. Climate* **1**, 91-95.

Old Farmer's Almanac, 1966: Vol. 174, 46 pp.
Parker, D.E., 1988: Stratospheric aerosols and sea surface temperature. *J. Climate* **1**, 87-90.
Pollack, J.B., Toon, O.B., Sagan, C., Summers, A., Baldwin, B., and VanCamp, A., 1976a: Stratospheric aerosols and climatic changes. *Nature* **263**, 551-555.
Pollack, J.B., Toon, O.B., Sagan, C., Summers, A., Baldwin, B., and VanCamp, A., 1976b: Volcanic explosions and climatic change: A theoretical assessment. *J. Geophys. Res.* **82**, 1071-1083.
Pollack, J.B., Toon, O.B., and Wiedman, D., 1980: Radiative properties of the background stratospheric aerosols and implications for perturbed conditions. *Geophys. Res. Lett.* **8**, 26-28.
Pollack, J.B., Toon, O.B., Danielsen, E.F., Hofmann, D.J., and Rosen, J.M., 1983: The El Chichón volcanic cloud - an introduction. *Geophys. Res. Lett.* **10**, 989-992.
Rampino, M.R., and Self, S., 1982: Historic eruptions of Tambora (1815), Krakatao (1883), and Agung (1963)—Their stratospheric aerosols and climatic impact. *Quart. Res.* **18** (2), 127-143.
Rampino, M.R., and Self, S., 1984: Sulphur-rich volcanic eruptions and stratospheric aerosols. *Nature* **310**, 677-679.
di Sarra, A., Cariani, M., Di Girolamo, P., Franco, G., Fuá, D., Knudsen B., Larsen, N., and Joergensen, T.S., 1992: Observation of correlated behavior of stratospheric ozone and aerosol at Thule during winter 1991-92. *Geophys. Res. Lett.* **19** (18), pp. 1823-1826.
Sedlacek, W.A., Mroz, E.J., Lazrus, A.L., and Gandrud, B.W., 1983: A decade of stratospheric sulfate measurements compared with observations of volcanic eruptions. *J. Geophys. Res.* **88**, 3741-3776.
Settle, M., 1979: Formation and deposition of volcanic sulphate aerosols on Mars. *J. Geophys. Res.* **B84**, 8343-8354.
Sigurdsson, H., 1982: Volcanic pollution and climate – The 1783 Laki eruption. *EOS* **36**, 601-602.
Sigurdsson, H., 1990a: Assessment of the atmospheric impact of volcanic eruptions. *Geol. Soc. Amer.*, Special Paper **247**, 99-110.
Sigurdsson, H., 1990b: Evidence of volcanic loading of the atmosphere and climate response. *Paleogeogr., Paleoclimatol., Paleoecol.* **89**, 277-299.
Simkin, T., and Fiske, R.S., 1983: Krakatau 1883 - a classic geophysical event. *EOS.,* **64**. 513-514.
Stith, J.L., Hobbs, P.V., and Radke, L.F., 1978: Airborne particle and gas measurements in the emissions from six volcanoes. *J. Geophys. Res.* **C83**, 4009-4017.
Stolarski, R.S., and Butler, D.M., 1979: Possible effects of volcanic eruptions

on stratospheric minor constituents chemistry. PAGEOPH 117, 486-497.
Stothers, R.B., 1984: The great Tambora eruption in 1815 and its aftermath. *Science* **224**, 1191-1198.
Thorarinsson, S., 1969: The Lakagigar eruption of 1783. *Bull. Volcanol.* **33**, 910-927.
Thorarinsson, S., 1979: On the damage caused by volcanic eruptions with special reference to tephra and gases. In *Volcanic Activity and Human Ecology*, P.D. Sheetsand and D.K. Graysson (Eds.) 125-159, Academic Press, N.Y.
Tilling, R.L., 1989a: Volcanic hazards and their mitigation: Progress and problems. *Revs. of Geophys.* **27** (2), 237-269.
Tilling, R.L., (Ed.), 1989b: *Volcanic Hazards: Short Course in Geology. Vol. 1.* Amer. Geophys. Union, Washington, D.C., 123 pp.
Tilling, R.L., and Dvorak, J.J., 1993: Anatomy of a basaltic volcano. *Nature* **363**, 125-133.
Tilling, R.L., and Lipman, P.W., 1993: Lessons in reducing volcano risk. *Nature* **364**, 277-280.
Tilling, R.I., Heliker, Ch., and Wright, T.L., 1987: *Eruptions of Hawaiian volcanoes: Past, Present, and Future.* U.S. Dept. of Interior/U.S. Geological Survey, 55 pp.
Tilling, R.I., Topinka, L., and Swanson, D.A., 1990: *Eruptions of Mount St. Helens: Past, Present, and Future.* U.S. Dept. of Interior/ U.S. Geological Survey, 57 pp.
Toon, O.B., 1982: Volcanoes and climate. In *Atmospheric Effects and Potential Climatic Impact of the 1980 Eruptions of Mount St. Helens*, NASA Conference Publication 2240, Proc. of Symp. held in Washington, D.C., Nov. 18-19, 1980, A. Deepak (Ed.), 15-36.
Toon, O.B., and Pollack, J.B., 1976: A global average model of atmospheric aerosols for radiative transfer calculations. *J. Appl. Meteorol.* **15**, pp. 225.
Toon, O.B., and Pollack, J.B., 1982: Stratospheric Aerosols and Climate. In *Topics in Current Physics– The stratospheric aerosol layer.* R.C. Whitten (Ed.), Springer Verlag, 121-147.
Turco, R.P., Hamill, P., Toon, O.B., Whitten, R.C., and Kiang, C.S., 1979a: A one-dimensional model describing aerosol formation and evolution in the stratosphere, I. Physical processes and mathematical analogs. *J. Atmos. Sci.* **36**, 699-717.
Turco, R.P., Hamill, P., Toon, O.B., Whitten, R.C., and Kiang, C.S., 1979b: A one-dimensional model describing aerosol formation and evolution in the stratosphere, II. Sensitivity studies and comparison with observations. *J. Atmos. Sci.* **36**, 718-736.
Wigley, T.M.L., 1991: Climate variability on the 10-100 year time scale:

Observations and possible causes. In *Global Change of the Past.* R.S. Bradley (Ed.), 83-101.

Wilson, J.C., Blackshear, E.D., Hyun, J.H., 1973: Changes in the sub-2.5µ diameter aerosol observed at 20 km altitude after the eruption of El Chichon. *Geophys Res. Lett.* **10** (N11), 1029-1032.

Wright, T.L., and Pierson, T.C., 1992: *Living with Volcanoes. The U.S. Geological Survey's Volcano Hazards Program.* U.S. Geological Survey Circular 1073, 57 pp.

CHAPTER 3

RECENT MAJOR EXPLOSIVE VOLCANIC ERUPTIONS

The most persuasive evidence of the climatic impact of volcanic eruptions was obtained on the basis of the analysis of observational data after three recent major eruptions: Mount St. Helens, El Chichón and Pinatubo. We shall discuss relevant results.

3.1. The Mount St. Helens Volcano Eruption

At about 08:30 in the morning on May 18, 1980, the explosive eruption of the Mount St. Helens volcano located in the southern part of the State of Washington (USA) took place accompanied by strong ejections of dust, gases and heat into the troposphere and stratosphere (Deepak, 1982, Newell and Deepak, 1982). It is important to emphasize that Mount St. Helens was the first major volcanic eruption observed from the ground, from the air, and from space.

Under the impact of wind the volcanic dust cloud moved to the east, steadily covering most of the United States. At noon on May 20 the cloud reached the central part of Illinois. Soon after the eruption various measures were undertaken to monitor the impact of the eruption on the atmosphere as completely as possible. For instance regular lidar soundings (589.0-µm wavelength) were started close to midnight on May 20 in Urbana, Illinois, and continued for 2½ weeks (Gardner et al., 1980). The ceiling of the lidar aerosol soundings reached 25 km with altitude resolution of 300 m. The analysis of the results revealed a presence of time dependent multilayered distribution of the aerosol. Apparently, by June 3 the vertical aerosol profile was close to what was usually observed in Urbana.

The May 18, 1980 eruption of the Mount St. Helens volcano was unusually strong, although the amount of ejected magma reached only 0.5 km^3.

Rose and Hoffman (1982) have pointed out that from the viewpoint of the magma amount there have been about 30 similar eruptions since 1900, but as far as the scale of ash ejections is concerned, not more than 10 to 20. A series of photographs obtained during the first few minutes of the eruption allowed detailed monitoring of the volcanological development of the eruption. Rose and Hoffman (1982) write: " ...the chain of events began with a large earthquake (magnitude 5.1). The north slope of the St. Helens dome which had been growing for 2 months at a rate of 1.5 m/day, detached itself and began to slide away. Ultimately the more than 2 km^3 of Mount St. Helens cone that was removed formed a 15-km-long hummocky debris flow in the the North Fork of the Toutle River. Phreatic eruptions, similar to those which had taken place beginning on March 27, then began from the summit crater. Within seconds, however, the character of the eruption changed dramatically. A strong lateral blast (pyroclastic surge) developed from the dome, directed mainly to the north. This pyroclastic surge moved more than 20 km to the north of the dome and devastated about 700 km^2 of forest land. The surge deposit contained mainly material from the old St. Helens dome but also contained dacite rocks which represented new magma. It is one of the most dramatic historic examples of pyroclastic surge. After the surge a huge plinian eruption column rose to a height of more than 20 km. The ashes erupted in this plinian phase were preserved partly as hot pyroclastic flows, restricted to the northern slopes of Mount St. Helens itself, and partly as an extensive ashfall deposit which is found to the east of of the volcano."

The plinian eruption of May 18, 1980 lasted about 9 hours (08:40–17:40). Radar data demonstrate that the column height for this period was consistently greater than 13 km. There were four periods of increased ash loading of the atmosphere during the eruption as shown by the radar. These periods occurred beginning at 08:45; 10:30; 13:30 and 16:00. The first ash pulse was the highest and moved eastward most rapidly (150 km/hr).

Some unusual features of the eruption (intensity, strong lateral blast, multimodal size distribution of ash particles and their mainly nonmagmatic composition) could be due to its phreatomagmatic origin.

About 6 hours after the Mount St. Helens eruption aircraft observations (NASA flying laboratory S-141) in the State of Utah were made of the volcanic cloud thermal emission in the 2–40 μm wavelength region. At the time of observations (May 21 19:00 LT, aircraft altitude 12.1 km) the volcanic cloud in the stratosphere was stretched along the distance of about 1300 km from the volcano to Utah. Kuhn et al. (1981) have processed observational data to get information on the volcanic cloud IR transparency. During the flight from the clear atmosphere to below the cloud area atmospheric thermal emission increased from 0.25 to 0.58 W/m^2.sr (upwelling IR irradiance was equal to

8.38 W/m².sr). The assessment of the IR transparency gave 0.936 which corresponds to an AOT equal to 0.073 while the background AOT value was 0.06.

On the basis of AOT = 0.073 approximate estimates of mean particle radius ($\approx$ 0.35 μm) and longwave radiative heating due to aerosols were obtained. Such a heating should be, apparently, compensated through cooling due to backscattering of shortwave radiation by the cloud of which the AOT in the visible region is twice as high. Thus, even the local impact of the volcanic cloud on the surface temperature should be negligible (such a conclusion requires, however, further confirmation).

A broad program of various observations which was started in the USA and other countries after the Mount St. Helens eruption opened perspectives of studies in the following directions:

(i) Environmental impacts of the eruption (in particular, climate, ecosystems, and water basins);
(ii) In situ measurements of ejecta (gaseous and aerosol components);
(iii) Ground-based, aircraft and satellite remote sensing studies;
(iv) Transport and diffusion of eruption products;
(v) Chemistry of volcanic products (including processes of gas-to-particle conversion);
(vi) Potential volcanic impact on weather and climate (paleoclimatic data, radiation budget changes, etc.).

As has been mentioned earlier these and other problems have been discussed in detail (Deepak, 1982, Newell and Deepak, 1982). The principal conclusions may be summarized as follows:

The volume of magma erupted at Mount St. Helens on May 18 was in the range from 0.3 to 0.6 km³. Eruptions of comparable volume occur on average every 8 to 10 years (in fact El Chichón and Pinatubo eruptions took place in 1982 and 1991). The ash ejecta was very fine grained with the majority in the form of particles whose diameter was less than 200 μm.

NOAA-6 satellite data showed that the ash cloud over Montana and Wyoming on May 19 increased the albedo by a factor of two. On May 20 ground-based AOT values as high as 0.48 at 0.875 μm were recorded at Boulder, Colorado (a typical background value is 0.053).

Assessment of SO_2 volcanic emission showed that more than $2 \cdot 10^8$ kg of sulfur was erupted on May 18, but the amount entering the stratosphere is not known. Since this sulfur mass is relatively small (partially because the magma was poor in sulfur) climatic impact of the eruption could not be significant. It is also clear that explosivity alone is a poor criterion for judging the global atmospheric impact.

VOLCANIC ACTIVITY AND CLIMATE

At about 12 km, the plume moved rapidly eastward, then turned to the southeast and later northeast, passed the United States east coast about 3 days after the eruption, and circuited the globe in about 16 days. At 27 km the debris moved toward the west with an average speed of about 8 m/s, corresponding to the circuit time of about 56 days. Reliable predictions of the cloud trajectory at the early stage of its development allowed aircraft to penetrate the cloud a day or two after the eruption. The analysis of aircraft measurement results discovered the presence of sulfuric acid in the stratospheric part of the cloud as well as gaseous sulfur in the form of SO_2, OCS and CS_2. H_2S was the dominant sulfur species in the troposphere.

Water vapor concentration in the plume on May 19 was found to be 10 or more times the background concentration. Enhanced levels of oxydized nitrogen compounds (NO, NO_2, and HNO_3) were also observed. Satellite remote sounding SAGE data and ground-based lidar soundings in various points of the globe allowed monitoring of spatial structure and temporal change of the plume, as well as drift in the stratosphere. Lidar scattering ratios (SR), for the material near 14 km were as large as 100 at 694.3 μm (background SR values at this wavelength are close to 1.1). Extinction coefficients as high as 0.01 km^{-1} were measured at 1.0 μm.

AOT values (SAGE data) for the NH in August varied from 0.001 to greater than 0.005 at 1 μm. The assessment of the stratospheric aerosol mass in the NH on the basis of lidar soundings at NASA Langley Research Center show an increase from $3 \cdot 10^8$ to $7 \cdot 10^8$ kg, while the global SAGE data yielded about $3 \cdot 10^8$ kg. Thus, there was doubling of the aerosol mass in the NH in comparison with the background level. If one can assume that the thickness of the stratospheric aerosol layer was 4 km, then the equivalent average mass density for the NH is about 0.3 $μg/m^3$, that is about four times higher than pre-erupted values.

Assessment of climatic consequences taking into account aerosol physical properties (single scattering albedo exceeded 0.98), led to the conclusion that the long-term radiative effects will be small in comparison to the normal variability of the atmosphere. Qualitative conclusions that observed physical properties of aerosols have to result in an increase of the planetary albedo (hence, tropospheric cooling) and in the lower stratosphere warming.

The results mentioned point to a necessity of more complete and regular observations. Aerosol sampling needs to be maintained so that composition measurements can be made in a volcanic plume as soon as possible after an eruption (probably within 12 hours) and as often as is logistically practical during the first 10 days after the eruption. A capability to numerically simulate 3-D plume trajectories in real time is necessary for use in making aircraft flight plans. To assess gas-to-particle transformation conversion rates in the volcanic cloud of the sulfur and the nitrogen species should be measured from an aircraft.

Satellite remote sensing monitoring should be applied to obtain data on global climatology of stratospheric aerosol. Lidar soundings in various parts of the globe as well as solar and longwave radiation measurements (total and spectral fluxes) should be accomplished at the surface and in the free atmosphere on a broader scale (especially in the SH, at equatorial and high latitudes). Reference observations at certain points for calibration purposes are needed. A very important problem is further development of 3-D models, including detailed consideration of photochemical processes and aerosol cloud kinetics. Paleoclimatological research should be a part of the whole program.

A very informative presentation of stratospheric aerosol satellite results monitoring in 1979–1980 on the basis of Nimbus-7 and SAGE has been published by McCormick (1981; 1982; 1983). Aerosol extinction coefficient (AEC) serves as an indicator of aerosol content in the stratosphere. Typical AEC background value for the layer between the tropopause and 20-km altitude is about $(1-2) \cdot 10^{-4}$ km^{-1}, (*in situ* measurements show that background aerosol is characterized by lognormal size distribution with the range of 0.1–0.6 µm for particle radius and number concentration of about 0.5 cm^{-3}).

Satellite data for May 23, 1980 clearly demonstrate an increase of AEC up to 10^{-3} cm^{-1} after the Mount St. Helens eruption. At altitudes below 15 km aerosol was moving eastward and at the 12-km level completed the first revolution around the Earth in about 2 weeks (with an average speed close to 25 m/s). Above 25 km the aerosol was slowly shifting westward with the speed about 6–8 m/s and circled the globe in 60–80 days. The analysis of mean zonal AEC cross section for September 1980 shows the concentration of aerosol mass mostly in the NH, especially in high latitudes. The hemispheric stratospheric aerosol mass had approximately doubled. Data for the winter in the SH (August 2–3, 1980) have revealed the presence of PSC with average background stratospheric aerosol concentration (outside of PSC) about $1.2 \cdot 10^{-4}$ cm^{-3}.

The analysis of SAGE data made by Kent (1982) shows that during the first period of observations (May 21–31, 1980) enhancements of stratospheric extinction were confined to latitudes between about 55° N and 25° N and longitudes between 10° W and 140° W. Individual layers were observed up to altitudes of 23 km. The values of AEC after the eruption were from 10^{-3} to 10^{-2} km^{-1}, that is, up to two orders of magnitude higher than the background (~ 10^{-4} km^{-1}). During the second period of observations (July 19–August 12, 1980) the volcanic aerosol was found to be widely distributed over the NH, the maximum concentration being north of 50° N. The aerosol showed considerable inhomogeneity and had reached as far south as 15° N but little, if any, had crossed the equator into the SH. Individual layers at different heights were still distinguishable. The total stratospheric aerosol loading on this occasion appeared to be greater than in May and corresponded to an increase in global stratospheric mass of between 50 and 100 percent.

Hirose et al. (1982) have discussed results of the chemical analysis of aerosol samples which had been taken regularly during the time period from June to August 1982 at the field station of the Meteorological Research Institute in Tsukuba, Japan (close to the Pacific coast). The results of X-ray fluorescence analysis of samples on glass filters show strong increase of Zn, Sb, Se and Cl concentrations in samples for June 4, 1980. This is especially true for Zn whose concentration had increased up to 7 µg/m^3, far exceeding background level (0.04 – 0.35 µg/m^3). Concentration ratios Zn/Se and Zn/Sb were close to those observed over the USA territory in June 1982 in the volcanic plume produced after the El Chichón eruption.

Concentration maxima mentioned coincided with stratospheric aerosol enhancement observed at lidar sounding stations in Tsukuba, Fukuoka and Nagoya on July 4. The eruption cloud circled the globe in 17 days. The whole totality of observational data indicates that aerosol chemical composition changes, observed in Tsukuba, were due to the Mount St. Helens eruption. Apparently stratospheric aerosol transport to the surface layer of the atmosphere was connected with the stratosphere-troposphere air exchange in the jet stream zone where aerosol is being transported.

Products of the Mount St. Helens volcanic ejecta were soon discovered after the eruption over the vast USA territory and beyond. Chung et al. (1981) have analyzed transport and temporal change of the volcanic plume using NOAA-6 satellite images in the visible spectral region and surface synoptic maps. The presence of volcanic aerosol was identified on the basis of meteorological visibility surface observations. In 1–2 days after the eruption the visibility dropped to 3–7 km. In 56 hours the volcanic cloud moved along the path of 1700 km (towards the U.S. northern border) with the mean speed about 40 km/hr, and its width reached 680–750 km. At the 200 hPa level (~ 10 km) the cloud shifted by 2700 km in 2 days with average speed of 25 m/s. Correlation spectrometer SO_2 concentration measurements discovered very high SO_2 concentration over Toronto during the passage of the cloud.

By May 20 the volcanic cloud reached the State of Michigan. Atmospheric turbidity measurements conducted during these days in Ann Arbor with the use of a modernized Volz photometer (Ryzner et al., 1981) were used to calculate Ångstrom turbidity coefficient (ATC). During the passage of the cloud there was an increase of ATC from 0.10 (before the arrival of the cloud and after its passage) to 0.25 as well as a decrease of parameter α in the relationship for the aerosol optical thickness: $\tau = \beta\lambda^{-\alpha}$ (λ -wavelength) from 1.8 to 0.6 which is an indication of weakened spectral dependence of turbidity. The aerosol optical thickness at the 0.5-µm wavelength had increased from 0.6 to 1.1. Direct solar and global radiation had decreased while there was an increase of diffuse radiation.

An important source of information on volcanic cloud dynamics were geostationary visible and infrared satellite images (Laver, 1982). Favorable observing conditions (small amount of clouds) made it possible to monitor dynamics of the Mount St. Helens eruption from the very beginning, when the initial explosion, shock wave, and visible horizontal dust distribution during the following weeks were readily apparent. Meteorological wind and height fields (on a daily basis from well over 100 stations in the USA, Canada and México) permitted the inference of the vertical distribution of volcanic dust as well as an explanation of the atmospheric behavior that caused the visible and nonvisible dust distribution.

A GOES picture taken at 16:45 Greenwich Meridional Time (GMT), approximately 25 minutes after the May 18, 1980, eruption shows ash spreading more than 90 km east of the volcano site. At 18:45 GMT, on the same day, a second eruption occurred, and a 20:15 GMT photograph indicates a new mushroom-shaped cloud which probably corresponds to this second eruption. The 23:45 GMT picture clearly shows that the leading edge of volcanic dust spread into western Montana at 9 to 12 km height by nightfall on May 18, 1980.

By the evening of May 19, horizontal dispersion had significantly increased. High cloudiness obscured some of the dust at and below about 700 hPa (3 km) over Montana and southern Canada. The dust between 500 hPa and 200 hPa (5 to 12 km) had dispersed across Wyoming and Colorado with its leading edge covering the Texas panhandle. Coincidently, ash at 100 hPa through 70 hPa (16 to 18 km) remained over Idaho and Washington. In the vertical the maximum concentration of volcanic ash, approximately 32 hours after the first Sunday morning eruption, lay in an upward-spiraling clockwise path skewed southeastward at jet stream levels over the Central Plains.

During May 20–21, visible GOES pictures clearly show the upper tropospheric ash shift from a southeast trajectory to a northeast one implying rotation about a large upper-air trough. From May 23 to May 28 the closed low aloft dissipated slowly, but there is evidence that the layers of dust may have lingered near the Mid-Atlantic States for up to 10 days after the initial eruption. At the same level (12 km), dust caught in the jet stream very possibly had travelled three-quarters of the way around the globe across China and over the western Pacific Ocean (Laver, 1982).

The major eruption of Mount St. Helens on May 18, 1980, had sufficient energy to traverse the troposphere (9 km above the mountain top) and to penetrate an additional 10 km into the stratosphere (Danielsen, 1982). This plume, initially quasi-vertical, rapidly acquired the horizontal momentum on the environmental winds and suffered differential rotation due to a positive speed shear in the troposphere and a negative shear in the stratosphere. Advecting rapidly eastward by the undulatory jet stream, the lower stratospheric portion of the plume circled the globe at an average speed ~ 25 m/s reentering North

America over California in early June. During the same period, the uppermost portion slowly lay over the northwestern United States and then moved westward over the northern Pacific Ocean. This plume dispersion was initiated by the vertical shears of the horizontal winds which converted a nearly vertical plume to a thin, quasi-horizontal, quasi-zonal lamina. Horizontal shear then dispersed the lamina meridionally while small scale, wave turbulent motions spread it slowly vertically.

Cheng (1982) has accomplished microscopical investigation of volcanic ash collected from the ground stations during the eruption which revealed a distinct bimodal size distribution with high concentrations of particle range at (1) 200 to 100 µm and (2) 20 µm to 0.1 µm. Most larger particles are solidified magma particles of porous pumice with numerous gas bubbles in the interiors the smaller ones are glassy fragments without any detectable gas bubbles. Very few of the fragments are below 0.1 µm and no evidence could be found to indicate that the small particles are formed by the condensation process.

Elemental analysis discussed by Cheng (1982) demonstrates that the fine fragments all have a composition similar to that of the larger pumice particles. Laboratory experiments suggest that the formation of the fine fragments occurs by the bursting of glassy bubbles from a partially solidified surface of a crystallizing molten magma particle. The production of the gas bubbles is due to the release of absorbed gases in molten magma particles when solubility decreases during phase transition. Diffusion cloud chamber experiments strongly indicate that submicron volcanic fragments are highly hygroscopic and extremely active as cloud condensation nuclei. Ice crystals also are evidently formed on those fragments in a supercooled (-20 °C) cloud chamber.

It has been reported earlier that charge generation by ocean volcanic eruptions is due to contact of molten lava with sea water. Cheng (1982) believes, however, that it seems to be insufficient to explain the observed rapid and intense lightning activities over the Mount St. Helens eruption. He has suggested therefore a hypothesis that highly electrically charged fine solid fragments are ejected by bursting of gas bubbles from the surface of a crystallizing molten magma particle. The charge-separation process by thermal electric effect is taking place within this crystallizing particle, and their polarities are determined by a temperature gradient between the colder surface of the solid glassy bubbles and the interior of the particle with a much higher temperature. After the bursting, positively charged fine fragments are ejected and carried by convection to the top of a volcanic plume and negatively charged, and heavier solidified plumic particles gradually fall by gravity to the base of the plume and eventually reach the ground. It has been suggested by Cheng (1982) that the fragmentation of volcanic ash and its accompanying electrification may play an important role in the generation of electric fields which may become strong enough to initiate lightning during a volcanic eruption and also to have a definite impact on climatological effects in the atmosphere.

Cadle and Heidt (1982) have conducted a comparison of constituents of Mount St. Helens eruption clouds with those of some other volcanoes. Table 3.1.1 contains the results of gas chromatographic analysis of samples taken from aircraft during the flights through, across, or along the eruption cloud (the St. Helens data were obtained during the time period April 3–10, 1980). Coincidentally, samples of particles were collected from the eruption clouds with a single stage impactor on Parlodian-covered electron microscope grids and examined by transmission and scanning with the same instrument and by energy dispersive X-ray analysis.

Table 3.1.1. Values or range of values for four constituents of eruption clouds from Mount St. Helens, Guatemalan Volcanoes, and Hawaiian Volcanos. After Cadle and Heidt (1982).

Volcano	Value in ppmv[a]		Value in ppbv[a]	
	CO_2	H_2	CH_4	COS
Pacaya (1978)	16	nd[b]	0.3	0.27
Pacaya (1980)	<3	nd	nd	0.03–0.087
Fuego (1978)	9–27	nd–9.6	nd–0.1	0.26–9.4
Fuego (1980)	<3–3	c	nd–0.27	–
Santiaguito (1978)	17	0.31	nd	0.015
Santiaguito (1979)	<3–7	nd–0.8	nd–0.3	–
Mauna Loa (1979)	557–2.1x10^4	0.8–20	0.8–2.9	0.09–51
Kilauea (1979) (Sulphur Bank)	1.7x10^4–1.5x10^5	0.9–1.2	nd–0.45	4.1–10
Mount St. Helens (April 1980)	<3–19	0.12–0.23	nd–0.59	nd–0.22

[a]Net, after background concentrations subtracted.
[b]Denotes not detected.
[c]Indicates that the analysis was not performed.

The most interesting feature of the results is the similarity in the magnitude of the values for the phreatic eruptions of Mount St. Helens and those for the magmatic eruptions of Fuego and Santiaguito in Guatemala. Subjectively, the eruption clouds penetrated by the aircraft contained similar quantities of ash. One might have expected much lower concentrations of volcanic gases in the Mount St. Helens clouds than in the others, and these results suggest that magmatic gases may have played a greater role in the phreatic eruptions than had generally been believed.

The concentrations of volcanic gases in the samples collected in Hawaii were much larger than those observed elsewhere because the former were collected directly from fumaroles. The samples collected in Hawaii had much larger ratios of CO_2 to the other gases than did any of the other samples. Ash particles collected from the Mount St. Helens eruption clouds during the early phreatic stages were much larger than those collected from the clouds of the Central American volcanoes.

A very representative data bank of the St. Helens eruption aerosols including the available worldwide lidar data as well as a series of airborne lidar flights has been discussed by McCormick (1982). These data show the dispersion of material at different altitudes during the early global circuits. The material in the lower stratosphere and upper troposphere was very patchy in horizontal extent with backscattering ratio values over the east coast of the United States greater than 100 at the ruby wavelength 694.5 nm. Two wavelength ratios and depolarization values for the material in the lower stratosphere (12 to 18 km) appear to have returned to the pre-May 18 values within a month after the eruption and this indicates a rapid conversion to spherical shapes and normal indices of refraction. The material above 20 km moved slowly westward while most of the ejecta moved eastward at various speeds and directions which varied considerably with altitude. The westward material was detected first by the Japanese lidar system and subsequently by the European and American ground-based systems. It circuited the globe in about 80 days. An airborne lidar flight in early September across the continental United States showed the layers to have homogenized considerably as one broad layer between about 14 and 21 km peaking at 18 to 19 km and another more intermittent thin layer between 21 and 22 km. The ruby peak backscattering signal was mainly due to an impact of sulfuric acid aerosol. The backscattering values have been used to provide an estimate of the increase in aerosol mass loading produced by the Mount St. Helens eruption. The assessment gave the value of the sulfate aerosol mass of about $5 \cdot 10^5$ tonnes which may be compared with a prevolcanic total for the whole globe of approximately the same value. This loading is smaller than that of other recent eruption: $3 \cdot 10^7$ tonnes (Mount Agung, 1963) and $5 \cdot 10^6$ tonnes (Mount Fuego, 1974).

Laulainen (1982) studied the effects of volcanic ash suspended in the atmosphere on the incident solar radiation which was monitored at the Hanford Meteorological Station in Washington subsequent to the major Mount St. Helens eruption on May 18, 1980 (May 16–31, 1980). Passage of the ash plume over Hanford (located close to the volcano) resulted in a very dramatic decrease of solar radiation intensity to zero. A reduction of visibility to less than 1 km was observed. Ash loading in the atmosphere remained very high for several days after the eruption, primarily as a result of resuspension from the surface.

Laulainen (1982) has obtained values of the aerosol optical thickness τ of the atmosphere from the following empirical relationship with the ratio of diffuse-direct solar radiation F:

$$F = 0.52\tau + 0.14 \tag{3.1}$$

The diffuse-to-direct ratio of solar radiation provided a useful index for estimating volcanic ash loading of the atmosphere. Aerosol mass concentration varied between 150–600 µg/m^3 (May 19) and 1–6 µg/m^3 (background values). Atmospheric clarity and visibility improved to near pre-eruption conditions following a period of rain showers.

After the major Mount St. Helens eruption on May 18, Lerfald (1982) made photometric and photographic observations at Boulder, Colorado (May 19–June 2) with the help of an 8-channel narrow band width solar photometer (0.5–1.1-µm wavelength region), a solar aureole photometer and two time-lapse camera systems. The solar photometer data revealed specific features of the atmospheric transparency and spectral dependence and showed that on May 20 the values of AOT for longer wavelengths exceeded background values up to 10 times. Photographs of the sky and circumsolar zone show spatial inhomogeneity of the eruption cloud.

Michalsky et al. (1982) have analyzed the data from a network of solar radiometers, 12 channel solar filter photometers operated on the North American continent (35°–61° N, 74°–120° W) for an average of 2 years before the first major eruption and later. The nearest station was located some 180 km east of the volcano (the event around July 22 was chosen because of the preponderance of clear and partially clear days before and after the event). The observational results showed the growth of the β turbidity coefficient of the Ångstrom expression for the 0.5 µm wavelength and the decrease of parameter α which characterizes the mean particle size, immediately following the eruption.

Aircraft measurements of the volcanic ash particle size distribution were conducted on May 18 and May 23 by Hobbs et al. (1983) in the eastern part of the State of Washington. Ash particles soon deposited on the surface and then (May 23) resuspended from the surface. For the resuspension of most of the small ash particles (this resulted in a rapid drop of visibility) it was enough to have wind speed of the order of a few meters per second. Measurements in the boundary layer of the atmosphere where most of the ash was located gave an average scattering coefficient value of about $4 \cdot 10^{-4}$ which is equivalent to the visibility of approximately 9 km (typical visibility in the region is close to 30 km). Due to measurements in two points median values of particle diameter were equal to 211 and 142 µm.

Ackerman and Lippens (1984) have discussed brightness changes of the atmosphere near the horizon on the basis of photometric processing of horizon zone photographs taken from balloons (35-km altitude) launched in France in May and June, 1980. The strong enhancement of aerosol forward scattering was registered on June 5 at altitudes lower than 16 km, as well as layered structure of volcanic aerosol distribution characterized by the presence of both vertical and horizontal inhomogeneity. Altitude dependence of aerosol optical properties was studied (including asymmetry ratio of forward and backscattering) as well as variations of color ratios.

Tephra and aerosols from the May 18 eruption were sampled in the lower stratosphere with a WB-57F aircraft by Sedlacek et al. (1982). The main body of the plume was intercepted over western Kansas on May 20, 48 hours after the eruption, at an altitude of 15.2 km and concentrations from filter samples were 26 ng of SO_4^{2-} per gram of air and 5799 ng of ash per gram of air. Angular glass pyroclasts ranged in size from 0.5 to 10 µm, with a mean grain size of 2 µm. Samples collected at altitudes of 16.7 and 12.5 km had only traces of SO_4^{2-} and ash.

A second flight was flown, 72 hours after the eruption, on May 21. From north Texas to central Wyoming, at an altitude 15.2 km, < 0.5 to 38 ng ash per gram of air and 1.0 to 2.2 ng of SO_4^{2-} per gram of air were sampled. At an altitude of 18.3 km, from central Wyoming to northwest New Mexico, the plume density and character were variable as described here:

Flight path (at 18.3 km)	Concentration (ng per gram of air)	
	SO_4^{2-}	Ash
NW Colorado	32.8	87.3
SW Colorado	0.6	11.6
NW New Mexico	36.1	29.6

Glassy pyroclasts similar to those sampled on the first flight ranged in diameter from 0.5 to 4 µm. Trace element analysis revealed some volatile element enrichment, but far less than previously observed in the plume from the St. Augustine volcano in 1976.

Follow-up observations of the NH SO_4^{2-} aerosol burden in late July and early August 1980 revealed an increase by a factor of 2.5 above the April 1980 measurements. Concentrations observed during April 1980 were increased by a factor of two above April 1975 (the lowest concentration seen in the previous three years). Quartz crystal micro-balance cascade impactor measurements and filter sampling on the July–August 1980 flights revealed almost no residual ash in the stratosphere. The SO_4^{2-} aerosol sizes centered between 0.6- and 0.3-µm aerodynamic diameter.

The results obtained by Sedlacek et al. (1982) show that the sulfur gases present in volcanic eruption clouds convert to sulfate aerosols within a few days instead of several months as would be predicted by an ozone oxidation mechanism. The principal contribution to the aerosol layer of the lower stratosphere is injection of volcanic material.

During the time period from June 4 to August 8, 1980, Lerzberg et al. (1982) conducted in the Great Lakes region 20 filter sampling flights of the

NASA Lewis F-106 aircraft at a level close to the local tropopause. These samples were analyzed for ^{7}Be as a stratospheric tracer. Quarter sections were analyzed for chloride by selective ion electrode. A filter sample taken above the tropopause on June 5 indicated a sulfate level of 50 times the baseline measurements, which is consistent with the trajectory predictions of the leading edge of the cloud on its second transit of the Earth.

Subsequent measurements over a period of 2 months showed an initial drop-off and formation of a persistent layer of sulfate above the tropopause with the concentration of 10 to 18 times previously measured background levels. Concentrations of nitrate above the tropopause exhibited considerable variability and showed some enhancement (2 to 10 times) compared with previously measured concentration levels. The source of the nitrate may also be volcanic ash evidenced by its temporal relationship to the sulfate concentration changes. Based on the null results of X-ray fluorescence measurements, there is no evidence of ash particle concentrations greater than 3.4 µg/m^3 persisting in the layer above the tropopause after the second transit of the cloud.

On the basis of routine measurements of ozone and SO_2 with Dobson and Brewer spectrophotometers of the Atmospheric Environment Service in Downsview, Ontario, Canada, Kerr et al. (1982) discussed the results for May 20 and 21, 1980, to show that large values of columnar SO_2 were observed with both instruments at the time of passage of the volcanic cloud. Enhanced SO_2 values were first observed at 18:00 GMT on May 20. The maximum column amount of SO_2 measured was 0.06 cm at 20:00 GMT. Typical SO_2 amounts due to pollution at the Downsview site are approximately 0.003 to 0.005 cm. At the time of maximum SO_2 enhancement, both Dobson and Brewer spectrophotometers measured a 0.040-cm decrease of total ozone. It is not clear whether the decrease of total ozone was caused by the volcanic cloud or natural ozone variability. Air mass trajectories indicate that the altitude of the debris cloud that passed over Downsview at the time, was between 10 and 12 km. Since the area of the volcanic cloud (due to satellite data) was about $5 \cdot 10^5$ km^2 and the average column amount for the area of SO_2 within the cloud was 0.40 cm, an assessment of the total volume and mass of the cloud gives $2 \cdot 10^8$ m^3 and $0.6 \cdot 10^6$ tonnes, respectively.

Soon after the Mount St. Helens eruptions (May 25 and June 13, 1980) during the time period from May 15 to June 17, Inn et al. (1982) took 19 aerosol samples at altitudes 13.1–20.7 km using a cryogenic device on board a U-2 stratospheric aircraft (five flights were made). Gas chromatographic analysis was made to determine concentrations of such sulfuric aerosol precursors as OCS, SO_2, and CS_2 as well as CH_3Cl, N_2O, CF_2Cl_2, $CFCl_3$. The results revealed a very strong increase of SO_2 concentration within the volcanic cloud (compared to pre-eruption conditions) as well as moderate growth of CS_2 and COS concentrations. At 14-km altitude SO_2 concentrations exceeded background values approximately 2000 times. Under nonvolcanic conditions the principal gas precursor of sulfuric acid aerosol was OCS; however, after the eruption the dominating contribution was made by SO_2 because during the eruption oxidation of gases under high temperature took place due to the presence of O, OH, O_2 and NO_2 in the volcanic cloud. Such conditions also stimulate rapid OCS and CS_2 oxidation up to SO_2.

Sulfur dioxide concentration was decaying rapidly with time. The asessment of the time constant of the decrease gave the value of approximately 1.3 days which is much shorter than the time constant of SO_2 transformation into H_2SO_4 vapor (about 1 year in the stratosphere). This indicates a significant role of mixing and diffusion. Measurements of Cl^-, SO_2^{2-} and NO_3 concentrations in water solutions of aerosol samples showed high mixing ratios of HCl as well as ($NO+NO_2+HNO_3$) and SO_2.

Farlow et al. (1982) made an analysis of stratospheric aerosol samples (ash particles and sulfuric acid droplets) which had been taken on board the U-2 aircraft using a wire impactor technique at various altitudes in the stratosphere during 6 months (spring and summer 1980) after the Mount St. Helens eruption. Samples had been taken in various geographical regions from Fairbanks (Alaska) down to Panama, but most of samples relate to California. The purpose of the analysis was to monitor temporal change of aerosol concentration, chemical composition, and size distribution. Table 3.1.2 data illustrate aerosol mineralogical composition which does not practically depend on particle size. All particles collected in one day after the eruption were absolutely dry, while ash particles of later samples were covered by the layer of sulfuric acid.

Farlow et al. (1982) have pointed out that ash grain sizes and composition depend on collection altitude, location within the drifting cloud, and days following their injection. Different volume percentage of major ash components such as glass, plagioclase, hornblende, and pyroxene emphasize the inhomogeneity of the clouds. Size distribution of ash particles vary with altitude. Generally small particles are depleted more rapidly at low altitudes (12 km) than at higher altitudes (17–18 km). Although samples collected 1 day after the first eruption of May 18 were dry, flow marks on the aircraft indicated that parts of the cloud contained heavy acid concentrations. Indeed, all other samples obtained within 1 to 4 days after later eruptions (May 25 and June 12) were covered with copious amounts of liquid ash. The proportion of liquid to ash varied considerably depending on sampling location and cloud age. Because the acid-coated ash globules were large, they fell rapidly from the stratosphere until, by late June 1980, only a residue of acid droplets remained. Size distributions and concentrations of these droplets have varied considerably at the later post-eruption stage.

Table 3.1.2. Characteristics of Mount St. Helens ash in the stratosphere. After Farlow et al. (1982).

Date Collected	Eruption Sampled	Days After Eruption	Altitude km	Glass	Mineralogy (volume percent)			
					Plagioclase	Hornblende	Pyroxene	Other
5/19/80	1	1	14	65	32	2.3	0.1	Sulfides biotite, chromite
5/19/80[a]	1	1	17	37	16	21	13	SiO_2[b]
5/22/80	1	4	18	60	13	5.0	22	c
5/27/80	2	2	17	80	20	c	c	c
5/29/80	2	4	12	64	18	4.6	c	SiO_2[b] Sulfides ilmenite
6/3/80	1	16	12	100	d	d	c	c
6/13/80	3	1	12	83	10	2	5	c

a Not included in size distribution data.
b Cristobalite?
c Not found.
d Questionably present.

Although stratospheric aerosol loading before the St. Helens eruption was fairly stable and characterized background conditions, Mathews et al. (1982) discovered an intrusion of volcanic dust over the northern Mojave Desert in April 1980, a month before the Mount St. Helens eruption. This intrusion was documented by particle size distributions, scanning electron microscope analysis of nucleopore filter samples, insolation measurements, observations by Navy and NASA aircraft, and meteorological data. The intrusions mentioned were due to a number of small but frequent eruptions which resulted in significant increases of dust loadings and drops of visibility. For instance, during the time period of April 14–18 there was a decrease of visibility up to 24–48 km in comparison with normal (under cloudless sky) visibility about 150 km. Visual observation from an aircraft on April 14 (close to noon time) discovered the presence of the haze layer at 4.6 to 7.6 km. Pyrheliometer surface observations of April 15 showed a decrease of solar radiation of about 2.5%.

Cadle (1980) has pointed out that observations made after the Mount St. Helens eruption opened the first opportunity to test the hypothesis of a gas-to-particle conversion mechanism of eruption clouds formation based on the analysis of SO_2 aircraft measurements in the troposphere and lower stratosphere over Europe before and after the eruption. The aircraft soundings data for May 24 show a typical SO_2 background vertical profile, but in few days the situation changed drastically: strong (~12 times) increase of SO_2 mixing ratio was

VOLCANIC ACTIVITY AND CLIMATE 181

registered near the tropopause (11 km) up to 0.460 ppbV (10.1 km) and 0.515 ppbV (11.3 km), whereas at higher altitudes (13.1 and 13.7 km) the SO_2 mixing ratio did not exceed 0.05 ppbV.

Lidar soundings in Garmisch-Partenkirchen confirmed the presence of the aerosol layer coincident with maximum SO_2 concentration. Particle number density (as well as the SO_2 mixing ratio) in the eruption cloud was an order of magnitude less than in the initial cloud in the vicinity of the volcano. This means that there was a joint motion of both particles and gases (the speed of the transport was maximum at the tropopause level).

The chemical analysis of gaseous volcanic emissions has discovered the presence of nitrogen oxides (which was unusual for volcanic ejecta). This finding may be significant for stratospheric photochemistry, especially from the viewpoint of an impact on the ozone layer.

Davis et al. (1981) have conducted mineralogical and chemical analyses of volcanic ash samples deposited in various parts of the United States after the Mount St. Helens eruption. The samples had been taken in the vicinity of the volcano (southwestern part of the State of Washington) and far from the volcano (Montana, South Dakota). Almost all samples contain ash from the May 18 eruption, but in three cases they were obtained after the June 12 weak eruption. The results show that principal ash components are amorphous glass and plagioclase as well as such minor components as quartz and some polymorphic SiO_2 compounds; other free SiO_2 was present in all the samples but its quantity never exceeded 8% of the total aerosol mass.

Observed in May maximum precipitation in the northern part of the Rockies coinciding with the trajectory of the volcanic cloud suggest, assuming the possibility of an impact of volcanic ash on enhancement of precipitation. This is, however, just a possibility which needs to be further studied.

An outstanding contribution to explorations of stratospheric volcanic aerosol and its climate implications has been made by a group of scientists from the University of Wyoming with the use of dustsondes (Hofmann and Rosen, 1981a; 1981b; 1982; 1983a; 1983b; 1983c; 1983d; 1984; 1987, Hofmann et al., 1983, Rosen and Hofmann, 1982, Rosen et al., 1992). Relevant references contain detailed descriptions of the instrumentation which was steadily improved; here we confine ourselves to a brief discussion of the principal results, obtained after the El Chichón eruption.

During the first 5 months after the eruption Rosen and Hofmann (1982) conducted numerous balloon (dustsonde) soundings of aerosol (with diameters ≥ 0.3 μm, and ≥ 0.5 μm) and CN concentrations over Laramie, Wyoming. On several occasions the volatility of particles was tested. Data processing shows that the particles in the initial cloud were relatively large and nonvolatile, but, in a relatively short time the aerosol began showing a dominant volatile component. Already the first sounding of May 19 showed the presence of a very pronounced layer of relatively large particles at about 13 km, which at that time was slightly above the jet stream core. Other than this feature, the aerosol profile was essentially identical to pre-eruption measurements. Later on the same day, a volatility test of the aerosol was made, which showed that particles were nonvolatile, whereas in the undisturbed regions, the usual volatile aerosol was observed. By May 26 no disturbed stratospheric aerosol conditions were observed: The approximate completion of the drift aerosol cloud passage took place. The observations strongly that the cloud was again seen over Laramie on June 20 after completing its second traversal.

The analysis of regular (once a week) soundings discovered the presence of four typical aerosol layers: close to the jet stream core (12.5–15 km); the main layer (> 15 km); the reversal layer which occurred in a region where the summer winds changed from westerly to easterly (about 15–20 km); the retrograde layer (this part of the cloud was first observed over Wyoming at about 24 km on July 10 moving towards the west and was reportedly detected by lidar over Europe about 10 days earlier). The layered structure disappeared about 4 to 5 months after the eruption and the broad main layer centered about 17.5 km emerged.

Rosen and Hofmann (1982) have pointed out that, although there were probably no CN in the original cloud due to expected very short coagulation lifetime, high concentrations of unusually small CN particles were observed about a month after the eruption. By the end of September the CN profiles and associated particle sizes were practically back to normal. A comparison of pre-eruption and post-eruption profiles indicates a total stratospheric aerosol loading increase of about a factor of 3. This material was essentially all volatile.

Later on Hofmann and Rosen (1984) have discussed the results of 36 dustsonde flights in Laramie for the time period from May 15, 1980 to May 13, 1981. The results show that after sedimentation of large ash particles an aerosol layer formed at the 18-km level consisting mainly of volatile particles, shown in Figure 3.1.1. An increase of aerosol concentration was registered up to an altitude of 26 km. The size distribution of equilibrium aerosol formed in 8 months after the eruption was log-normal in the radius range, $0.01 \leq r \leq 0.5$ µ, with maxima of number density at 0.7 µm and 1 µm. The mode of large particles $r \approx 1$ µm was also discovered.

The results of CN measurements show that CN coagulate and disappear 2 months after the eruption, shown in Figure 3.1.2. The sizes of CN of volcanic origin are usually smaller than the background stratospheric CN and they do not have an insoluble core, which is a characteristic feature of background CN. The transformation of the sulfuric acid size distribution during the first 3 months after the eruption is, apparently, due to coagulation of CN.

VOLCANIC ACTIVITY AND CLIMATE 183

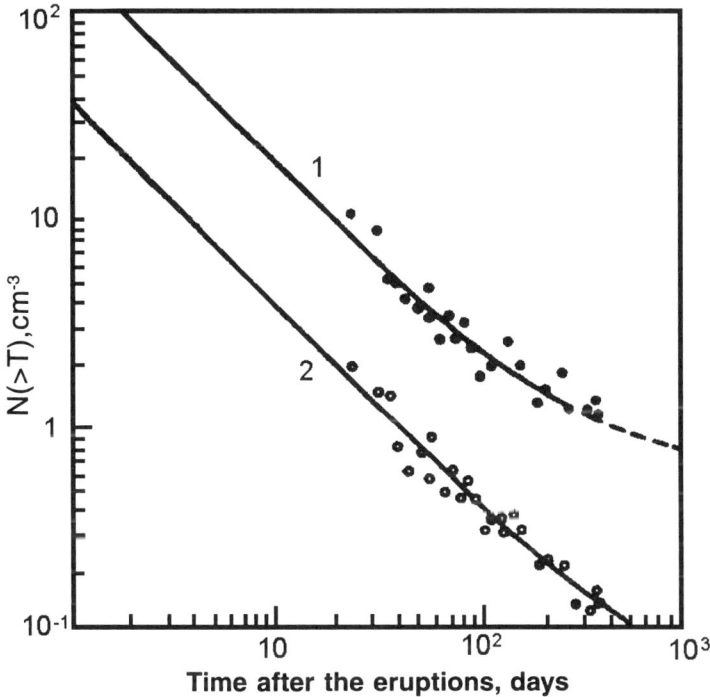

Figure 3.1.1. Dependence of total number particle concentration $N\ (>r)$ at 19 km altitude on time for $r \geq 0.15\ \mu m$ (1) and $r \geq 0.25\ \mu m$ (2) due to observations in Laramie. After Hofmann and Rosen (1983b).

The interpretation of data on the evolution of eruption aerosol during the first 1.5 months after the eruption demonstrates the presence of a continuously functioning mechanism of CN formation. The maximum growth of aerosol concentration was observed at 22 km altitude. As a rule, sizes of particles formed in the stratosphere after the Mount St. Helens eruption were less than after the Fuego eruption in 1977 and less than in the case of background aerosol observed in the beginning of 1979. Columnar mass of equilibrium eruption aerosol in 6 months after the eruption was about 10^{-7} g/cm^2. Assuming homogeneous aerosol distribution in the NH stratosphere, an assessment of the total aerosol mass gives approximately 0.25 Mt which is three times more than the background value, but twice less than the similar figure for the Fuego eruption.

Gadian and Davies (1981) have undertaken a preliminary numerical modeling of the effect of volcanic aerosols formed after the Mount St. Helens eruption on climate of the NH. They assumed (considering that the scale of

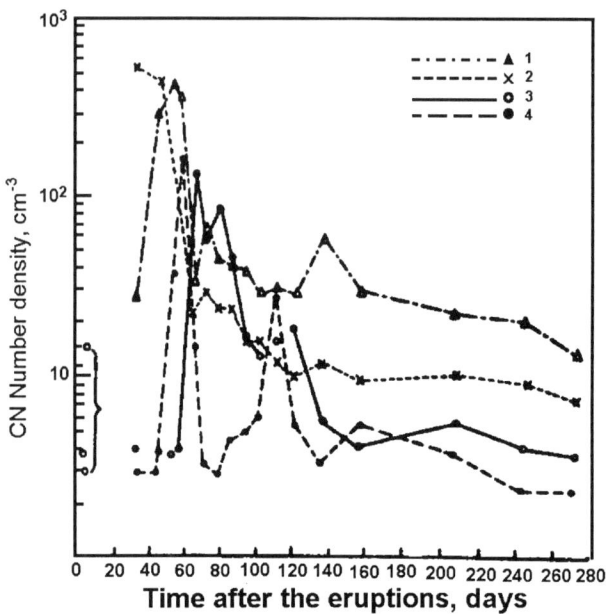

Figure 3.1.2. Dependence of CN concentration on time at altitudes 17 km (1), 15 km (2), 22 km (3), and 24 km (4) from observations in Laramie. The range of background values is given as the vertical axis. After Hofmann and Rosen (1983b).

ejecta was three times less than during the Krakatau eruption) that during the first 90 days there was a decrease of solar global radiation at the surface for latitudes $> 60°$ N of the order of 10%, and during the subsequent 5 months 5% for the whole NH (in comparison with climatic average values). Under such conditions the southern edge of the polar ice cover should move to the south by 2–3° latitude (in comparison with the control case) and then steadily return to the initial location about 1 year after normalization of the global radiation level.

The prediction of volcanically induced air surface temperature decrease should be 1.5–2.0 °C starting from the moment of autumnal equinox. Under conditions of the global radiation decrease during the first 90 days an enhancement of meridional temperature gradient in polar regions takes place. During this period the general atmospheric circulation which corresponds to the control case, is characterized by the presence of a summer type weak anticyclone in midlatitudes, which resulted, apparently, from the change of meridional temperature gradient. Probably these pressure minima led to the increase of heat transport towards the pole by about 50%.

Approaching the autumnal equinox a dominating feature of the general circulation appears as well developed anticyclones accompanied by strong jet streams which are more intense and form earlier than in the control case. During the winter months the intensity of jet streams is close to that observed in the control case, but the jet streams are shifted to the south by 3°–4° latitude (the same is true for cyclone trajectories), which agrees with the shift of the ice edge to the south. It should be pointed out the the assessments made by Gadian and Davies (1981) are based on overestimated values of the decrease of global radiation after the eruption. As a number of other similar studies show, climatic consequences of the Mount St. Helens eruption were insignificant because of the limited scale of sulfur dioxide emissions.

On the basis of a comprehensive one-dimensional time-dependent physico-chemical model of stratospheric sulfate aerosol layer formation Turco et al. (1979; 1982a; 1982b; 1983) conducted numerical modeling to simulate the evolution of physical and chemical properties of aerosol, which were formed after the Mount St. Helens eruptions on May 18, and 25 and June 13, 1982. The model takes into account the broad range of particle sizes (from 10 Å to 32 μm), multicomponent aerosol composition (volcanic ash and multimodal sulfur acid aerosol), the volcanic cloud dynamics, homogeneous and heterogeneous chemical reactions, as well as the chemical interaction between OH, SO_2 and H_2O. Table 3.1.3 data characterize typical gaseous ejections of the Mount St. Helens volcano. Turco et al. (1982a) have considered computational results.

The time histories of the gaseous species concentrations, sulfate aerosol size dispersions, and ash burdens in the eruption clouds are based on the long-term buildup of stratospheric aerosols in the NH and the persistent effects of injected chlorine and water vapor on ozone. It is concluded that SO_2, water vapor, and ash are the most important substances injected by the volcano into the stratosphere, with respect to both the widespread effects on composition and the impact on climate. Despite the seemingly large magnitude of the Mount St. Helens eruption (a three-fold aerosol enhancement over the NH in the fall and winter of 1980) the volcano probably had little influence on the climate (< 0.05 °C global surface temperature cooling) or on stratospheric ozone (<0.2% maximum hemispherical reduction). The chlorine injected above the tropopause by the eruption was only 1% to 10% of the sulfur injection.

The SO_2 and SO_4^{2-} data collected in the clouds provide circumstantial evidence for an initial SO_2 oxidation rate (including ash particles) of 15%/ha. Rapid removal of the ash may have terminated the reaction. Some measurements suggest that large quantities of SO_4^{2-} (relative to SO_2) were injected directly into the stratosphere, but such an assumption cannot be considered reliable.

For a number of reasons H_2O injections into the stratosphere are important because they

(i) Create a medium in which aqueous sulfur reactions could occur;
(ii) Generate excess OH radicals which affect the SO_2 oxidation rate via homogeneous chemistry;

(iii) Favor homogeneous nucleation;
(iv) Participate in the chemical cycles that control stratospheric ozone.

Table 3.1.3. Composition of the emissions of Mount St. Helens. After Turco et al. (1982a).

Emitted Species	Principal Roles	Ambient Air (1)	Fumarolic (2)	Stratospheric Plume (3)
			Relative Abundances	
H_2O	Cloud formation, initiate HO_x chemistry, washout of $SO_4^=$, Cl^- and NO_3^-	$3-5 \times 10^{-6}$	$\sim 10^{-2}$	$\sim 10^{-4}$
SO_2	⎱ Precursors of sulfate	$\sim 5 \times 10^{-11}$	⎱	$\sim 10^{-7}$
H_2S	⎰ aerosols consume	$\ll 10^{-9}$	⎰ $\sim 10^{-5}$	$\ll 10^{-9}$
OCS	OH radicals	$1-5 \times 10^{-10}$		$\sim 10^{-9}$
CS_2		$\leq 10^{-12}$		$\leq 10^{-10}$
HCl	⎱	$\sim 10^{-10}$	⎱	$\leq 10^{-9}$
CH_3Cl	⎰ Ozone catalysis	$\sim 10^{-10}$	⎰ $\sim 10^{-5}$	$\sim 10^{-9}$
CH_3Br		$\sim 10^{-11}$		-----
$S/Cl^{(4)}$		$\sim 0.1-1.0$	~ 1.0	$1.0-1000.0$
CO_2	Infrared transmission	3.6×10^{-4}	$\sim 3.6 \times 10^{-4}$	3.6×10^{-4}
CO	Tracer of OH	$\sim 10^{-8}$	?	$\sim 10^{-7}$
NH_3	Aerosol reactions	$\leq 10^{-10}$	?	-----
N_2O	Precursor of NO_x	$\sim 10^{-7}$		$\sim 10^{-7}$
NO_X	Active in ozone cycles	$\leq 10^{-8}$		$\sim 10^{-8}$

1. Air in the lower stratosphere.
2. Clouds near the volcano. Data sources are given in the text.
3. See the text for a discussion of the stratospheric data.
4. Molar ratio, for gases plus particles.

As Turco et al. (1982a) have pointed out the volcanic ash lofted into the stratosphere by the Mount St. Helens eruption consisted of numerous small particles (<3 µm), containing <10% of the mass, and a relatively few large particles (>3 µm), holding the rest of the mass. The large ash promptly fell out of the stratosphere (~ month), leaving acid aerosols, some with ash cores, as the dominant long-term volcanic particles. Even so, most of the fine volcanic ash particles were removed from the stratosphere by sedimentation and diffusion after ~ 3 months.

The size distribution of volcanic aerosols is characterized by the presence of at least four modes: < 0.01 µm (particles formed by homogeneous nucleation); 0.01–0.1 µm (ambient aerosols continuously mixed into the cloud); the volcanic ash: < 3 µm and > 3 µm.

Turco et al. (1982a) have emphasized the necessity to further observe the particles and gases (including such an oxidant as ozone) in specific eruption clouds. The measurements should begin within 1 day of the eruption and continue periodically for several weeks to 1 month. We shall see later on to what extent such suggestions have been actually realized.

3.2 The El Chichón Volcano Eruption

On March 28, 1982, at 23:30 local time an eruption of the El Chichón volcano in Chiapas (a southeastern state in México) started. El Chichón underwent three major eruptions in late March and early April 1982. This stratovolcano of the late Pliocene or early Pleistocene, located at the east end of the Mexican neovolcanic zone of the current Chiapas volcanic arc, has exhibited only solfataric activity over a long period. Since Chiapas is located at a junction of the American, Cocos, and Caribbean tectonic plates, its volcanic activity is supposedly caused by continuous shifting of the Cocos plate beneath the southeastern part of México.

As has been pointed out by Hofmann (1987,1988) and Hofmann and Rosen (1987) more was learned about the effects of volcanic aerosols on the stratosphere following the El Chichón eruption than from all previous eruptions combined, which was due mainly to the wealth of research techniques that could be brought to bear with little delay. For the first time, trace chemical species such as HCl and OH were observed to increase, and other species, possibly related to these, were observed to decrease (O_3, NO, and NO_2) following the eruption. Volcanic SO_2 was observed from space for the first time, and its conversion into H_2SO_4 vapor, condensation nuclei and finally sulfuric acid aerosol were monitored closely by both *in situ* and remote sensing techniques. At least three factors contributed to the severity of the eruption aftermath. These include the apparently large fraction of sulfur involved for an eruption of only moderate proportions, the latitude of the volcano, and the time of year of the eruption. The final eruption, on April 4, penetrated to altitudes in excess of 25 km and probably injected more sulfurous gases than any eruption in the previous 100 years. The assessment of the total amount of erupted SO_2 Total Ozone Mapping Spectrometer (TOMS) data was 3.3 Mt, and the lifetime, for loss to chemical conversion into H_2SO_4, was of the order of a month rather than a year as thought previously. This conclusion has been confirmed by the analyses of balloon data. The relevant sequence of reactions is

$$SO_2 + OH + M \rightarrow HSO_3 + M,$$
$$HSO_3 + O_2 \rightarrow SO_3 + H_2O,$$
$$SO_3 + H_2O \rightarrow H_2SO_4$$

Ozonesonde data for Hohenpeissenberg (48° N), which have been discussed by Bojkov (1987), indicate that the main O_3 reduction from late 1982 to early 1983 occurred in the 20-km region, although there was evidence for a reduction at all levels. The large O_3 reduction in 1983 is probably a result of both deficient ozone transport (QBO effect) and possibly a volcanic effect.

As Hoffer et al. (1982) noted, during the first 10 days of volcanic activity, when three large-scale eruptions occurred, about 0.3 km^3 of andesite pyroclastic matter was ejected (without lava streams). The initial product erupted was tephra enriched with a crystalline component, with a larger amount of silicon and alkalines compared to the pyroclastic matter of the second and third eruptions. The inital tephra consisted mainly of juvenile substances, and then the eruptive product contained both the juvenile and lithoid fractions.

First eruptions were characterized by large amounts of ejected ash, moderate amounts of pumice, and smaller amounts of lithoid fragments. During the first phase of the eruption that continued until April 2, large amounts of light gray ash were ejected which covered the adjacent area northeast of the volcano. The thickness of the ash layer reached 0.5 m at a distance of 15 km and decreased to 0.2 m at a distance of 75 km from the volcano. The ash layer in Villahermosa, the capital city of the adjacent State of Tabasco, was 0.1 m thick.

The second phase consisted of two large-scale eruptions (April 3, 19:33; April 4, 05:36), when brown-gray ash with a large percentage of lithic tephra was ejected and propagated mainly to the east of the volcano. On April 4 at 10:30 the rate of settling for ash near Teapa reached 0.33 g/m^{2}s, which inmersed this region in nearly total darkness, reducing visibility to 5 m. By 12:30 the rate of settling decreased by 0.5 g/m^{2}s. On this day, pyroclastic streams, consisting of hot ash and large blocks of pumice moved down the slopes of El Chichón. The thickness of the tephra layer measured on April 5 near Palenque (125 km east of the volcano) exceeded 0.4 m.

The chemical analyses of 30 samples of tephra collected on April 3 to 7 at different locations revealed two types of tephra:

(i) Light gray matter with a high percentage of silicon (59%, on average) ejected on March 28 – April 2, covered with tephra containing large amounts of lithic components (the April 3 and 4 eruptions);
(ii) Basalt tephra with a high percentage of iron, oxides of magnesium, and calcium.

As Robock (1983) noted, the stratospheric post-El Chichón eruptive cloud on April 4, 1982 was apparently the most powerful cloud in the last century (by that time). This cloud caused a local temperature decrease of more than 5 °C, and a further cooling down to 0.5 °C was expected near the surface in 1984–1985. The interpretation of the unique database of complex post-eruption ground-based balloon, aircraft, and satellite observations opened up the possibilities of verifying the models of gas-to-particle conversion as a source of the stratospheric H_2SO_4 aerosol, as well as calculating the transport and the gravitational settling of particles and the impact of aerosols on the radiative regime and climate.

Robock (1983) gave a brief review of the reports at two conferences held in late 1982, at which the following aspects of the problem were discussed: the global-scale diffusion of the eruptive cloud, variations in its gas and aerosol composition, and the possible impact of the eruption on weather and climate. The large size of the eruptive cloud permitted its continuous monitoring using satellite-derived imagery (following the Mount St. Helens eruption in 1980, the eruptive cloud could be identified only during the first 3 to 4 days).

The eruptive cloud from El Chichón circled the globe for 21 days at an average speed of 22 m/s. Lidar soundings in Hawaii, over which the thickest part of the cloud passed on April 9, gave values of the backscattering coefficient (determined with respect to the Rayleigh scattering) exceeding 200, which had never been observed before, with a maximum at the 26-km level. The eruptive cloud was stratified. Two months after the eruption the upper layer (with a maximum concentration at a height of 20 km) propagated apparently over the entire globe.

Air and aerosol samplings from aircraft at altitudes up to 20 km permitted the chemical analysis of the cloud to be made. Aerosol sounding data contained information about the concentration, size distribution, and (indirectly) the composition of the aerosol, which turned out to contain 95% of H_2SO_4. From data of observations on the Hawaiian Islands an additional eruption-caused attenuation of the direct solar radiation reached 20%, and of the global solar radiation 7%. An estimation of the cloud mass from its optical thickness was 15 million tons.

Rao and Takashima (1986) have discussed results of actinometric observations at several locations in the continental United States and Hawaii after the El Chichón eruption (1982–83). These results show that directly transmitted solar radiation was depleted by up to 25%; diffuse sky radiation was enhanced by a factor of up to 3; and solar global radiation was lowered by up to 5%. Radiative transfer computations for a realistic model of the atmosphere have simulated measured values reliably enough. Galindo (1992) has confirmed the above results for México using satellite-derived solar irradiance data showing a monthly maximum irradiance extinction of 28% during the winter of 1982–1983. Figures 3.2.1 and 3.2.2 show mean solar radiance extinction over México after the El Chichón eruption from May to December 1982. Note the effect of the easterly phase of the QBO that shifted the stratospheric volcanic aerosol cloud westward during summer 1982.

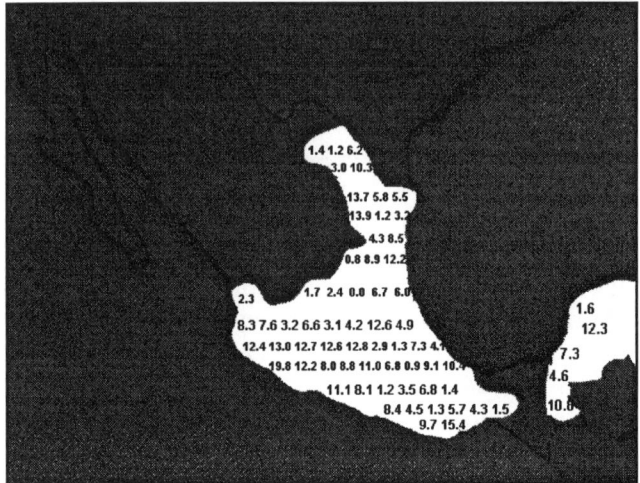

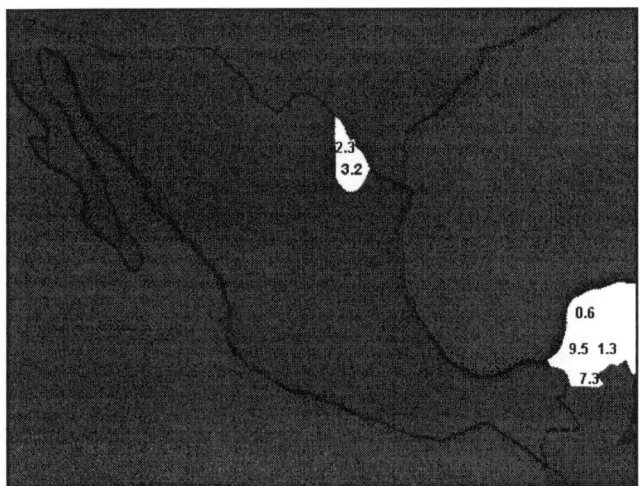

Figure 3.2.1. Mean monthly solar irradiance extinction over México under cloudless conditions due to El Chichón stratospheric aerosol layer. (1) May; (2) June, 1982.

VOLCANIC ACTIVITY AND CLIMATE

3

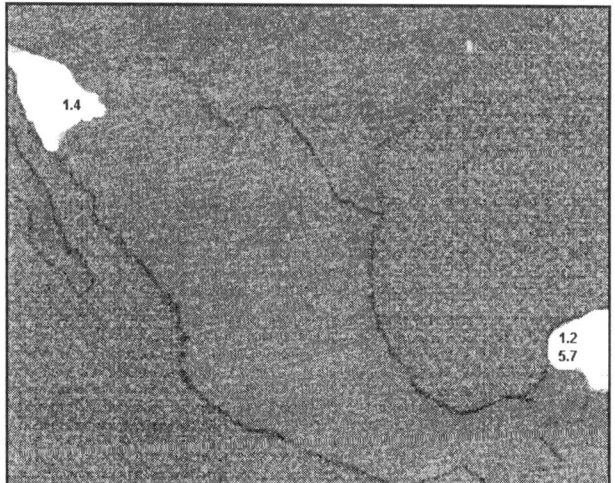

4

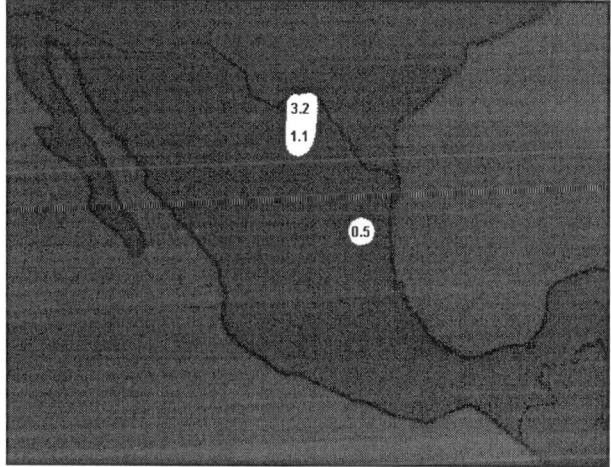

Figure 3.2.1. (Continued) Mean monthly solar irradiance extinction over México under cloudless conditions due to El Chichón stratospheric aerosol layer. (3) July; and (4) August 1982.

RECENT MAJOR ERUPTIONS

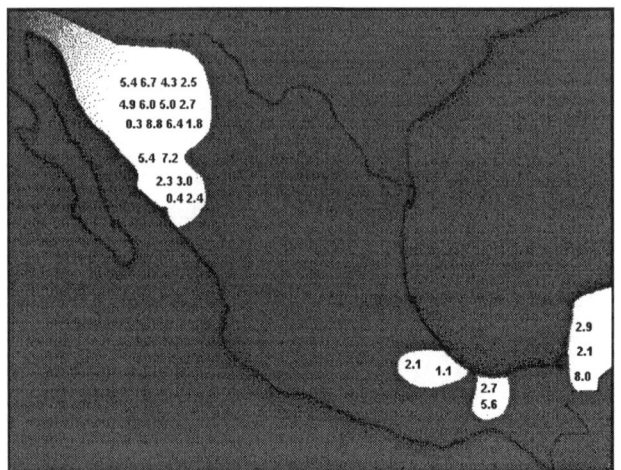

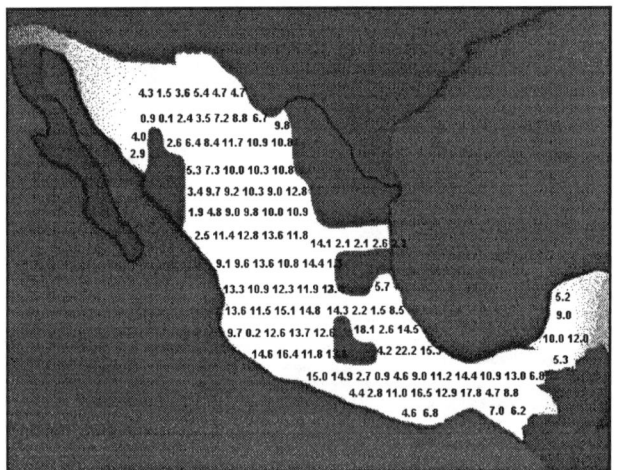

Figure 3.2.2. Mean monthly solar irradiance extinction over México under cloudless conditions due to El Chichón stratospheric aerosol layer. (1) September; (2) October 1982.

VOLCANIC ACTIVITY AND CLIMATE

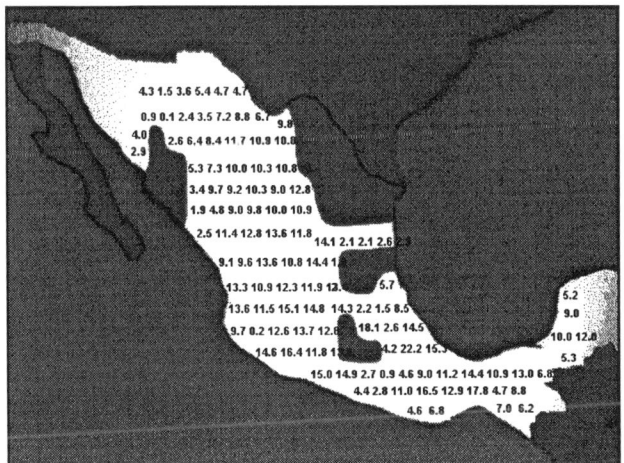

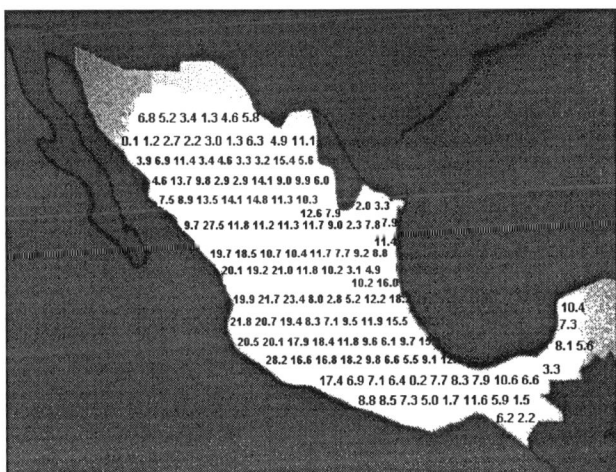

Figure 3.2.2. (Continued) Mean monthly solar irdiance extinction over México under cloudless conditions due to El Chichon stratospheric aerosol layer. (3) November, and (4) December 1982.

Wendler and Kodama (1986) have conducted detailed radiation measurements of 13 independent radiative fluxes at Fairbanks (central Alaska: 64°49' N, 147°52' W) during the time period from 1 July 1979 to 30 June 1984. These data showed that the El Chichón volcanic cloud had a major impact on the surface radiative regime, the maximum of which was observed in the winter of 1982/83, about 9 months after the eruption. The direct solar radiation was reduced by as much as 38% (3-month mean value), the ratio of diffuse over global radiation was increased by 91%, and the global radiation was reduced by about 5% (in general, the results for the global radiation are inconclusive). These values show that the volcanic cloud was a strong forward scatter, while relatively little energy was absorbed or reflected back to space. Data for other U.S. and Canadian stations show maximum drop of the radiation within 11%–33%. Substantial increases were discovered for the aerosol optical thickness and Linke's turbidity factor. Effects of the stratospheric aerosol cloud were seen all through 1983 and the spring of 1984. In the summer of 1984, however, the radiative values were back to normal. It is important to emphasize that all these results have been based on the analysis of observational data for clear sky conditions. Consideration of all the data (mean annual cloud amount in Fairbanks is 70%) show that with the exception of the diffuse radiation and, to a lesser extent, the ratio of diffuse to global radiation, no clear trends due to the volcanic aerosol cloud can be seen. In the agreement with some other observations the mean global radiation does not show a clear trend. This result may be of significance in the context of discussion between Dutton (1990) and Mass and Portman (1989), concerning latitudinal dependence of the volcanically induced depletion of global radiation.

The results of one-dimensional radiation transfer model calculations for the wavelength region 175–800 nm made by Michelangeli et al. (1989) show that the spectral global radiation increased by 8% within the aerosol layer (16–30 km); longward of 300 nm. At certain altitudes and wavelengths below 300 nm the global radiation decreased by 15%. Such changes in the level of spectral distribution of solar radiation have a direct impact on stratospheric photodissociation rate constants. O_3, NO_2, NO_3, $ClNO_3$, and $HOCl$ have absorption cross sections between 300 nm and 800 nm and are among the species whose photodissociation rate constants increase by 10%, while those of O_2, H_2O, NO, and HCl, absorbing below 300 nm, decrease by 15% at altitudes within aerosol clouds.

Volcanically induced radiation and temperature changes affect stratospheric chemical composition in various ways. For example, model calculations predict 7% O_3 decrease at 24 km (Michelangeli et al., 1989), in agreement with observations. This is due to both a 6% decrease in the photodissociation rate constant of O_3 and 7% increase in $[O]/[O_3]$ concentrations ratio. The latter results from a change in the partitioning between O and O_3 because of increased ozone photodissociation and the decreased rate of ozone formation reaction (from $O + O_2 + M$).

VOLCANIC ACTIVITY AND CLIMATE 195

Heterogeneous reactions such as those invoked to denitrify the polar stratosphere and precondition the early spring chemistry probably occurred on the volcanic aerosol particles. Of all heterogeneous reactions explored Michelangeli et al. (1989) favor the reaction $ClNO_3 \rightarrow HCl$. This reaction results in a decrease of NO and NO_2 and an increase of HCl concentration, consistent with observations. It is important to emphasize that one of the less understood aspects of the chemistry of the Earth's stratosphere is the effect of aerosols whose optical and chemical properties are not yet fully understood. Table 3.2.1 data compiled by Michelangeli et al. (1989) summarize observations and modeling results concerning volcanically induced change in the gaseous composition of the stratosphere after the El Chichón eruption.

Table 3.2.1. Comparison of observations of species after the El Chichón eruption and model results. After Michelangeli et al. (1989). [© American Geophysical Union]

Species	Observation		Model			
	H km	Change %	H km	Case A%	Case B%	Case C%
O_3	24-30	-6 -10	24	-3.0	-7	-12
HCl	25	-4.6				
	21.6-27.4	30 -40	24	8	10	102
	>12km	40	>12km	2.7	1.7	41
NO	26	~-50	26	-4	~1	44
	30	-75	30	-2	-5	-60
NO_2	26	~-50	26	-3.5	-3	-49
	25-32	-50				
$NO+NO_2$	>12km	-50	>12km	-3.03	-2.8	-28
HNO_3	>12km	~0	>12km	0.5	-2.9	-25
OH	>0 km	35	>0 km	-0.1	-0.8	6

Comments: In case A model calculations have been made taking into account only the perturbation of the radiation field; case B equivalent to case A plus temperature perturbation; case C equivalent to case B plus inclusion of heterogeneous reaction $ClNO_3 \rightarrow HCl$. All model results for 20° N, 1400 LT (solar zenith angle 45°)

The following heterogeneous reactions involving volcanic aerosols have been investigated:

$$ClNO_3 \rightarrow HCl^*,$$
$$ClNO_3 + HCl \rightarrow Cl_2 + (HNO_3),$$
$$ClNO_3 + H_2O \rightarrow HOCl + (HNO_3),$$
$$N_2O_5 + H_2O \rightarrow (2HN_3),$$
$$Cl \rightarrow HCl^*.$$

The asterisk denotes speculative reactions in order to increase HCl directly. Since it is uncertain whether or not the product HNO_3 would remain in the aerosol or escape into the gas phase HNO_3 is mentioned in parenthesis. Michelangeli et al. (1989) found that invoking a heterogeneous reaction of the form $ClNO_3 \rightarrow HCl$ on the surface of volcanic aerosols (the NO_3 is sequestered in the aerosol) provides a satisfactory explanation of the observed changes in HCl, NO, and NO_2. Combining the results due to radiative and thermal perturbations and a heterogeneous reaction, it is possible to explain all the observations summarized in Table 3.2.1 except for HNO_3 and OH (presumably, the observations are in error, but the HNO_3 observations require an alternative explanation).

The spatial propagation of the eruptive cloud was monitored in detail from satellite data (Kondratyev et al., 1983a, 1983b; Mojena and García, 1984). The first of four large-scale El Chichón eruptions happened on March 28, 1982 at 23:32 local time. An analysis of immediate GOES images in the visible (0.55 to 0.75 μm), spatial resolution (1 km) and infrared (10.5 to 12.5 μm; 8 km) spectral regions gave a minimum BT of -75.2 °C of the eruptive cloud at 20:30 on March 29. A comparison with radiosonde data showed that this temperature corresponded to the height 16.3 km (the height of the tropopause was 16.5 km), higher than the ceiling of the aircraft used for aerosol sampling; very important results were obtained, which characterized the properties of the secondary aerosol layers and the aerosol sedimentation from the major cloud. The following measurement instruments were used: the Knollenberg photoelectric counter, wire impactor, quartz cascade impactor, condensation nuclei counter, filter samples, cryogenic cells for air sampling, and cells to measure the water vapor concentration from the Lyman-alpha radiation absorption. An important contribution was made by aircraft lidar sounding in July and October 1982. At Mauna Loa observatory, observations were made with the use of the airborne photometer, the Eppley photometer and pyrheliometer, the Dobson spectrophotometer, and the ruby lidar.

An analysis of the aerosol samples taken on April 19 over the Gulf of California showed a prevailing contribution by silicate particles covered with an H_2SO_4 layer, with a maximum number density at the diameter 2 μm, and a mass concentration of 1 to 4 mg/m³. As a rule, the particles were clusters of several particles. On April 15, the highest concentration of sulfates of about 160 ppb was recorded (compared with the background concentration of 1 to 2 ppb, by volume). The SO_2 concentration in April was low ($50 \cdot 10^{-12}$), but in July at a height above 21.8 the concentration exceeded $100 \cdot 10^{-12}$.

An unusually high COS concentration ($350 \cdot 10^{-12}$) was registered on May 5, at a height of 18 km. The water vapor concentration at a heights of 19 to 20 km was 5 ppm more than the background concentration (4.4 ± 0.4 ppm). The total SO_2 content in the whole eruptive cloud from the April 4, Nimbus-7 data was estimated at $(3-4) \cdot 10^6$ tons. During the period April 9 to 20, the cloud's optical thickness at $\lambda = 0.425$ μm varied between 0.6 and 0.75 and decreased to 0.25 in July.

The Solar Mesosphere Explorer (SME) satellite carried three devices to study the mesosphere: IR radiometer, and visible and IR spectrometers (both of them covering the wavelength interval 0.3 to 2.4 μm). They measured the concentration of various minor optically active components of the stratosphere and mesosphere. In particular, two IR scanning radiometers measured the distribution of atmospheric thermal emission at the limb in the bands of water vapor (6.3 μm) and ozone (9.6 μm).

After the April 4, 1982 El Chichón eruption, the Earth's radiation field was disturbed by the eruptive aerosol cloud, which made it possible to use the SME data to monitor the process of gas-to-particle conversion and propagation of the eruptive aerosol. Most informative were data for channels 6.3 μm and 9.6 μm of the radiometer and 1.9 μm of the spectrometer. There was practically no molecular scattering and therefore the aerosol scattering was clearly manifested. The vertical resolution of the observational data constituted 3.5 km, with a vertical reference error of 2 km.

Barth et al. (1983) performed an analysis of data for three orbits for the latitudinal belt 30°–120° W (the shift of successive orbits was 24° longitude, and the latitudinal resolution 5°). This analysis showed that the combined effect of the 21-day westward round-the-globe transport and mixing had determined the formation of an Earth-encircling aerosol cloud by the first week of June in the latitudinal band from the equator to 30° N. According to the available data, about $(3-4) \cdot 10^{12}$ g of sulfur dioxide were ejected into the stratosphere (up to 30 km) on April 4, 1982. In the process of westward round-the-globe motion SO_2 reached odd hydrogen and odd oxygen residing in the stratosphere, giving H_2SO_4 vapors and then H_2SO_4 water droplets (through homogeneous nucleation of condensation on nuclei). The growing size of droplets intensified the gravitational settling until its effects exceeded the rate of growth of particles (at a height of 27 km this equilibrium state was reached 8 weeks after the eruption).

Apparently, an initial maximum of the eruptive aerosol cloud concentration at a height of 27 km was determined by SO_2 ejected to this level, but it could also have been governed by rapidly intensifying photodissociation of SO_2 above 27 km. A slower rate of the gas-to-particle conversion at lower altitudes could be connected both with slow formation of H_2SO_4 due to a lower concentration of odd hydrogen as well as with the specific initial vertical profile of SO_2 concentration. From observational data, by the end of May, SO_2 ejected to the stratosphere was mainly used in the formation of H_2SO_4 aerosol, whose concentration started decreasing and the level of its maximum lowered, reaching a height of 20 km in 24 weeks (see also Vedder et al., 1983).

Patterson et al. (1983) performed laboratory measurements of the optical properties of three samples of El Chichón ash falling out at a distance of 12 to 80 km from the volcano. The ash consisted of pumice (in the form of nonmineral and glassy particles) and rock particles. The content of glass in the pumice reached 80% (the phenocrysts that constitute glass are mainly andeside and hornblende with apatite, augite, and anhydrite). Rock particles are characterized by the following proportions: 20% glass, about 50% plagioclase, and 30% hornblende. Basic components of the insoluble part of ash are SiO_2 (59%) and Al_2O_3 (18%).

The chemical composition of ash particles by different eruptions of El Chichón turned out to be similar. The samples examined had ions Ca^{2+}, SO_4^{2-}, Na^+, K^+, Mg^{2+}, HCO_3^-, and Cl^-.

Measurements of the imaginary part of the complex refractive index n_{im} gave very small values of the order of 0.001 at a wavelength of about 0.5 μm and revealed a weak wavelength dependence. The optical parameters of sample AN 101 taken at a maximum distance from the volcano (80 km) happened to be close to the respective parameters of stratospheric aerosols. The results obtained show that the complex refractive index for stratospheric silicate ash is $1.53 + 0.001i$.

Recent observations suggest that the erupted gases and particles are a major source of an abnormal enrichment of the background aerosol in remote regions (e.g., in the Antarctic). In this regard, on November 3, 1982, Kotra et al. (1983) carried out measurements of concentrations of various gaseous components and aerosols in the eruptive plume from El Chichón using the NASA flying laboratory—the Electra-429. In the period of observations hydrogen sulfide was a major gaseous sulfur compound in the plume.

Using the neutron activation technique, 29 elements were found in the filter and impactor aerosol samples. The visually observed plume reached a height of 0.7 km, width of 3 km, and a length of 1 km, extending in the wind direction.

These observational data were used to estimate the injections into the atmosphere of minor gaseous components, normalized against total injections of sulfur compounds. The results obtained revealed an enrichment with such volatiles as sulfur, chloride, arsenic selenium, bromine, antimony, iodine, tungsten, and mercury, with respect to the basic pyroclastic, by a factor of 60,000 to 20,000. Arsenic, antimony, and selenium were concentrated mainly in small particles less than 3 μm in diameter. Calcium and natrium occurred mainly in large particles, and the distribution of aluminum and magnesium was bimodal. The composition of ash particles ejected to the atmosphere and reaching the stratosphere during a large-scale eruption (the aircraft-derived May 1982 data for a height of 18 km) was characterized by a 10-to-30-fold increase of the content of some elements as compared to ash ejected to the atmosphere during the quiet period of volcanic activity.

In the period before and after the eruptions of El Chichón (March 23–December 17, 1982) Wilson et al. (1983) measured the size distribution of stratospheric aerosols at altitudes from 19.6 to 21.6 km with the condensation nuclei counter (an estimation of the number density of particles larger than 0.01 µm in diameter), wire impactor, and Knollenberg photoelectric counter (the diameter range of 0.1 to 3 µm) carried by the U-2 aircraft. The aerosol concentration field in the regions not affected by the eruption (the flights were made along the Pacific coastline of the United States) turned out to be homogeneous to a large extent. The largest contribution to the aerosol number density before the eruption (six to seven particles per cubic centimeter) had been made by particles less than 0.1 µm in diameter.

Since the characteristic time for coagulation of such particles is about 9 months, they were probably a product of the December 23, 1981 Nyamuragira (Zaire) volcano eruption, which is considered to be a major factor in the appearance of the "mysterious" volcanic cloud. A high concentration of post-eruption submicron particles (April–May) reflected the effect of the post-eruption process of gas-to-particle conversion. Measurements made in November and December revealed a decrease in content of particles less than 0.1 µm in diameter compared with the pre-eruption conditions. Results of simultaneous measurements of the SO_2 volume concentration and aerosol near the lower boundary of the eruptive cloud (2 weeks after the eruption) made it possible to estimate the rate of transformation of SO_2 to the H_2SO_4 aerosol within the range 15 to 1200 molecules/cm^3.

During the summer, fall, and winter following the eruption of El Chichón, Gandrud et al. (1983) obtained filter samples of sulfate aerosols using the high altitude U-2 aircraft (in the region of the U.S. western coastline) and balloons (near 33° N) at altitudes of 15 to 28 km. An analysis of samples showed that the values of the post-eruption sulfates mixing ratio is about two orders of magnitude greater than the background levels. The processing of aircraft filter samples gave results consistent with those of synchronous impactor sampling, and calculations of the total content of sulfates in an air column agreed with results of aircraft lidar sounding, Nimbus-7 data, and results of balloon-borne photoelectric counter sampling.

From data of the first series of the July 23, 1982 aircraft sounding (25–36° N, near 129° W) and August 5 and 17 balloon sounding, a maximum sulfate mixing ratio was observed at a height of about 23.5 km and constituted 78 ppb (by mass). The final series of aircraft (January 20–21, 35°–50° N) and balloon (January 23) sounding gave a maximum of 500 ppb at a height of 20 km. The total content of sulfates in the layer from 15 to 29 km varied (from aircraft data) between 0.017 and 0.24 g/m^2, whereas the background value in

1976 was 0.00096 g/m², the post-eruption content of sulfates thus increasing 18 to 25 times. Taking into account available data, the content of sulfates in the spring of 1982 at 33° N latitude averaged 0.020 to 0.021 g/m².

Vedder et al. (1983) discussed results of the gas chromatographic analysis of the SO_2 content in air samples taken from U-2 and ER-2 aircraft between April 16 and December 13, 1982 at altitudes of 15 to 22 km. These flights had been planned to take into account the evolution of the eruptive cloud from El Chichón which covered the region 23°–52° N, 108°–130° W. Measured values of the SO_2 mixing ratio varied within 8 to 132 ppb (by volume), with an error of ±23%. Mixing ratios registered before the eruption constituted (20–170)•10^{-12}. The level 50•10^{-13} was considered as the background level. Data of aerosol measurements indicate that the results of the early flights refer to the periphery of the volcanic cloud, and later flights refer to the central part of the cloud, but in the period of its evolution, when most of the erupted SO_2 was transformed into the H_2SO_4 aerosol.

From December 4, 1982 until January 4, 1983, using the flying laboratory Convair–990, Dutton and DeLuisi (1983a; 1983b; 1987), performed measurements of the spectral transparency of the atmosphere above the aircraft (against the sun) with two calibrated filter photometers operating at 0.368, 0.500, 0.675, and 0.778 µm (the first photometer), and 0.380 and 0.500 µm (the second photometer). The aircraft flew at a level of 250 to 200 mb along the 120° W meridian in the 5° S – 55° N latitudinal belt.

The use of observational results to calculate the AOT, determined as a vertically integrated product of the mass coefficient of aerosol extinction by the mass aerosol concentration, revealed a relative maximum near 5° N and a relative minimum near 25° N. The AOT increased after a relative constancy in low latitudes). A volcanically induced averaged (over wavelengths and in time) AOT increase reached 0.10 ± 0.01 (the background aerosol was practically absent, constituting only 0.005).

Estimates show that such an AOT increase must cause an 18% drop of direct solar radiation extinction at the level of the tropopause (for the representative solar zenith angle 60°). Data on the AOT latitudinal and spectral variability point to the existence of two specific latitudinal zones characterized by different spectral signatures of aerosol due to different size distributions of aerosol in the southern and northern latitudinal bands caused mainly by the aerosol transformation in the process of its transport.

Increases of atmospheric optical depth (AOD) were observed in both Vancouver (Hay and Darby, 1984) and México City (Galindo et al., 1996) from May to July 1982. This AOD increase can principally be ascribed to short lifetime large ash particles (silicates) rather than to stratospheric enhanced acid aerosol. A second AOD increase was observed in Vancouver from October 1982 to September 1983, shown in Figure 3.2.3. For México City, the second AOD increase was of shorter duration, extending from March to September 1983, as shown in Figure 3.2.4. Acid stratospheric aerosols are considered responsible for AOD increases in both cities because colored twilights coincide with the El Chichón eruptions (Galindo et al., 1996). Notice that the AOD increases in México City can be seen starting January 1982. These AOD increases might be related to the Nyamuragira eruption of December 1981 (Fig. 3.2.4).

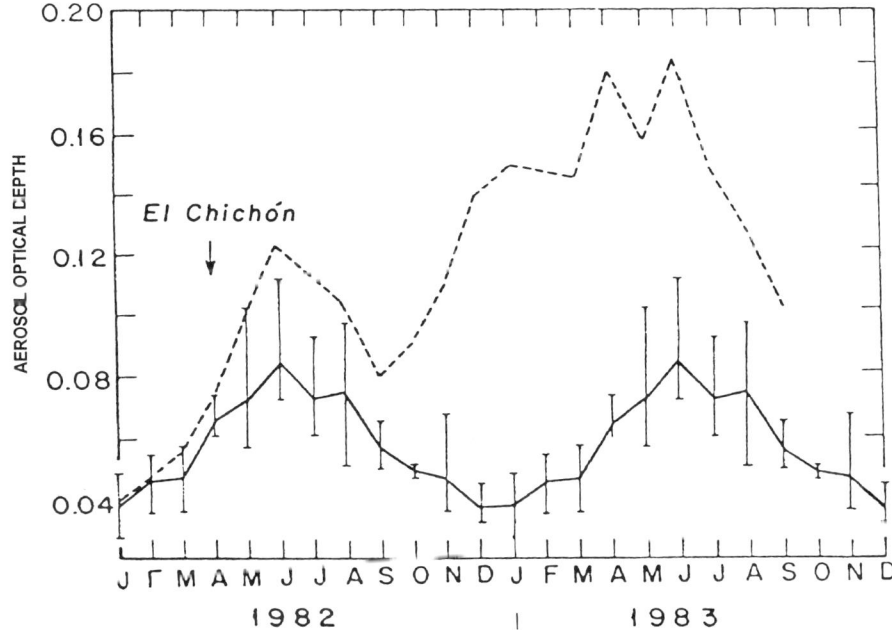

Figure 3.2.3. Aerosol optical depth at Vancouver for 1982 and 1983, and the monthly mean aerosol optical depths (and absolute maximum and minimum values) for the 5 years prior to the eruption of El Chichón. After Hay and Darby (1984).

Wittenborn et al. (1983) performed measurements of the IR atmospheric transparency (against the sun) in the wavelength interval 8 to 13.5 µm from an altitude of about 11 km using the Convair–990 spectrometer with a wedge interference filter. (The flights were made in December 1982 in the latitudinal band of 5° S–50° N.) An analysis of registered transmission spectra revealed not only the CO_2 and O_3 bands but also an absorption band near 8.5 µm that can be identified as that of the erupted H_2SO_4 aerosol. The share of absorbed radiation in the 10–41° N latitudinal band constituted 0.02 per unit airmass, about half this absorption being determined by a contribution from aerosols. An estimate of the ratio of optical thickness of the eruptive cloud at 8.5 µm to that of 0.5-µm wavelengths gave 0.14 at 20° N in December 1982.

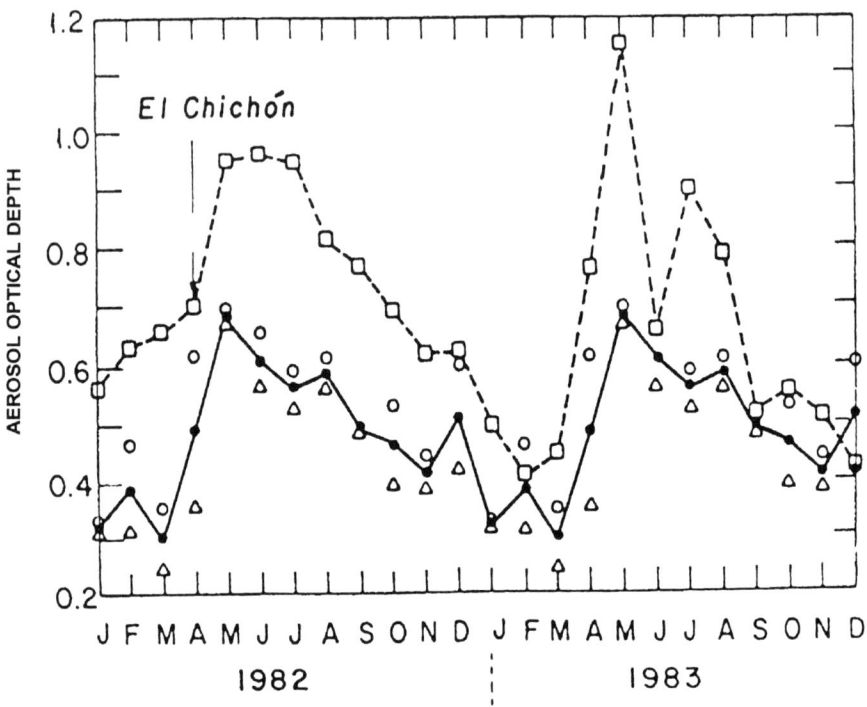

Figure 3.2.4. Aerosol optical depth at México City for 1982 and 1983, and the monthly mean aerosol optical depths (and absolute maximum (O) and minimum (△) values) for the 5 years prior to the eruption of El Chichón. After Galindo et al. (1996).

Pollack et al. (1991) have also analyzed the Convair–990 IR transmission data and those of a second flight performed during April/May 1983. Observations obtained during the second flight covered a wide range of latitudes in both hemispheres. Values of AOT for the wavelength 8.4 µm were calculated as well as the ratio (R) for infrared to visible (0.5 µm, Dutton and DeLuisi, 1983a; 1983b; 1987). An average value $R = 0.25 \pm 0.05$ has been obtained (the data implied that concentrated sulfuric acid was the dominant source of aerosol extinction in the 8- to 12-µm spectral atmospheric window and that the volume modal particle radius equalled 0.6 ± 0.1 µm at the time of the observations). Due to such optical properties the volcanic cloud produced a significant warming of the NH stratosphere, enhanced the planetary albedo, and caused a significant net radiative cooling of much of the NH. An important exception was, however, high northern latitudes during the winter of 1982, where a net radiative warming was expected. Similar effects should have characterized the SH, but with a much reduced magnitude, due to the cloud's reduced opacity there.

As Yue et al. (1991) have pointed out, the aerosol extinction coefficient at several wavelengths measured by the SAGE II satellite system beginning at the end of 1984 provided a unique opportunity to investigate the decay of the El Chichón aerosol perturbation on a global scale. In this context the monthly averaged SAGE II AOT values at 1.02 µm, 0.525 µm, and 0.453 µm over 10° latitude bands for the period from December 1984 to December 1988 have been calculated and analyzed.

Yue et al. (1991) have demonstrated that the temporal AOT variation at latitude bands higher than 30° in both hemispheres is the result of two components: an exponential component and a sinusoidal component. In general, the exponential component is latitude dependent, with large AOT values at higher latitudes. The sinusoidal or seasonal component has a maximum at local winter and a minimum at local summer, its amplitude being decreased each year from 1985 to 1988. In 1985, 1987 and 1988 the AOT exponential component was decreasing with time, but in 1986 the decrease was interrupted by the eruption of the Nevado del Ruíz volcano which took place at the end of 1985.

The gradual increase of the ratio of AOT values at 0.525-µm to 1.02-µm wavelengths from about 2 at the beginning of 1985 to about 3.5 at the end of 1988 was an indication of aerosol size decrease due to the diminishing effect of El Chichón (again, the eruption of Nevado del Ruíz temporarily changed the trend).

Numerical modeling performed by Yue et al. (1991) show that the annual variation of the tropopause height and the growth and evaporation of aerosol particles due to the annual variation of ambient temperature cannot satisfactorily explain the presence of the relatively large sinusoidal AOT component. This component was discovered by SAGE II observations at higher latitudes. The transport of aerosol from the low latitudes during the latter part of the year and the incursions of Arctic air at midlatitudes in the early part of the year suggested by Hofmann and Rosen (1982) may play a role in the generation of the sinusoidal AOT component observed in the NH.

Kent and Yue (1991) have analyzed a potential of CO_2 lidar sounding from the viewpoint of the retrieval of stratospheric aerosol properties. They show (through model calculations) that the backscattering coefficient at wavelength 10.6 µm varies by a factor of the order of 100 between background and immediate postvolcanic conditions.

After a series of El Chichón eruptions on March 28 and April 3 and 4, 1982, 18 aerosol sounding were made by Hofmann and Rosen (1983a, 1983b, 1983c, 1983d) in Laramie, Wyoming (41° N) and in southern Texas (27°–29° N) to study the mechanisms of gas-to-particle conversions that determine the transformation of ejected SO_2 into concentrated H_2SO_4 solution droplets.

Previous observations made after the eruptions of Mount St. Helens (1980) and Alaid (1981) revealed a continuous formation of both submicron droplets (r ~ 0.02 µm) through gas-to-particle conversions during about 1.5 months after the eruption and very large droplets (r ~ 1 µm). The latter made the largest contribution to the formation of the backscattering signal during lidar sounding.

The observational results were obtained using various instruments that measured the number density of condensation nuclei (r ~ 0.01 µm; devices are used beginning in 1973), particles with radii r ≥ 0.15 µm and r ≥ 0.25 µm (aerosol radiosonde; used beginning in 1971), and large particles (r ≥ 0.25; 0.95; 1.2; 1.8 µm); this sensor, used from 1981, could measure the concentration of particles down to minimum values 1 cm^{-3}.

Since the post-El Chichón stratospheric aerosol at altitudes above 20 km during the first 4 to 5 months was located in the band 0°–30° N, along with aerosol sounding in Laramie soundings were made in southern Texas on May 18–19 in Laredo (27.3° N), August 21 in Sinton (27.8° N), and October 23 in Del Rio (29.2° N). In the latter case all four devices were launched in one balloon and an experiment was made with the samples warmed to determine the boiling point for the aerosol.

Analysis of observational data showed that there were always two major aerosol layers, the lower layer located between the tropopause (heights 12 to 16 km) and an approximate level of 21 km, and the center of the upper 5-km layer near 25 km, as shown in Figure 3.2.5. Typical values of the number density of particles in these layers constituted about 10 cm^{-3} (r > 0.15 µm) during the entire period from May to October, but the concentration of large particles (r ≥ 1 µm) decreased at the time from ~ 1 to 0.1 cm^{-3}.

The aerosol layers were always divided by an unusually pure atmospheric layer, without any vertical mixing between them. The horizontal development of the upper layer was limited, but the lower layer had extended to Laramie by June. The May 18–19 sounding data revealed high concentrations of both CN and large particles in both layers, which points to the process of continuous formation of new particles with a radius (r) of 0.01 µm as a result of gas-to-particle conversion, since the time for coagulation of such particles (their concentration constitutes about 750 cm^{-3}) is less than 5 days.

Ground-based measurements of the atmospheric spectral transparency made by Lockwood (1982) in Flagstaff, Arizona, during a week in mid-May in the wavelength interval 330 to 850 nm gave maximum values of the optical thickness of the aerosol cloud in the visible of about 0.3 (this means that soundings in Laredo refer to conditions of thickest aerosol layers in the stratosphere). Forty-five days after the eruption the aerosol size distribution in the layer 4.5 to 25.5 km was bimodal with modes, to which radii of about 0.02 µm (new particles) correspond, as well as 0.7 µm (enlarged particles due to the condensation of H_2SO_4 vapors existing before the eruption, when their modal radii constituted about 0.08 µm).

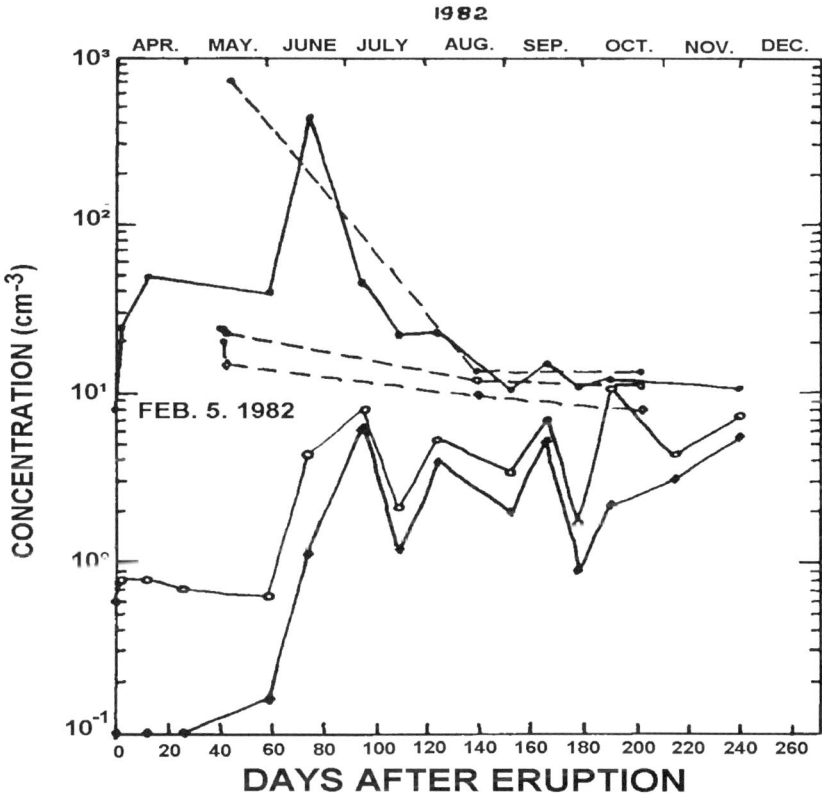

Figure 3.2.5. Dependence of aerosol number density near a level of maximum concentration above 20 km from observations at Laramie (solid lines) and in southern Texas (dashed lines) after the eruption of El Chichón. ●, $r \geq 0.01$ μm; ○, $r \geq 0.15$ μm; ◊, $r \geq 0.25$ μm. After Hofmann and Rosen (1982). [© by American Geophysical Union]

Although the number density of submicron aerosols is much higher than that of large particles, the latter contribute most to the total mass concentration of aerosols (35 μm/cm^3 compared to 1 μg/cm^3). The August 21 data of the sounding in Texas revealed a relatively low concentration of the smallest particles in both cases, indicating that the process of formation of new particles had ceased by that time. These data revealed the existence of a thin aerosol layer at altitudes of 31 to 33 km.

Results of the October 23 sounding in Texas were similar to those for August 21. Since from the 25-km height large particles (~ 1 μm) drop by 1 km every 10 days, the persistent concentration of these particles points to continuing growth of particles due to the accretion of H_2SO_4 vapors, which requires a concentration of H_2SO_4 not lower than $(3-4) \cdot 10^{-7}$ cm^{-3} (about 0.5 ppb). According to data of the soundings in Laramie, the formation of the gas-to-particle conversion aerosols (r ~ 0.01 μm) ceased after June, when a rapid decrease of particle concentration began, typical of the process of coagulation. By December 1982 the aerosol concentration observed in Laramie was approximately the same as in the latitudinal belt 27°–29° N and was characterized by a relatively larger contribution by particles with a radius of r ≥ 0.25 μm than by particles with r ≥ 0.15 μm.

Calculations of the aerosol mass in an air column above the prescribed level were made on the supposition that particles are droplets of a 75% H_2SO_4 water solution (density 1.65 g/cm^3). These revealed a rapid initial decrease of the aerosol mass in Texas due to the transport and gravitational settling of particles, and at Laramie a slow growth of the aerosol was observed as erupted material was absorbed into the atmosphere, as indicated in Figure 3.2.6.

In May, when according to satellite data the spreading of aerosols was confined to the altitude belt 0°–30° N, the aerosol mass in an air column above 15 km at 27.3° N constituted 0.18 g/cm^2 (about 100 times greater than before the eruption), and in the entire atmosphere within 0°–30° N it reached nearly 23 Mt (apparently, this was an upper limit, since aerosols were nonuniformly distributed).

Based on data of aircraft lidar soundings, by October the upper aerosol layer covered the latitudinal belt at least from 10° S to 40° N, and the lower layer from 20° S to 50° N, with the aerosol mass in the atmosphere above 20 km reaching 5.5 and 2.3 Mt, respectively (5.1 Mt for the lower layer at altitudes above 15 km). The total aerosol mass in both layers is 8.3 Mt, from which it follows that the initial mass apparently exceeded 10 Mt (but was below 20 Mt). By December the effect of gravitational settling had still not dominated the process of particle growth, making it possible to estimate the lifetime of erupted stratospheric aerosol.

The heating of the input aperture of the photoelectric counter up to 150 °C (performed in October) at a slow descent of the balloon (29° N) showed that ≥ 99% of the upper layer particles and 98% of the lower layer particles were volatile (or had a nonevaporating nucleus with r < 0.15 μm). This finding suggests that the bulk of the upper layer is composed of large droplets of an 80% H_2SO_4 water solution, and that of the lower layer, droplets with an H_2SO_4 concentration of about 60% to 65% (various concentrations are explained by the different temperatures of the layers). These estimates are in agreement with the theory of stratospheric sulfuric acid aerosol formation (Hamill et al., 1988, Turco et al., 1979; 1982a; 1982b; 1983).

VOLCANIC ACTIVITY AND CLIMATE 207

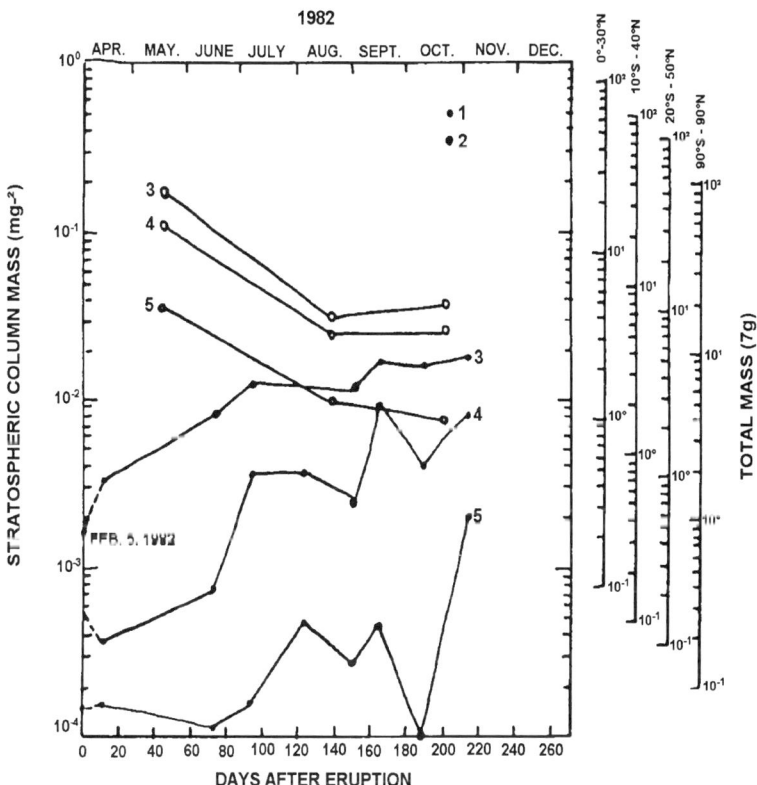

Figure 3.2.6. Temporal variations of the aerosol mass in the vertical air column above three altitudes, from observations at Laramie (●) and in southern Texas (○) after the eruption of El Chichón. The scale on the right makes it possible to estimate the aerosol mass (1 Tg = 10^{12}g) in different latitudinal belts: 3, r>15 μm; 4, r>20 μm; 5, r>25 μm. After Lockwood (1982). [© by American Geophysical Union]

From observations at Mauna Loa 2 days after the April 4 eruption the maximum SO_2 content in the atmosphere was equivalent to a mixing ratio of about 5 ppm, assuming that the SO_2 was concentrated in the 5-km layer near 25-km altitude. Faster growth of particles at 25 km than at 18 km (although in the latter case less H_2SO_4 is required) can be explained by the accelerated SO_2 oxidation reaction resulting from a higher temperature at a 25-km height and (or) higher concentrations of H_2SO_4 in the upper layer. Probably, the existence of two aerosol layers is explained by the view that the lower one formed as a result of weaker ejections of sulfur on March 28 and April 3, and the upper one resulted from the most powerful April 4 eruption. This conclusion does not contradict the results of analysis of heights and motion of eruptive clouds using satellite imagery.

Aerosol soundings were made by Hofmann and Rosen (1983d) in Laredo (26.3° N) and Del Rio (29.2° N) on May 19 and October 23 (1.5 and 7.5 months, respectively, after the April 4 El Chichón eruption). An analysis of the data by Hofmann and Rosen (1983d) showed that the distribution of aerosol by radius n(r) in the layer of maximum concentration (24.5 to 25.5 km) is bimodal and well approximated by the lognormal function

$$n(r)dr = \sum (N_i/\sqrt{2\pi})\exp(-\alpha_i^2/2)d\alpha_i. \quad (3.2)$$

Here N_i is the total number density of the ith mode, $\alpha_i = - \ln (r/r_i)\ln \sigma_i$ where r_i is the ith mode (median) radius and σ_i the ith (nondimensional) modal width. The value of $\int n(r)dr$ is the total number of particle N(r) with a radius exceeding r.

From the May 19 observations, the aerosol of fine (large) size is characterized by the following parameters:

$N_o(N_1) = 150$ (4 cm^{-3}); $\sigma_o(\sigma_1) = 2.80$ (1.77); $r_o(r_1) = 0.02$ (0.72 μm); $m_o(m_1) = 1$ (35 μg/m^3).

On October 23, 7.5 months after the eruption, the following values were registered:

$N_o(N_1) = 10$ (0.2 cm^{-3}); $\sigma_o(\sigma_1) = 1.50$ (1.10); $r_o(r_1) = 0.27$ (1.0 μm), and $m_o(m_1) = 2.9$ (1.5 μg/m^3).

Thus modal radii of volcanic aerosol particles are 0.02 and 0.7 μm, the concentration of large-size particles being nearly the same as before the eruption. Hence the conclusion can be drawn that these particles resulted from the growth of old particles, whereas fine-sized particles formed after the eruption are a product of gas-to-particle conversion. Regular aerosol soundings at Laramie showed that there was a gradual post-eruption growth of the aerosol layer at a height of about 18 km, and the upper (25 km) aerosol layer started to appear sporadically in July and reached some stability by the end of 1982.

The October 23 sounding in Del Rio revealed an almost complete absence (ceased formation) of smallest particles (r ~ 0.01 μm), which caused an increase of the modal radius of the small-size fraction up to 0.3 μm. Simultaneously, large-size particles grew, which manifested through an enlargement of the modal radius from 0.7 to 1.0 μm and a 20-fold decrease of their number density (probably because of gravitational settling). Stratospheric aerosol mass concentration decreased eight times, with the main contribution to mass from the 0.2-μm mode particles.

Six to eight months after the Mount St. Helens (1980) and Alaid (1981) eruptions, a secondary 1-μm mode was also observed, with the main mode remaining consistent with the pre eruption one of ~ 0.08 μm. This means that following the eruption of El Chichón, the aerosols formed consisted of much larger particles, and hence the total mass reached about 10 Mt in 6 months (as opposed to a total aerosol mass following the Mount St. Helens eruption of 0.25 Mt).

Recent theoretical studies and observations have led to the conclusion that the atmosphere at altitudes of 25 to 30 km is characterized by an almost saturating concentration of water vapor of about 10^5 to 10^6 mol/cm^3, preserved in volcanically quiet periods due to photodissociation of carbonyl sulfide in the stratosphere. After eruptions, when the stratosphere gets large amounts of SO_2, a strong supersaturation occurs with respect to H_2SO_4, and favorable conditions are created for the formation of the sulfuric acid aerosol:

(i) Provided that the supersaturation is large enough (the sulfuric acid concentration at 25 km exceeds 10^7 molecules/cm^3), a homogeneous nucleation and formation of H_2SO_4-H_2O droplets takes place without participation of condensation nuclei; and

(ii) If there is an accretion of vapor on already existing particles, which stimulates their growth. An intermediate process (with respect to those mentioned) is associated with possible condensation on numerous ions and ion clusters.

Since the final product (small droplets of the sulfuric acid solution) does not suggest its mechanism of formation (either homogeneous or ion nucleation), the complex of aforementioned processes is usually considered as one process. The process of formation of new droplets in supersaturation conditions is very fast and intensive, which determines a high initial concentration of particles ($\geq 10^3$ cm^{-3}). On the other hand, however, coagulation of particles determined by their square-number density is also very intensive. The coagulation-induced growth of droplets causes a rapid decrease of their number density, but in this case the lifetime of droplets increases. The lognormal size distribution of aerosols is formed.

Consideration of the growth of particle observed from May to October (from 0.08 to 0.72 μm) gave the H_2SO_4 concentration $2 \cdot 10^7$ molecules/cm^2, which exceeds the background concentration at a 25-km height at least 100-fold. It follows that the rate of H_2SO_4-H_2O homogeneous nucleations at a temperature of -85 °C varies within 10^3 to 10^{-3} cm^{-3}s^{-1}, with the water vapor mixing ratio growing from 3.5 to 10 ppm (by volume). This is equivalent to the formation of 10^2 to 10^6 particles/cm^3 per day and corresponds to the observed maximum concentration of particles with r ~ 0.01 μm of 750 cm^{-3}.

The time constant that determines the loss of H_2SO_4 vapors spent in the formation of particles was estimated at 5 min. This is approximately the same time constant that characterizes the gas-phase reactions of continuous transformation of SO_2 into H_2SO_4 vapors. Estimates for previous eruptions gave a time constant for the eruptive aerosol layers after the eruption of El Chichón; one may expect that climatic consequences of this eruption would be more substantial than those observed earlier.

Using data of numerous balloon soundings at Laramie (41° N) and soundings on four occasions in southern Texas (27°–32° N), Hofmann and Rosen (1983b) analyzed the mechanisms of SO_2 to sulfuric acid droplet conversion in the stratosphere following the eruption of El Chichón.

Data on condensation nuclei at Laramie show two periods of new aerosol nucleation during the first 100 days after the eruption and in early 1983. The first period is probably due to sulfuric acid vapor supersaturation. As for the other period, the mechanism of particle production is connected with thermal nucleation of aerosols in polar regions from eruptive H_2SO_4 vapors ejected to high latitudes and supersaturated, being rapidly cooled.

The initial pronounced bimodality of aerosol in the layer at altitudes of about 25 km (modes 0.02 and 0.7 μm) can be considered to be a result of new particle nucleation from the gas phase and the intensive growth of pre-eruption stratospheric particles.

Formation of new particles ceases approximately 3 months after the eruption, but the growth of particles still continued after 5 months. By December 1982 the stratospheric layer of volcanic aerosol reached the Laramie latitude and two earlier layers of volcanic aerosol and 25 km merged into one layer located between the tropopause and 30 km. Data on particle growth dynamics show that 40 days after the eruption the sulfuric acid vapor concentration was about 10^7 molecules/cm^3 and the lifetime ~ 22–45 days. The assessment of the global sulfuric acid aerosol mass 9 months after the eruption gave a value of the order of 10 Tg, but soon after the eruption it reached up to 20 Tg (these values correspond to the SO_2 mass equal to 5 and 10 Tg). Weakening of the eruptive aerosol layer started in April 1983. Initial assessment gave the following values of decay times (e-times decrease): 8.5 months ($r \geq 0.15$ μm) and 7.5 months ($r \geq 0.25$ μm and total aerosol mass).

The analysis of observation data made by Hofmann and Rosen (1987) led to the following conclusions concerning succession of volcanic stratospheric aerosol formation processes:

1. After initial ejection of SO_2 into the stratosphere gas reactions of H_2SO_4 vapor formation begin; in about 1 day H_2SO_4 vapor concentration becomes high enough for nucleation processes (gas-to-particle conversion) which leads to the formation of high number concentration (~ 10^3 cm^{-1}) of tiny H_2SO_4 aerosol droplets.
2. In a certain time interval Δt, when attenuation of the strength of

SO$_2$ source starts, and the impact of diffusion and coagulation increases, the droplet number concentration decreases up to 100 cm^{-3}. After that the growth of droplets dominates due to condensation of H$_2$SO$_4$ vapor on droplets, the number concentration of droplets being controlled by gas-to-particle conversion depending on altitude (this condensation process controls aerosol size distribution).

Comparing the rate of decay of aerosol following the eruption of Fuego in 1974 and El Chichón in 1982 with the same balloon-borne particle counters led Hofmann and Rosen (1987) to conclude that simple sedimentation and diffusion are not the only factors affecting the decay of stratospheric aerosol from volcanic eruptions. El Chichón aerosol was characterized by a much larger decay time that may be attributed to the combined impact of the magnitude and altitude of the eruption as well as stratospheric warming in the polar regions, which vaporize sulfuric acid aerosol at high altitudes. The resulting vapor then diffuses to the 20-km region where it takes part in aerosol growth, thus prolonging the decay and maintaining the relatively larger aerosol size. Even more than 5 years after the El Chichón eruption, aerosol properties had not returned to pre-eruption conditions. An important result of the post-El Chichón observations was the discovery of the interrelationship between the growth of eruptive aerosols and the drop of extratropical total ozone content in the atmosphere (Bojkov, 1987, Bojkov et al., 1990, Zerefos et al., 1992).

The most important aspect of the climatic impact of volcanic eruptions is a volcanically induced tropospheric temperature change and a possibility to identify volcanic signal (VS) from temperature observations. Angell (1988) has persuasively shown that this is not a simple problem at all, especially in the El Chichón case when overlapping of such strong forcing as the eruption and the ENSO 1982–83 event took place. Therefore there was unambiguous evidence of low stratospheric warming after (and due to) El Chichón, but not of tropospheric or surface cooling (this subject will be discussed in more detail in Chapter 5).

Analysis of the magnitude of lagged correlations between El Niño dynamics—SST variations in the eastern equatorial Pacific, and tropospheric temperature variations in the tropics (a global 63-station radiosonde network)—has led Angell (1988) to the conclusion that the reason why tropospheric temperatures cooled following the Agung eruption in 1963 but warmed following the El Chichón eruption in 1982 is the occurrence of a very strong El Niño after El Chichón but only a relatively weak El Niño after Agung. It is possible, however, to filter out the El Niño contribution through the use of empirical correlation between tropical tropospheric temperatures and equatorial SST (this may be done by displacing the tropospheric temperatures two seasons earlier with respect to SST and dividing the SST deviations from the mean by 2 before subtraction from the tropospheric temperature deviations). In this case the magnitude of the tropical tropospheric cooling following El Chichón becomes similar to that following Agung (about 0.5 °C) but is of shorter duration.

After Angell (1988) applied a similar adjustment procedure for the five major volcanic episodes since 1880, the eruptions of the tropical volcanoes of El Chichón, Agung, Santa Maria, Pelée, Soufrière, and Krakatau are all indicated to have brought about a decrease in NH continental surface temperatures of about 0.3 °C, whereas the eruption of Katmai in Alaska is indicated to have decreased this temperature by only about 0.1 °C, as shown in Table 3.2.2. Thus, it may be concluded that the reason why the evidence for volcanically induced cooling of the Earth's surface has been uncertain and controversial in the past is that such cooling may or may not be observed, depending on the extent of sea surface warming in the eastern equatorial Pacific (El Niño) shortly after the volcanic eruption.

Table 3.2.2. Surface air temperature (°C) change between the 2-year interval immediately before and the 2-year interval immediately after the five largest volcanic eruptions since 1980. After Angell (1988). [© by American Geophysical Union]

Volcano	Northern Hemisphere		Southern Hemisphere		Adjustment *
	Continental	Marine	Continental	Marine	
Krakatau (6° S)	-0.17	-0.07	-0.25	-0.15	-0.12
Santa Maria (14° N)	-0.35	-0.38	-0.34	-0.25	0.08
Katmai (58° N)	0.05	-0.08	-0.22	0.15	-0.16
Agung (8° S)	-0.27	-0.08	-0.21	-0.18	-0.03
El Chichón (17° N)	-0.14		-0.01		-0.21

* Estimates of the values to be added to SAT changes in order to determine the contribution of the volcanic eruptions to those changes based on the variation in equatorial SST during the 4 years.

VOLCANIC ACTIVITY AND CLIMATE

3.3 The Mount Pinatubo Volcano Eruption

Powerful explosive volcanic eruptions are a unique type of natural calamity with consequences of great importance not only on a regional but also on a global scale. Of special interest are impacts of volcanic eruptions on global climate change and the ozone layer (Kondratyev, 1988, 1989, 1992, 1993; Kondratyev et al., 1983a; Kondratyev and Moskalenko, 1984; Kondratyev and Cracknell, 1996). Every explosive volcanic eruption is a large scale "experiment" nature which opens up a possibility to study the sensitivity of nature (e.g., climatic system, ozone layer) to external disturbances. The El Chichón volcanic eruption in 1982 was in fact, the first "experiment" on when it was possible to undertake a comparatively complete program of observations (including satellite remote sensing) to monitor relevant environmental consequences.

Various observational means were further developed by 1991 when the Pinatubo volcanic eruption took place on Luzon Island (Philippines). By that time a very broad complex of satellite, aircraft, balloon and ground-based instrumentation for monitoring environmental dynamics had been developed and applied for various purposes. It had become possible to not only monitor propagation of volcanic products over the globe but also observe (with the use of both *in situ* and remote sensing techniques) various properties of eruptive aerosols, as well as their impact on the Earth's radiation budget and climate.

The principal purpose of this section is to survey briefly some preliminary results of observations and numerical modeling. As far as the observations are concerned it is very important to emphasize the unique combination of *in situ* and remote sensing measurements which has been achieved.

3.3.1. Basic Information Concerning the Mount Pinatubo Eruption

The explosive eruption of Mount Pinatubo (Luzon Island, Philippines: 15.14° N; 120.35° E) of the andesite type took place on June 1991 and had a more powerful impact on the atmosphere than all earlier eruptions monitored with the use of satellite information (the previous Pinatubo eruption happened 635 years ago). Gigantic ejections of SO_2 and other volcanic products led to the formation of a stratospheric aerosol layer up to 30-km altitude and above which has been documented by satellite remote sensing data (McCormick, 1992; McCormick and Veiga, 1992). Above 20-km altitude the aerosol cloud was quickly moving to the west and its front edge made a complete revolution around the Earth in about 3 weeks.

Boville et al. (1991) have undertaken numerical modeling to simulate global transport and dispersion of the Pinatubo aerosol cloud with the use of a high resolution, 35 level (about 300-km horizontal resolutions), stratospheric version of the NCAR Community Climate Model (CCM2) with an annual cycle. Calculations have been made under the assumption that a passive tracer was injected into the model stratosphere over the Philippine Islands (at 27 points) on model day June 15, and the transport was simulated for 180 days using an accurate semi-Lagrangian 3-dimensional advection scheme.

In accordance with observations the simulated volcanic aerosol cloud initially (for about the first 10 days) drifted westward and expanded in longitude and latitude (by July 15 the main tracer cloud had circled the globe about 1.5 times). The bulk of the aerosol cloud dispersed zonally to form a continuous belt in longitude, and remained confined to the tropics (30° N – 25° S) centered near the 20 hPa level for the entire 180 day model run, although a small amount was transported episodically into the upper troposphere in association with convective disturbances. Aerosol transported to the troposphere was dispersed within a few weeks into the NH extratropics. In the SH the aerosol was mixed into the regions equatorward of the core of the polar night jet during the first 50 days, but penetration into Southern Polar latitudes was delayed until the final warming in November 1991. As we shall see later, these predictions are basically in agreement with observations.

Franceschini et al. (1991) have applied a 2-dimensional model of the troposphere and lower stratosphere with steady transport to simulate sulfur chemistry and the formation of sulfate aerosol after volcanic eruptions. The chemistry includes the main families (NO_x, NO_y, HO_x) plus the sulfur compounds, while the heterogeneous processes are modeled with a microphysics code which takes into account nucleation, condensation and coagulation. The comparison with available observational data confirmed, in general, the adequacy of the simulation as far as aerosol concentration and size distribution is concerned but revealed strong (two orders of magnitude) underestimation of predicted sulfuric acid concentration in the troposphere. (This may be explained by too large a flux of condensation nuclei and partly by the absence of sulfate production from methane sulfuric acid in the model). The calculated concentration of aerosol particles in the stratosphere is larger by a factor of three than that observed and the vertical and meridional distribution is slightly different with respect to observations.

As has been pointed out by Pitari (1992), the injection of large amounts of SO_2 in the tropical lower stratosphere following the eruption of Mount Pinatubo can produce a not negligible perturbation in the stratospheric dynamics. Sulfate aerosols formed by nucleation H_2SO_4 vapor, which is in turn produced from the SO_2 plume, are responsible for a net heating of the order of

0.15 °C/day in the equatorial stratosphere, located where the thickest portion of the aerosol cloud is found. Pitari (1992) has assessed the dynamical response of the middle atmosphere to such diabatic forcing on the basis of a 3D stratospheric model. The calculations show that in a few months after the eruption, a change in the diabatic circulation larger than 10% is found in the lower stratosphere along with a stronger planetary wave activity during the 1991/92 winter season. The model predicts a temperature anomaly in the tropical lower stratosphere of about 2 °C that closely resembles the observations at the 30-hPa level.

In 3 weeks after the eruption (July 7, 1991) aircraft measurements of various eruption cloud characteristics were started, which covered (during the time period of six flights from July 7 to 14) the region of 4.5°–37° N; 45°–80° W. The aircraft instrumentation included nadir-directed lidar and correlation spectrometers; a side-looking interferometer (Fourier) spectrometer, and a multichannel radiometer to measure upward and downward fluxes of direct solar and global radiation (McCormick, 1992).

A representative synchronous database of Nimbus 7 satellite TOMS observations were used to retrieve TOC and total SO_2 content (TSC) in the atmosphere. An important contribution of satellite observational data was also made due to the AVHRR and SAGE occultation data.

An initial analysis of observational data (McCormick, 1992) has led to the conclusion that the total mass of volcanic effluent reached 20 Mt (1 Mt = 10^{12} grams), and that it was 3 times higher than after the El Chichón eruption in 1982. During the first few months after the eruption the basic portion of volcanic aerosols stayed within the tropical belt 30° N – 20° S.

Due to optical measurement data, the total mass of eruptive sulfuric acid aerosols reached ~ 20–30 Mt (these assessments agree with SO_2 data). The AOT of the stratosphere for the wavelength 0.5 µm varied within 0.2–0.4. Additional absorption of solar radiation due to volcanic aerosols resulted in significant stratospheric warming. For instance, in September the temperature rise at the 30 hPa level within 0° – 30° N belt reached 3.5 °C. A drop of total NO_2 content in the atmosphere was observed over New Zealand which supports the assumption concerning stimulation of ozone destruction processes in the stratosphere under the impact of heterogeneous reactions on surfaces of aerosol particles.

Dutton and Christy (1992) have pointed out that the Pinatubo eruption produced about 1.7 times stronger global solar radiation forcing than El Chichón. (The average total aerosol optical thickness for the first 10 months after the eruption is 1.7 times greater than that observed following the 1982 eruption of El Chichón.) Direct solar radiation was observed to decrease by as much as 25%–30% at four remote locations widely distributed in latitude, while monthly-mean clear-sky total solar radiation at Mauna Loa, Hawaii, decreased by as much as 5% and averaged 2.4% and 2.7% in the first 10 months after the

El Chichón and Pinatubo eruptions, respectively. By September 1992 the global and NH lower tropospheric temperatures had decreased 0.5 °C and 0.7 °C, respectively, compared to the pre-Pinatubo levels.

As Robock and Mao (1992) noted a typical result of modeling volcanic climatic signal in the NH is the warming over both Eurasia and North America and the cooling over the Middle East, which agree with observations (Groisman, 1992). Therefore it may be concluded that winter warming over NH midlatitudes and high latitudes and subtropical cooling are indeed volcanic effects to be expected in the winter following large tropical volcanoes, in the first or second winter following midlatitude volcanoes, and in the second winter following high latitude volcanic eruptions from either hemisphere. Robock and Mao (1992) have pointed out that an enhanced zonal wind driven by heating of the tropical stratosphere by the volcanic aerosols is responsible for the regions of warming, while the cooling is caused by blocking of incoming sunlight.

3.3.2 Satellite Observations

Existing satellite remote sensing observations allow complete enough monitoring of volcanic eruptions and their environmental impact. These are not only digitized images in the visible and infrared wavelength regions, which are regularly available from meteorological satellites, but also various data on minor gas and aerosol components in the stratosphere (TOMS and SAGE instrumentation), AOT (AVHRR data) and others.

Processing Nimbus-7 TOMS data has opened opportunities to retrieve both TOC and TSC (in the latter case with an accuracy of about ± 30%). Initial data on SO_2 content relate to a few days (June 11–14, 1991) preceding the cataclysm, when several weak eruption precursors took place (Bluth et al., 1992).

The first eruption cloud of the size ~100 km² has been discovered on June 12 and contained 25 kt (25×10^3 tonnes) SO_2. By June 13 the cloud shifted 1100 km to the west (being found over Vietnam) and its mass reached 110 kt. On June 14 the cloud was hardly distinguishable over the Indian Ocean (15 °N) and later disappeared. On June 13 a second similar aerosol cloud (9100 km², 15 kt) was discovered over the west edge of Luzon Island but it was observed for only 1 day. On June 15 the third cloud was documented (7500 km², 450 kt) elongated toward the west of Pinatubo up to South Vietnam (along the distance of 1600 km).

Cataclysmic eruption started on June 15 before noontime and continued until the next morning. This eruption was discovered through TOMS data analysis only on June 16 (by this day the satellite was over the zone of eruption). An assessment of the total mass of eruptive SO_2 at this moment gave an estimate of 15.5 kt (this value has probably been underestimated because of certain instrumentation malfunction). By June 23 the eruptive cloud covered a region measuring 15×10^6 km² stretched over a distance of the order of 10^4 km from Indonesia to Central Africa, being fairly homogeneous in space.

Two weeks after the eruption (June 30) the length of the cloud, located within the latitude zone 10° S – 20° N, had increased to 16×10^3 km and the SO_2 mass reached 12 Mt. Thus, 2 weeks after the eruption the cloud contained 60% of the initially ejected SO_2. This gaseous cloud was distinguishable during its movement around the whole globe (22 days).

Table 3.3.2 contains comparative data which characterize different eruptions. Due to the great power of the Pinatubo eruption, serious climate consequences were to be expected.

Table 3.3.1. Comparative characteristics of different volcanic eruptions (TOMS data).

Eruption characteristics	Kevarupta/ Katmai, 1912	El Chichón 1982	Nevado del Ruíz, 1985	Pinatubo 1991
SO_2 content, kg	$(5.2-20)10^3$	$7.0.10^3$	750	$2.0.10^4$
Ash ejections, g	$3.4.10^{16}$	$3.0.10^{15}$	$4.8.10^{13}$	$1.0.10^{16}$
Ratio of gas to ash ejections	$(2-6).10^{-4}$	$2.3.10^{-3}$	0.015	$1.9.10^{-3}$
Volcanic explosivity index, VEI	6	4-5	3	-6

The continuous functioning of SAGE II occultation instrumentation on board ERBS gave an opportunity to obtain a long series of observations for the atmospheric content of a number of minor gaseous components (MGC) and aerosols, beginning in 1984.

On the basis of aerosol extinction data for the wavelength 1020 nm in July–August 1991, McCormick and Veiga (1992) have analyzed the evolution of eruptive aerosol global spatial distribution (including both sulfuric acid and solid aerosol components). About 1 month after the eruption these observations covered mainly middle and tropical latitudes. During this time period and later aerosols were observed up to 29 km, being basically concentrated in the 20- to 25-km layer. In the beginning of the period considered AOT values reached maximum in the latitude band 10° S – 30° N, but by the end of July high AOT values took place at least up to 70° N, the maximum to the north of 30° N in the layer below 20 km.

In both hemispheres the correlation has been observed between anticyclonic high pressure systems and advection of volcanic material from subtropics located at the 21-km (16 km) level into the midlatitude zones of jet streams in the Southern (Northern) Hemisphere. By August in the whole atmosphere of the SH a 10-times enhancement of AOT took place in comparison with the beginning of June mainly due to the increase of aerosol content above 20 km. Assessments of the total volcanic aerosol mass gave values within 20–30 Mt, which significantly exceeds similar values (~12 Mt) for the El Chichón eruption.

Successful development of the AOT retrieval technique to process NOAA Satellite AVHRR data (retrieval errors are within 0.03–0.05) has opened the possibility to start regular AOT mapping over the world ocean beginning from July 1987. Availability of such information allows monitoring the dynamics of such phenomena as dust outbreaks from the Sahara and the Arabian Peninsula. Aiming to analyze the variability of aerosol optical properties regularly after the Pinatubo eruption, Stowe et al. (1992) undertook weekly aerosol mapping for 12 weeks after the eruption which demonstrated how weak desert dust outbreaks are in comparison with the eruption. It should be noted that the technique to retrieve AOT is based on the intercomparison between measured outgoing shortwave radiance (OSR) and numerical modeling results for a broad range of optical atmospheric models for pixels of 1x4 km^2 size. A multispectral (0.63- and 3.7-μm channels) algorithm used to identify clear sky cases is similar to the algorithm which was applied to retrieve sea surface temperature (in the case of AOT mapping a 1° latitude x 1° longitude resolution was used).

Analysis of a series of AOT maps has shown that volcanic aerosols travel around the globe in 21 days. The spatial distribution of aerosols is characterized by the presence of a horizontal inhomogeneity even 2 months after the eruption (AOT maxima were observed in August near Kamchatka and over the Bering Sea due to forest fires in Siberia). If one can apply as a criterion for volcanic aerosol recognition the threshold AOT value equal to 0.1 then it appears that in 2 months volcanic aerosol covered 42% of the Earth's surface (this is two times more than after the El Chichón eruption). By the end of the first 2 months the aerosol layer was limited by the belt 20° S – 30° N with the presence of separate patches in higher latitudes.

The maximum AOT value observed on August 23 was equal to 0.31. The total global SO$_2$ mass equal to 13.6 Mt (the mass of sulfuric acid aerosol should be a factor of 1.8 larger) corresponds to observed AOT values.

The reduction of the ERB, resulting from the growth of the Earth's albedo, due to scattering by volcanic aerosols, reached 2.5 W/m^2 (~1.3%). This

VOLCANIC ACTIVITY AND CLIMATE

is equivalent to a decrease of the global average SAT of not less than 0.5 °C in the case of a homogeneous aerosol distribution over the globe in the subsequent 2–4 years. (The reduction of ERB is a result of the increase of OSR by 4.3 W/m² and the decrease of OLR by 1.8 W/m².) Of course relevant global climate cooling may not happen under real conditions because of the multifactor nature of SAT variability (e.g., ENSO impact). The important consequences of the growth of the columnar aerosol content are its influence on the global ozone layer (due to enhancement of ozone-destroying heterogeneous chemical reactions on surfaces of aerosol particles), as well as certain enhancement of precipitation acidity (as a result of sulfur acid aerosol deposition into the troposphere).

For a number of reasons a very important parameter is the amount of volcanic SO_2 ejected into the atmosphere. McPeters (1993) has undertaken the retrieval of the amount of SO_2 in the atmosphere using NOAA-11 SBUV/2 spectrometer data, since the UV spectrum of backscattered solar radiation has enough representative SO_2 signatures in the wavelength interval 300–310 nm. Four SO_2 absorption bands are located within this interval, being centered at 302.3, 304.3, 306.5 and 308.7 nm. The SO_2 amount is calculated using the max-to-min ratios of the SO_2 structure. The max-to-min band ratio (BR) was calculated for each of the four bands. The BR is calculated as the ratio of the SO_2 absorption maximum (albedo minimum) to the adjacent absorption minima (albedo maxima). The SO_2 amount is derived from pre-computed tables of BR values as a function of solar zenith angle and SO_2 amount. For the 302.3-nm band, for instance, the band ratio is

$$BR_{302} = \frac{(R_{301} + R_{303})/2}{R_{302}} \qquad (3.3)$$

where R is the ratio of pre-volcano to post-volcano zonal average albedo spectra. The use of the band ratio technique to infer SO_2 amounts permits better accuracy (~10%–20%) and sensitivity (about 0.5 milli.atm.cm of SO_2) than the TOMS retrieval, but with relatively poor spatial coverage because the measurement is in nadir only.

As McPeters (1993) has pointed out there were only seven scans on June 19 with identifiable SO_2 signatures because at this time the volcanic cloud was still very localized. On July 1 there were 29 scans between 35° N and 12° S with the then highest concentration of SO_2 over the Atlantic. On July 17, SO_2 was detected in 30 scans around the world but in decreased concentration. Estimates of the total SO_2 budget made after the cloud had spread sufficiently for the sparse SBUV/2 sampling to be adequate, indicated there were $8.14 \cdot 10^6$ metric tonnes of SO_2 in the stratosphere on July 1, 1991, and 4.1 Mt remaining on July 17. These estimates correspond to an e-folding time of about 24 days for the conversion of SO_2 to aerosol, and are consistent with an initial injection into the stratosphere of 12–15 Mt SO_2. According to TOMS data, the SO_2 amount on June 17 was 18.5 Mt (Bluth et al., 1992). One can conclude therefore that TOMS appears to overestimate SO_2 by about 50%.

Future perspectives of satellite monitoring of global scale transport and climatic impact of volcanic aerosols are connected, first of all with the beginning of the EOS functioning at the end of the century (Kondratyev, 1992). The volcanic component of the EOS includes monitoring surface phenomena and atmospheric phenomena which is characterized by Table 3.3.2 data (Mouginis-Mark et al., 1991). Special emphasis will be made on climate impacts of volcanic eruptions.

From a volcanological viewpoint, available observational means will permit the mapping of volcanic terrains, the assessment of volcanic hazards, and the analysis of both lava-producing and explosive eruptions. Three modes of data collection have been planned: long-term monitoring (once per year, approximately 200 volcanoes); short term monitoring (once every 16 days); and analysis of active eruptions (as frequently as possible).

Table 3.3.2. EOS instruments and their applications to volcanology. After Mouginis-Mark et al., 1991. [© Elsevier Science Inc. Reprinted by permission.]

Instrument	Application
(A) Surface phenomena and plumes	
GLRS	Deformation rate of volcanoes via laser ranging, topography of eruption plumes, plume height via altimeter.
HIRIS	Visible/near infrared (0.4–2.5 μm), 30 m pixel. Used to determine energy output from extremely high temperature features such as lava flows and lava lakes.
ITIR	Analysis of thermal properties of summit lakes, eruption plumes, and fumaroles, and investigation of volcano lithology. Topography of plumes.
MISR	Surface bidirectional reflectance data stereo images provide topography of surface, time series views of moving flows give velocity of rapidly moving flows (e.g., mud flows, pyroclastic flows).
MODIS	Eruption detection will be provided by low resolutions (1 km/pixel) thermal-IR (7.3–12.0 μm) measurements to search for thermal and SO_2 (3.96–4.05 μm) anomalies. Multitemporal MODIS data used to determine eruption plume dispersal rates over time periods of days to weeks.
SAR	Will fly on a separate platform from other EOS instruments. Radar interferometry will be used to determine volcano topography, and the

VOLCANIC ACTIVITY AND CLIMATE

temporal change in morphology of a volcano due to eruptions. Multitemporal SAR images permit studying the rate of advance of lava flows under any weather conditions, day or night, or studying the volcano morphology in areas prone to frequent cloud cover (such as Indonesia or the Aleutians).

(B) Volcanic emissions

GOMR	Stratospheric SO_2 from 0.3050 μm; 0.3125 μm, and 0.3175 μm channels (NOAA satellites).
MLS	Vertical profile of SO_2 (15–30 km altitude), HCl (15–60 km), H_2O (10–90 km); 100 km x 6 km spatial resolution (EOS platform B).
TES	Tropospheric SO_2, CO, OCS, HCl (daytime only), H_2S (in high concentrations only), and quantitative amounts of H_2O and CO_2 (Platform B).

(C) Volcanic aerosols

MISR	Four bands: 0.44 μm; 0.55 μm; 0.67 μm, and 0.86 μm at four look angles for, and aft at 28°, 46°, 60°, and 72°. Provides aerosol data and cloud angular reflectance. Local mode spatial resolution 240 m, global mode 1.9 km.
SAGE III	Tropospheric and stratospheric aerosols, H_2O, 1–2-km vertical resolution.

Abbreviations:
GIRS	Geodynamic Laser Ranging System
HIRIS	High-Resolution Imaging Spectrometer
ITIR	Intermediate Thermal Infrared Radiometer
MISR	Multiangle Imaging Spectro-Radiometers
MODIS	Moderate Resolution Imaging Spectrometer
SAR	Synthetic Aperture Radar
GOMR	Global Ozone Monitoring Radiometer
MLS	Microwave Limb /Sounder
TES	Tropospheric Emission Spectrometer
SAGE	Stratospheric Aerosol and Gas Experiment

3.3.3 Aircraft Observations

A broad program of aircraft observations ("Electra" flying laboratory) of the plume of eruption products was started by NASA scientists 3 weeks after the eruption (Valero and Pilewskie, 1992). Six flights conducted during the time period June 7–14, 1991, covered the region 34° N – 4.5° S; 76° W – 46° W. Since altitude of the aircraft could not exceed 5.5–7.0 km, in a number of cases it created serious complications for remote sounding of the stratosphere throughout the upper troposphere.

The complex of onboard instrumentation included, in particular, instrumentation to measure total fluxes of direct solar, diffuse and solar global radiation; multichannel radiometer (with oscillating shadow-ring to measure direct solar radiation); and multichannel flux radiometer (MFR). All types of spectral instrumentation had similar spectral channels (380, 412, 500, 675, 785, 865, and 1064 nm) and were directed to zenith and nadir. Data of solar radiation measurements were used to calculate AOT. (The Rayleigh component was calculated with the use of observed atmospheric pressure values; extraterrestrial spectral distribution of the solar radiation was found through the extrapolation of measured values.) The errors of irradiance measurements did not exceed 1–2%.

Analysis of observational data discovered the growth of AOT with decreased distance from the volcano (15.14° N), but maximum AOT values were found in the vicinity of 10° N, which reflects specific features of the propagation of the volcanic cloud. The maximum AOT was observed on July 13 at wavelength 412 nm, being equal to 0.49. The spectral measurement data revealed a rapid AOT decrease with growing wavelength in the near IR and allowed retrieval of the particle effective radius as a parameter of aerosol size distribution. The analysis of relevant data led to the conclusion about the existence of two aerosol types: monomodal aerosol (July 7, 8, 12 and 13 data) and bimodal aerosol (July 10 and 14 data).

The aerosol particles effective radius r_{eff} was equal to 0.18 and 0.35 µm (July 8 and 13 data), and the columnar mass was 35 and 80 mg/m^2, respectively. Bimodal size distribution observed on July 10 was characterized by r_{eff} values equal to 0.1 and 0.8 µm, while the columnar mass reached 40 mg/m^2. Apparently, the existence of a bimodal aerosol size distribution resulted from gas-to-particle conversion processes of submicron sulfuric acid aerosol in the presence of large volcanic ash particles. Calculations of the surface-atmosphere system albedo (under various assumptions about aerosol properties) gave the maximum value 0.12 for the surface albedo 0.05, solar height 10° and AOT equal to 0.5.

Regular mapping of AOT variations on the basis of AVHRR data continued until 1993. It led to the conclusion that in the high northern latitudes after July 1993 (no measurements were possible in the high southern latitudes because of weak solar illumination) the Pinatubo volcanic aerosol cloud had essentially become undetectable (< 0.02 AOT, NOAA 13 AVHRR data). It is possible, however, that more sensitive instruments such as lidar and SAGE will be able to track dissipation of the aerosol cloud down to background levels of 0.003 AOT. For instance, lidar data for Garmisch-Partenkirchen (Germany) during July to early September 1993 reveal the continued presence of an aerosol layer that peaks at about 16- to 19-km altitude (Jäger, 1992). The backscattering ratio for Nd-YAG wavelength has decreased to about 1.9 from 3.7 at the same

VOLCANIC ACTIVITY AND CLIMATE

time in 1992; the corresponding decrease at the ruby wavelength was 3.2 from 7.6 in August 1992. Lidar data for the French Antarctic Station Dumont D'Urville (66.67° S, 140.02° E) show that integrated backscattering values increased steadily throughout most of 1992, maintaining a peak through most of the month of August before beginning a rapid decline by 1993 to values prior to those at the start of 1992.

Aircraft lidar sounding conducted in July 1991 (Nd:YAG lidar, 532-μm wavelength) revealed multilayered structure and horizontal inhomogeneity of the aerosol plume in the stratosphere in 3–4 weeks of the Pinatubo eruption (Winker and Osborn, 1992). As an indicator of aerosol content the SR was chosen, being defined as a ratio of the sum of aerosol backscattering and Rayleigh scattering coefficients to the coefficient of Rayleigh scattering (with the reference to the SR value at the tropopause level, which was 16 km).

The thickness of aerosol layers located within the 17- to 26-km altitude level varied between < 1 km and several kilometers. The layers located to the north of 20° N were relatively weaker (SR values about 3 and smaller), but south of 15° N values of SR, as a rule, exceeded 10. Maximum SR values of the order of 80 and higher were observed during July 12–13 at an altitude ~25 km, but north of 11° N a sudden disappearance of the powerful aerosol layer took place.

Assessments of column aerosol mass made under assumption of pure sulfuric acid aerosol led to the value of 95 mg/m², which is equivalent to a global aerosol mass equal to 8 Mt. Such an estimate is rather arbitrary, however. Presumably, by the moment of observation only about half of the ejected SO_2 was transformed into sulfuric acid aerosol, and eventually the sulfate aerosol mass could reach about 16 Mt (after the El Chichón eruption about 12 Mt of aerosol was formed).

Polarization lidar installed on board "Electra" aircraft produced a linearly polarized signal at the 532-nm wavelength. The backscattering signal due to stratospheric volcanic aerosol was registered for two orthogonal polarizations with subsequent calculations of both SR and depolarization ratio (DR) defined as a ratio of signals for perpendicular and parallel polarization components. (These were the first aircraft DR observations.) Lidar sounding data with 15-m resolution were smoothed up to 150-m resolution.

The analysis of DR data discovered the presence of aerosol layers with significantly different polarization signatures, which was the result of variable chemical composition, physical characteristics and shape of particles (Winker and Osborn, 1992). As is well known, depolarization of backscattered light indicates the presence of scattering by anisotropic particles. Possible causes of the depolarization in the case considered could be the existence of crystallized sulfuric acid droplets, aggregates of frozen droplets and silicate particles of irregular shape. June 12–13 data demonstrate the presence of a layer with high depolarization near 25 km. Apparently, in the latter case almost pure sulfuric acid aerosol with a small admixture of ash and other nonspherical particles took place while in the first case mixed aerosols were observed. For reliable interpretation of these data (explanation of depolarization effect) *in situ* measurements of aerosol size distribution, chemical composition and particle shape are necessary.

Since the assessments of global SO_2 emissions after the Pinatubo eruption, made on the basis of TOMS data, gave the value of the order of 20 Mt, gas-to-particle conversion had to result in the formation of 41 Mt of sulfuric acid aerosol. In this context aircraft measurements with the help of upward directed UV correlation spectrometer COSPEC V are of interest (Hoff, 1992). The seven-channel COSPEC allowed measurement of scattered radiance within the wavelength interval 297–315 nm with the purpose of retrieving columnar SO_2 content.

Processing observational results for various atmospheric masses (solar zenith angles) gave estimates of reduced SO_2 content within 8.9 to 64.2 µm.atm.m. Maximum values that correspond to minimum masses (minimum solar zenith angles) were observed near 14° N. Probably, the most reliable value is somewhere between the average value 24.6 µm.atm.m (for all data) and noontime value 40.9 µm.atm.m. An increase of SO_2 content during the time period July 12–14 was observed, which may be explained by the arrival of the most powerful plume of eruption products at the 26-km altitude, discovered by lidar sounding. The assessment of the global SO_2 mass gave values from 8.8 Mt (for average global columnar SO_2 content equal to 25 µm.atm.m) to 14.4 Mt (41 µm.atm.m). By the time of observation about 25 Mt of volcanic aerosol had been formed.

With the use of an aircraft IR Fourier spectrometer for spectral measurements of direct solar radiation for solar heights from 0° to 15° after the Pinatubo eruption Mankin et al. (1992) processed observational data for wavenumbers 1354–1368 cm^{-1}; 2926 and 1141 cm^{-1}) to retrieve the columnar content of SO_2, HCl and O_3 in the stratosphere. Spectral resolution was equal to 0.06 cm^{-1}, and the technique of retrieval was based on an intercomparison of measured and calibrated atmospheric absorption spectra. Retrieval errors for SO_2, HCl and O_3 are 25%, 20% and 16%, respectively. The results obtained show that ozone content values do not differ from those measured with the application of conventional techniques. In the case of HCl, observations did not reveal any significant increase of HCl content as was discovered after the El Chichón eruption. Thus, the Pinatubo eruption did not result in enhancement of Cl content in the stratosphere. Sulfur dioxide content values are within those limits that correspond to COSPEC V measurement results. Maximum SO_2 content was observed near 15° N, but significant quantities of SO_2 were also found within the 15°–30° N belt.

VOLCANIC ACTIVITY AND CLIMATE 225

If one can assume homogeneous global SO_2 distribution in the 20- to 30-km layer then its average mixing ratio is equal to 32 ppbV. If it is also taken into account that the SO_2 plume occupies the 20° S – 20° N latitude belt, then the SO_2 mass in the plume 4 weeks after the eruption should be equal to 5.5 Mt. Of course, such estimates may be considered only as very approximate. They show, however, that even in 4 months, a significant amount of volcanic SO_2 still remained in the troposphere. Aircraft observations from middle tropospheric altitudes did not permit assessment of stratospheric water vapor dynamics.

3.3.4 Balloon Observations

Balloon observations were very important for studies of consequences of the Pinatubo eruption. During the July 17–August 28 time period, seven dustsonde (DS) flights were accomplished by the University of Wyoming experts from Laramie (41° N). These launchings were made both before and after the arrival of the volcanic aerosol cloud to the region of Laramie. Apart from standard instrumentation to measure atmospheric pressure, temperature and ozone concentration, the DS instrumentation complex included the following basic sensors: CN; photoelectric counter (particle radius $r > 0.01$ µm); fast response eight-channel aerosol particle photoelectric counter ($r > 0.15$; 0.25; 0.50; 1.0; 2.0; 3.0, and 10.0 µm) as well as special two-channel ($r > 0.15$ and 0.25 µm) particle counter to investigate particle volatility. For a number of flights a three stage cascade impactor was used to obtain aerosol samples for subsequent electron-microscopic analysis in the laboratory. Maximum flight altitude varied within 35.5–8.8 km.

Deshler et al. (1992) accomplished the analysis of all measured CN vertical profiles, which showed that the first arrival of the volcanic aerosol cloud took place July 16. (A contribution to the increase of CN concentration at the tropopause level was, apparently, the result of oil fields fires in Kuwait.) By July 26 aerosol number density below the level of westerly transport in the upper stratosphere reached up to the record level of about 50 cm^{-3} at 17-km altitude. This stable lower aerosol layer due to a rapid transport of particles to northern latitudes was also discovered during subsequent DS flights.

The early evidences of the layer, formed above 20 km (within the westerlies) were observed as a thin layer at the 23-km altitude. This layer was also identified through lidar sounding data in Boulder, Colorado, during the time period July 29 – August 2. By August the layer had ceased to be noticed but appeared again over Laramie on August 29 with lowered aerosol concentration after the eruption cloud made a full revolution around the Earth. The end of August was characterized by the presence of an extended (in the vertical direction) aerosol layer at altitudes from the tropopause up to 22 km, and in the middle of September up to 30 km.

The CN concentration at 15- to 20-km altitudes was so high (due to gas-to-particle conversion processes of the sulfate aerosol formation) that the CN countersaturation threshold was overcome. This occurrence means that CN concentration exceeded 500 cm^{-3}, i.e., was approximately 2 orders of magnitude more than the background values for 18- to 20-km altitudes.

Experiments with heating of aerosol particles up to 150 °C showed that 95%–98% of aerosols were volatile. This finding allows identification of aerosol particles as droplets of a high concentration sulfuric acid water solution and assessment of the characteristic time of SO_2 to H_2SO_4 transformation as equal to 38 days. (Earlier assessments for smoke aerosols in Kuwait arrived at the estimate of about 50% of volatile aerosols.)

Aerosol size distribution was characterized by the presence of large particles above the level of wind direction reversal in the stratosphere (20- to 23-km altitude) in comparison with 16- to 17-km altitude. The presence of aerosol particles with modal radius ~0.07 μm was typical at altitudes below 20 km, while above this altitude the radius increased to 0.35 μm (in case of single-modal lognormal size distribution). In the lower aerosol layer and at the edge of the upper layer the aerosol size distribution was initially bimodal (August 2 data). On July 30 at the 23 km altitude measured values of mass concentration, mass mixing ratio and total surface of particles were equal to 28 μg/m^3, 0.48 ppm and 84 μm^2/cm^3, respectively, which are much higher than similar parameters obtained after the El Chichón eruption. Since observations discussed led to the discovery of a high concentration of small size sulfuric acid droplets it is assumed that such droplets resulted either from the process of homogeneous nucleation or through condensation of small ions.

During the flights of the DS in Laramie on July 26, July 30 and August 8, samples of volcanic aerosols at altitudes from 15.5 km to 37.5 km were taken with the help of a three-stage cascade impactor, which allows separation of particles with diameters > 4, 1, and 0.25 μm (Sheridan et al., 1992). These samples on thin substrate films were subject to electron microscopic analysis that showed that almost all small-size aerosol (> 99%) consisted of submicron droplets of a water solution of sulfuric acid, which later (in the course of processing) transformed into ammonium sulfate (($NH_4)_2SO_4$) particles. The remaining particles of volcanic aerosols with sizes up to 10 μm had sulfate and sulfite composition, with certain admixture of rock components.

Dustsonde optical counter data showed that maximum aerosol particles number density for particles with r > 0.15 μm in the stratosphere was higher in July than in August. This fact has also been reflected in the 20%–30% thicker layer of deposited submicron stratospheric aerosol in July than in August.

Detailed analysis of submicron sulfate aerosol did not discover any presence of solid or dissolved CN particles that consisted mainly (98%–99%) of sulfuric acid. It is therefore possible to assume that the aerosol was formed through homogeneous nucleation. This assumption is in agreement with the results of the experiments to study volatility of particles, which demonstrated that 95%–98% of volcanic aerosols consist of a water solution of sulfuric acid. The samples taken on July 30 contained a large quantity > 1 μm soil particles that were covered by the layer of sulfuric acid, and July 26 data revealed that, as a rule, submicron sulfate particles were surrounded by one or more rings of sulfur-containing particles of smaller sizes. (It is possible that such particles resulted from the deposition of small droplets on the film.)

During the time from June 1991 to January 1993, Deshler et al. (1993) conducted 29 flights of the balloonborne aerosol instruments in Laramie to measure the number concentration of CN (r > 0.01 μm) and optical aerosol (r > 0.15 μm to 10.0 μm, or with radius > 0.15 to 2.0 μm, in eight size classes). The minimum detectable concentrations are approximately 0.0006 cm^{-3} (optical aerosol) and 0.007 cm^{-3} (CN). The analysis of data on aerosol vertical profiles, size distribution and volatility shows that aerosol size distributions were best represented when bimodal lognormal distributions were fit to the data.

For the first 2 months, following the Pinatubo eruption the vertical aerosol distribution was very structured and inhomogeneous (Deshler et al., 1992). After that time the measurements can be grouped into three general periods:

(i) Aerosol filling the stratosphere between the tropopause and 27 km, but maintaining a highly layered structure, indicating that vertical mixing was not complete (September through November 1991);
(ii) The layered vertical structure was replaced by a fairly homogeneous and static aerosol layer in early 1992;
(iii) Slowly decreasing concentrations of aerosol both at 0.15 and 1.0 μm (beginning in the summer of 1992).

In May 1992 two flights were performed using an optical aerosol counter (r > 0.15 μm) with its intake heated to 150 °C, similar to flights made soon after the eruption (Deshler et al., 1992). The results were similar: the concentrations of particles > 0.15 μm were reduced by a factor of over 50, which reflects the domination of sulfuric acid aerosol.

After an early short-lived intense aerosol layer, the stratospheric maximum surface area and mass (40 μm^2/cm^3, 16 ppbm) was observed to occur approximately 180 days after eruption. The aerosol was then observed to remain relatively homogeneous both in altitude and time during 1992, with the maximum surface area and mass remaining relatively constant between 20 and 30 $\mu m^2/cm^3$ and 30 and 60 ppbm. Although the Pinatubo aerosol was decaying, and the size distributions were changing due to coagulation and gravitational settlement, the effects of these processes on changing the stratospheric maximum aerosol surface area and mass were minimal up to January 1993.

Rosen et al. (1992) have used a new instrumentation (backscattersonde) to monitor the penetration of Pinatubo aerosol into the North Polar Vortex. It has been discovered that some volcanic debris had arrived at far northerly latitudes below 20 km by October and was apparently incorporated into the initial vortex. A significant amount of material was transported northward by the end of the winter.

3.3.5 Remote Sensing Observations from the Surface

Surface observation results were a very important component of the whole totality of measurement data, for instance, lidar sounding at the Mauna Loa Observatory (Hawaii: 19.53° N, 155.58° W, altitude 3400 m) detected on the July 1, 1991 arrival of the volcanic plume formed after the Pinatubo eruption. In this context Goldman et al. (1992) used solar spectra measured with the help of a Fourier spectrometer to retrieve SO_2 columnar content in the atmosphere, summarized in Table 3.3.3. The retrieval was accomplished through intercomparison of measured and calculated absorption spectra for the v_3 absorption band of SO_2 centered at 1362.05 cm^{-1}. The stronger band v_1 band at 1157.71 cm^{-1} could not be used because of the strong overlapping with H_2O and CH_4 bands. As shown in Table 3.3.3, in August SO_2 content decreased up to the sensitivity threshold of the spectrometer (~1.10^{13} molecules/cm^2), which may be interpreted to mean that by this time practically all volcanic SO_2 was transformed into sulfuric acid aerosol.

Table 3.3.3. Summary of the data on SO_2 columnar content. After Goldman et al. (1992). [© by American Geophysical Union]

Observation Date of 1991	Spectral resolution (cm^{-1})	SO_2 Vertical Column Amount (in 10^{16} molecules cm^{-2})
May 11	0.004	< 0.9
July 9	0.010	(5.1 ± 0.5)
August 10–24	0.004	< 0.9

It is quite natural that the most broadly applied observational technique to monitor volcanic aerosol dynamics was lidar remote sounding. Since there were no strong volcanic eruptions between 1986 and 1990 it was possible to assume that by 1991 (before the Pinatubo eruption) stratospheric aerosol was of a background type. As has been mentioned, lidar sounding at Mauna Loa (ruby laser) showed that the volcanic plume in the stratosphere reached Hawaii on July 1. The layer was concentrated at altitudes of 21.5 to 22.8 km, where maximum SR (SR is defined as the ratio of the sum of aerosol and Rayleigh backscattering coefficient to Rayleigh backscattering) increased up to 22, from SR < 1.5 observed in June. At this time the tropopause was located at 14.2-km altitude. The analysis of subsequent sounding (July–August) revealed (De Foor et al. 1992) that in most cases the plume was located below 26–27 km (the tropopause height varied within 14–16 km), while after the El Chichón eruption volcanic aerosol reached the level of 30 km. In the course of the evolution of the plume an enhancement of aerosol backscattering coefficient (ABC) was observed, as well as its cyclic variability, due, apparently, to the motion of the plume around the globe with the average zonal velocity equal to 70–80 km/hr.

The growth of integrated (for the 15.8–33.0-km layer) backscattering coefficients (NRBS) in the July–August period could be approximated by the following dependence on the year D with the reference to the value of NRBS when D = 180

$$\text{NRBS} = 0.374 \ (\pm 0.87)(D/180) + 1.16 \ (\pm 0.76) \cdot 10^{-3} \quad (3.4)$$

The related correlation coefficient amounts to 0.67. The volcanically induced decrease of atmospheric transmittance for solar radiation was about 13%, which is approximately equal to that observed by Galindo (1992) after the El Chichón eruption using satellite derived insolation.

Twenty days after the Mt. Pinatubo eruption the volcanic aerosol cloud was detected by lidar at 14 km from Frascati, Italy (Gobbi et al., 1992). Three and a half months after the eruption, the aerosol perturbation generated by Pinatubo reached and exceeded the maximum loads recorded 11 months after the El Chichón eruption.

Jäger (1992) made an analysis of lidar sounding results (Nd:YAG laser) for Garmisch-Partenkirchen (Germany: 47.6° N, 11.1° E) and discovered the first arrival of volcanic plume to this point on July 1, 1991, when aerosol (ABC) data exceeded the background level, but the next day such an increase was not observed. During the next few months very variable ABC vertical profiles were registered. The layered structure of the aerosol vertical distribution may be explained by specific features of the general atmospheric circulation. Aerosol that was concentrated in the 7- to15-km layer at the beginning of July was moving along the shortest trajectory. The double-layered aerosol distribution observed in August agrees well with the presence of typical (for middle latitudes) west to east transport at altitudes less than 20 km, and reversal of the wind direction above 20 km. After the transition to the typical winter circulation regime (beginning from September) with prevailing west to east transport in the whole thickness of volcanic aerosol layer, a single powerful aerosol layer was formed between the troposphere level and 28-km altitude with the enhancement of the transport of volcanic ejecta to the north. This is quite similar to what was observed after the El Chichón eruption (Galindo, 1992; Galindo et al., 1996).

Lidar aerosol sounding of the troposphere and stratosphere at three points around Boulder, Colorado, were made with the use of three lidars at the wavelengths 10.591 µm (CO_2 laser); 0.694 µm (ruby laser); 0.574 µm (Post et al., 1992). The first arrival of the volcano plume was observed on July 27 (it is quite probable that stable stratus cloudiness which prevailed during the previous 5 days prevented earlier detection of volcanic aerosols). In 1982 the first identification of the volcanic plume in Boulder was made 4 months after the El Chichón eruption while the Pinatubo ejecta reached the same point in 1.4 months, in spite of a much larger distance between the volcano and the observation point. Undoubtedly, such differences may be explained by specific features of stratospheric circulation and/or volcanic emissions. There was also such a difference that an enhancement of backscattering by the atmosphere was observed soon after the Pinatubo eruption, while in the case of El Chichón it took place 8 months after the eruption.

A specific feature of the Pinatubo volcanic cloud was its strong variability, which required frequent observations, especially at the initial stage of the cloud evolution. Calculations of the ratio of integral SR (for the layer between the tropopause and 30 km) for the CO_2 lidar and in the visible spectral region show a decrease of this ratio with time, which points to the decrease of particle effective size.

Avdyushin et al. (1993) have discussed lidar aerosol sounding obtained during the time period from July 1991 to April 1992 using a lidar network consisting of five Rayleigh-Mie lidars in the wavelength range 532–589 nm. Two Russian lidar stations operated by the Roscomhydromet were onboard the ships RV *Professor Zubov* (doubled-yttrium-aluminum laser, 539. 5 nm) and RV *Professor Vize* (dye laser RGW, 589 nm). Three French lidar stations operated by the Centre National de la Recherche Scientifique (CNRS) were located at the Observatory of Haute-Provence (OHP: 44° N, 6° E), at the Centre d'Essai des Landes et Biscarosse (CEL: 44° N, 1° W) and onboard the military ship *Henri Poincare* (LIMA). The observations covered the western sector of the NH (8° S – 60° N, 80° W – 6° E).

The data show that there is a great deal of variability in the structure of the volcanic aerosol cloud, formed after the Pinatubo eruption. Aerosol extinction coefficient (AEC) data indicate the presence of aerosols primarily in the altitude range of 17–25 km, though they have on occasion been measured as high as 32 km. Comparison of the generally low latitude measurements of *Zubov* and *Vize* with those of the midlatitude region of OHP, CEL, and LIMA shows that at low latitudes there is a presence of a strong high altitude cloud at 17 km, whereas at midlatitudes there is the persistent presence of only the lower altitude cloud near and below 20 km with weaker, higher altitude features that pass through sporadically.

As Avdyushin et al. (1993) have pointed out the average (AOT) value for the 8°– 22° N latitude belt was 0.1 and reached values of almost 0.2 on a regular, nearly periodic basis. The maximum AEC was 0.08 km^{-1} at 23–24 km measured by *Zubov* in July. At midlatitudes (40° – 44° N) the first evidence of aerosols appeared in early July at 0.0024 km^{-1} (CEL). The first clear evidence of elevated aerosol concentrations above 18 km appeared at CEL and LIMA on September 21, with extinction values of 0.0025 km^{-1}. High AEC values at this altitude were first detected at all three midlatitude sites during the second week of October when AECs of up to 0.017 km^{-1} were measured at CEL. On average, the AOT at midlatitudes was near 0.05. However, on one occasion (December 8–11) both CEL and OHP measured high AOT values of up to 0.125 with peak AEC of 0.021 km^{-1} at 23 km (CEL). On this same occasion, higher than normal AEC values (up to 0.001 km^{-1} were measured at record high altitudes (near 30 km) by all three midlatitude sites.

The use of 530-nm lidar onboard RV *Professor Zubov* has offered a unique opportunity to monitor spatial and temporal changes of the aerosol cloud, formed after the Pinatubo eruption (Nardi et al., 1993). The analysis of vertical SR profiles shows that during the whole period of observation (July 11– September 21, 1991) strong variability of the volcanic cloud was observed within the latitude belt 8° – 43° N .

These data demonstrate the aerosol cloud to be primarily in the 16–17-km altitude region, with a high degree of variability in the vertical structure and in the longitudinal structure. Two dominant aerosol layers were observed at approximately 20 km and 24 km, with maximum observed SR of 19.5 and 51, respectively. The altitude of these aerosol peaks did not appear to rise appreciably over the duration of the measurements. The variability of the vertical structure is, apparently, due to the advection of zonal structure over a local observation point. The latitudinal dispersion rate of the volcanic aerosol cloud is estimated at ~5° per month. A clear periodic variability was observed in the altitude structure during the entire data set primarily within the altitude region of the layer, from which a mean zonal wind profile was produced by a spectral analysis of aerosol profiles. Estimates of horizontal and vertical diffusion coefficients from the data of cloud smearing effects gave, roughly, $1.0 \cdot 10^5$ m^2/s and $2.9 \cdot 10^{-3}$ m^2/s, respectively.

Ansmann et al. (1993) have conducted volcanic aerosol monitoring with a combined Raman elastic-backscatter lidar that allows determination of both the aerosol extinction and of the extinction-to-backscatter ratio (lidar ratio) vertical profiles. A XeCl eximer laser was used to generate light at the 308-nm wavelength. Signals, elastically backscattered from air molecules and particles and inelastically (Raman) backscattered from nitrogen and water vapor at 332 and 374 nm, were detected. The observations were made at Geesthacht (53.5° N, 105° E), 40 km southeast of Hamburg and continued for about one year (measurement began on August 12, 1991). Each observation lasted 2 hours during which 1.5 million laser shots were transmitted. The particle extinction coefficient is derived from the nitrogen Raman signal profile, while the backscattering coefficient is obtained from the ratio of the Rayleigh and particle elastic backscatter signal to the nitrogen Raman backscatter signal.

As Ansmann et al. (1993) have pointed out, the Pinatubo aerosol layer was located between the tropopause and about 24 km, but traces of stratospheric aerosol were frequently found down to 5 km. The AOT of the stratosphere in 1992 was of the order of 0.15. The ratio of AOT to column integrated backscatter varied between about 15 and 60 sr (most values were found between 20 and 30 sr).

The determined extinction-to-backscatter ratio contains information about the size range of the scattering sulfuric acid droplets. Lidar ratios of 15, 20, 35 and 50 indicate effective radii between 0.75 and 1.40, 0.30 and 0.75, 0.15 and 0.30, and < 0.2 µm, respectively. (These are found from Mie calculations of particle extinction and backscattering coefficients for a large number of different stratospheric bimodal lognormal size distributions). Assessments of column aerosol mass and particle surface area have given values of 0.05 g/m^2 and $2.5 \cdot 10^{11}$ µm^2/m^2, respectively, in the first half of 1992.

Raman lidar measurements over southeastern Kansas of stratospheric aerosol produced by the June 1991 eruption of Pinatubo were made by Ferrare et al. (1992) on 10 nights during November and December 1991. Both aerosol backscattering and extinction profiles were derived simultaneously. Peak aerosol concentrations were located between 19 and 22 km and varied significantly from night to night. Observational data agree with the estimates of particle mode radii between 0.3 and 0.5 µm. Aerosol optical thickness at 351 nm varied between 0.04 and 0.06.

Johnston et al. (1992) started daily observations in December 1980 of the twilight sky brightness at the zenith during sunrises and sunsets in Lauder (New Zealand: 45.5° S, 170° E) to retrieve columnar NO_2 content in the atmosphere (CNC) with the use of brightness data for the wavelength about 450 nm where the strong NO_2 absorption band is located. The analysis of CNC data over 11 years revealed that, apart from regular daily and annual variations,

VOLCANIC ACTIVITY AND CLIMATE 233

there was a long-term variability apparently due to volcanic impact (El Chichón, 1982) and solar activity (11 year cycle). On the whole, CNC values in the middle of the 1980s were less than at the beginning and end of that decade. During the entire observational period except 1991, minimum CNC values in the annual course took place in June. However in July–August 1991, a gradual decrease of mean monthly "morning" CNC values was registered that was not observed earlier. Although this kind of trend was not observed in the case of CNC "evening" values, by October 1991 average values (taking into account both morning and evening values) reached only 75% in comparison with minimum CNC values of Octobers previously observed.

Since a part of the maximum CNC drop in October—34%–45% (sunrises) and 30%–40% (sunsets) in comparison with 1981, 1989 and 1990 data with similar levels of solar activity—could be due to the contribution of added signal scattering by volcanic aerosols, it is reasonable to believe that the actual CNC drop exceeded 20% and the cause of this drop could be the impact of the heterogeneous chemical reactions with the participation of aerosols. As it is known, the reaction

$$N_2O_5 + H_2O \rightleftharpoons 2HNO_3$$

on surfaces of aerosol particles must lead to the decrease of NO_2 and O_3 contents, but to the growth of HNO_3 content. Since there was no decrease of ozone content after the Pinatubo eruption, it is necessary to find out whether there was a growth of HNO_3 content or not.

3.3.6 Climatic Impact

The climatic implications of volcanic eruptions have recently been of great concern. As we have seen, an extensive set of observations of volcanic effluents and their propagation in the atmosphere has been carried out by conventional means (for example, ground-based, aircraft, balloon) as well as satellites. Substantial progress has been reached in the development of the techniques for numerical modeling of climate and large-scale transport (Kondratyev and Moskalenko, 1984, Kolomeev et al, 1993).

Of principal interest are attempts to reveal a volcanic signal in observed climatic changes. The difficulty of the solution of this problem is that the interannual variability of SAT is determined by various processes that can be classified as connected with the external forcings, internal regular oscillations and random processes, appearing, largely, due to the dynamic instability of the atmosphere. The most probable external factors are large-scale volcanic eruptions and increased CO_2 concentration in the atmosphere (to date, there is not sufficient evidence for the effect of solar activity on climate). Regular changes in the AGC in different parts of the world have been related to ENSO. A considerable contribution belongs to atmospheric internal variability. With the varying intensity and location of cyclones and jet streams a random variability of heat fluxes occurs in the atmosphere, leading to a random variability in air temperature.

The continuing numerical modeling to detect a VS from SAT observations has not, so far, given definite results: information about the level and statistical significance, the geographical distribution and duration of VS remains contradictory, which is partly explained by inadequate data on both explosive eruptions and temperature.

In this connection, Bradley (1988) undertook an analysis of data of some catalogs of eruptions to assess the reliability of their information content, as well as of a new homogeneous data series on SAT for the NH conditions. The analysis took into consideration that powerful explosive eruptions (to which a volcanic explosivity index, VEI, of about 5–6 corresponds) simultaneously affect the SAT, with a maximum response 20.3 months after the eruption and subsequent gradual attenuation during 2–3 years. The volcanic signal ΔT is strongest in the summer and in the fall, but is almost zero in the winter. On the whole, the ΔT decrease over the NH continents reaches about 0.4 °C. Apparently, there are also secondary maxima of the SAT anomaly a year after the primary maximum, and the third, weaker, maximum one year later.

The latitude at which an eruption took place is an important factor that determines the VS. If the high-latitude eruption signals are the strongest in high and midlatitude eruptions, then the effect of low-latitude eruptions is confined, largely, to the tropical latitudes and midlatitudes. Thus it turns out that the midlatitudes are much more subject to the effect of both low- and high-latitude explosive eruptions. The eruptions with VEI ~ 4 somehow affect the SAT field only during the first post-eruptive months.

Depending on the season, the temperature decrease varies between 0.05 °C and 0.1 °C. The effect of large explosive eruptions on the SAT decreases over the land, for the last 100 years (except for the strongest five eruptions), and has been characterized by the fast formation and short lifetime of VS.

For the substantial manifestation of VS in simultaneous variability of temperature, the explosive eruptions of the intensity considered should repeat regularly almost every 3 years but during the last 100 years there was no such repeatability. The conclusion can be drawn about the unjustified assumption (which has been made sometimes) with respect to the leading role of eruptions in the formation of the temperature change through the last century.

Groisman (1992) has conducted an analysis of observational data on regional scale SAT changes after the Pinatubo eruption in comparison with similar data from previous eruptions. The principal conclusion of the analysis is that on a regional scale, even the sign of temperature change can be different, as has been observed after volcanic eruptions in European Russia in winter during the last 200 years.

As is well known, the most distinct climatic impact of volcanic eruptions is stratospheric warming (Angell, 1993; Kondratyev, 1988, 1989, 1993; Kondratyev and Cracknell, 1996). In this connection Labitzke and McCormick (1992) (see also McCormick, 1992) have undertaken a comparison of aerological data on stratospheric temperatures in the NH for 30 and 50 hPa levels, beginning from June 1991, with monthly mean temperatures for 20 years (1965–1984) and 26 years (1964–1989). This comparison has led to the conclusion about significant temperature rises in July, August, September and October 1991 within the 0° – 30° N latitude belt.

In September and October increases in temperature at the 30 hPa level took place, which reached (in a number of points) 3.5 °C at the 2–3σ level of statistical significance (in June the temperature was below and in July close to the mean values). Mean monthly zonal temperatures at the 30 hPa level near 20° N in September and October were about 2.5 °C higher than mean values for 26 years in the presence of the rise of mean daily values of about 3 °C. An even more significant stratospheric temperature rise was observed to the south of 20° N. It is assumed that all these changes were due to the enhancement of solar radiation absorption in the stratosphere by volcanic aerosols, (although the role of upwelling longwave radiation absorption cannot be ignored). It is to be expected also that simultaneous stratospheric warming took place in the SH, especially in the 0° – 20° S latitude belt. Satellite and lidar sounding data on volcanic aerosol spatial distribution point to the maximum stratospheric warming at 23- to 26-km altitudes in the 35° N – 20° S latitude zone.

Angell (1993) has compared the low stratospheric warming (100–50-hPa or 16- to 20-km layer) following Agung (1963, 8° S), El Chichón (1982, 17° N) and Pinatubo (1991, 15° N) eruptions on the basis of a 63-station radiosonde network and using nine-season-average temperatures to minimize the influence of the QBO. Some of the earlier studies have used the peak temperature following a volcanic eruption as a measure of stratospheric warming, but as Angell (1993) has noted, the peak temperature is sensitive to the phase of the QBO in which the eruption occurs. For instance, the eruption of both Agung and El Chichón occurred in the easterly wind phase of the QBO in the low tropical stratosphere so that volcanic warming was augmented by warming associated with the QBO. However, the Pinatubo eruption occurred in the transition from the west to the east wind phase of the QBO so that, initially, volcanic warming was diminished by cooling associated with the QBO. At least in the tropics, comparison of the peak temperatures following the eruptions is unlikely to provide an assessment of the relative impact of these eruptions on stratospheric temperature.

The mean period of oscillations of the QBO is about nine seasons (varying from 7 to 11 seasons) and the typical "volcano interval" (the duration of the warming due to major volcanic eruptions) is also about nine seasons (for Pinatubo data for only five seasons were available, which determines the preliminary nature of relevant conclusions), Thus, Angell (1993) defined the "volcanic warming signal" as the average of four temperature differences between temperatures for the volcanic interval and for two nine-season intervals preceding the following volcanic interval.

The results of data processing show that the lower stratospheric (16–20 km) warming following the Agung and El Chichón eruptions was largest in the equatorial zone and smallest in the polar zones. The warming was not, however, symmetrical with respect to the equator, being larger south than north of the equator. The warming following El Chichón was slightly larger than following Agung in all climatic zones except the south polar zone. Pinatubo data demonstrate that in the north extratropics and tropics the warming following this eruption was comparable to the warming following Agung and El Chichón, but in SH temperatures and polar zones the warming is considerably greater following Pinatubo (may be due to a contribution from the eruption of Cerro Hudson in Chile, 46° S). Globally, therefore, the low stratospheric warming following Pinatubo is greater than following Agung or El Chichón. In all three cases the warming in the tropics was at its maximum at 20-km altitude. The altitude dependencies were such that (based on a 10-station tropical radiosonde network with data to greater heights) the warming following El Chichón exceeded the warming following Agung to a height of at least 31 km, the difference in warming increasing with height. The warming following Pinatubo is similar to that following Agung to a height of 24 km, but beyond that point it became less (an unexpected result to be reexamined as more data become available). There is no evidence of warming in the high-tropospheric 9–16 km layer following the three eruptions as well as (due to the rocketsonde data) no evidence of warming in the 36–45 km layer following El Chichón.

Kinne et al. (1992) have pointed out that stratospheric aerosol from Pinatubo heated the tropical lower stratosphere by about 0.3 K/day mainly due to absorption of upwelling thermal emission. An important consequence of this heating was upward motion and subsequent ozone losses by 10% to 30%. Such a situation suggests the presence of an automatic buffering link between heating rates, temperature and ozone concentration in the tropical lower stratosphere.

On the basis of a 3-D global climate model developed at the Goddard Institute for Space Studies (GISS) Hansen et al. (1992) made preliminary estimates of the climatic impacts of volcanic eruptions. To accomplish simulations three scenarios of volcanically induced stratospheric AOT have been prescribed:

(i) E1: Aerosol properties correspond to conditions of the El Chichón eruption (aerosol consisting of sulfuric acid droplets; AOT decreases from 1 immediately after the eruption to zero in 6 months; volcanic aerosols initially homogeneously distributed within the 0°–30° N latitude belt, then propagate over the globe with AOTs twice as high in the 30°–90° N zone as in the 30° N – 90° S zone.
(ii) 2xE1 doubled AOT is prescribed in comparison with E1;
(iii) P: the same temporal AOT variation as in the E1 and 2xE1 models (exponential decrease during 10 months after the eruption with the time constant equal to 1 year), but AOT is 1.7 times more than for E1, as well as more showing a realistic global distribution than earlier (hemispherically homogeneous aerosol distribution is reached by January 1992).

Numerical modeling results show that in the case of scenario P mean global ERB changes due to the Pinatubo eruption could reach about 4 W/m^2 by the beginning of 1992. It means that volcanically induced climate cooling could exceed greenhouse warming due to accumulation of all greenhouse gases since the beginning of the industrial revolution. The decrease of mean global SAT by the end of 1992 could reach 0.5 °C, being three times more than the MSD for the mean global SAT.

Calculations of SAT changes for scenarios E1 and 2xE1 have demonstrated that in these cases volcanically induced climate cooling was too low to change the 1991 climatological conditions—one of the warmest years in the current century—because of gigantic thermal inertia of the climatic systems and low level of initial forcing. However, in the case of the 2xE1 scenario (as in the case of scenario P) the SAT decrease reached 0.5 °C and if this estimate is realistic the predicted climate cooling has to be much larger than the ~0.2 °C warming, typical for ENSO impact. Although assessments of future regional-scale climate changes are not reliable, they do show probabilities. For instance, the probability, ignoring volcanic impact, of unusually cold winters in Moscow may be shown to decrease from 33% in 1950s to 15%–20% at the present, but, taking into account volcanic forcing, the probability may be shown to increase to 30%–50% during the 1992–1994 time period. As far as the level of confidence in numerical modeling results is concerned it is limited by uncertainties of the GISS model, which does not take into account, for example, interaction with deep layers of the ocean and thermal inertia of the climatic system. Much more adequate should be the parameterization of cloudiness-radiation interaction. Volcanic climatic impact will be discussed in more detail in chapter 5.

3.3.7 Impact on the Ozone Layer

Since the discovery of the Antarctic TOC spring minimum ("ozone hole") the problem of global ozone dynamics has became a key issue of environmental studies (Kondratyev, 1989; World Meteorological Organization, 1991), especially in the light of serious TOC depletions observed in highly populated midlatitude regions of the NH (Stolarski et al., 1992, Zerefos et al., 1992). It is now well known that TOC in the NH midlatitudes and polar latitudes has been declining in the past two decades, i.e., after the chlorine loading of the stratosphere exceeded certain levels (Bojkov et al., 1990; Stolarski, 1993; Stolarski et al., 1992; Zerefos et al., 1992). During the winter-spring seasons (December-January-February-March: DJFM) the long-term trend for the 1970s and 1980s was 3.1% ± 0.4% per decade over North America and Europe and slightly less over Asia and the Far East (Bojkov et al., 1990).

If, however, one extends the period to include the last three winter-spring seasons the cumulative ozone decline from December 1969 through March 1993 will be close to -14%, for North America, Europe and Siberia between 45° N and 65° N. Bojkov et al. (1990) have analyzed TOC anomalies for these continental-scale regions with 10, 29 and 11 ozone-measuring stations, respectively. Table 3.3.4 shows some of the results.

Table 3.3.4. Mean monthly and seasonal total ozone deviations from the long-term averages (in %). After Bojkov et al. (1990). [© by American Geophysical Union]

	Dec.	Jan.	Feb.	Mar.	Season	Year
N.America	-7.2	-9.5	-9.1*	-9.9*	-8.2*	1991/92
	-4.9	-15.3*	-15.0*	-11.1*	-12.1*	1992/93
Europe	-5.0	-16.4*	-9.8	-8.8	-10.3*	1991/92
	-5.4	-11.5	-16.7*	-13.9*	-12.8*	1992/93
Siberia	-7.0	-8.9	-13.3*	-16.6*	-12.1*	1991/92
	-8.3	-13.6	-10.9*	-19.4*	-14.1*	1992/93

*Denotes departures of significance at 95% level.

The total ozone deficiencies for the entire DJFM season over all the above mentioned regions were about 11% and 13% below the long-term trend normal during the two consecutive years (1991–92 and 1992–93, respectively).

These helped to reduce the cumulative ozone decline since the winter-spring of 1969–70 to be about 14% in the latitude belt of 45° N – 65° N. Frequencies of days with ozone values deviating below the long-term mean by more than 2 has been 10 times higher than the 35-year average. Ozone drops for 5-day intervals sometime exceeded 20% and in a number of cases reached 30%.

As far as causes of TOC decline are concerned they may be different, but primarily connected with varying synoptic situations and photochemical processes. Bojkov et al. (1990) have pointed out that the transport processes (transport of ozone-poor airmasses forced in addition by vertical motions) have played an important role, along with possible chemical destruction (when cold air from the Arctic, known to have excess of ClO content, has moved over the sunlighted latitudes) in the observed record-low total ozone values during the 1991–92 and 1992–93 seasons. Bojkov et al. (1990) have underlined, however, that the ozone deficiencies mentioned do not have similar rates of decline and did not reach even close to the extreme low values regularly observed during the Antarctic spring ozone hole problem.

McCormick and Veiga (1992) have analyzed SAGE I/II ozone data from the period 1973–1991 to derive global trends in both stratospheric column and as a function of latitude. The linear trend in TOC above 12-km altitude averaged between 65° S and 65° N is 0.30% ± 0.19% per year or -3.6% over the time period February 1979 through April 1991. The column trend above 12 km is nearly zero in the tropics and increases towards the high latitudes with values of -0.6% per year at 60° S and -0.35% per year at 60° N. The TOC losses occurr mainly below 25 km (mostly in the 17–20-km layer). Negative trend values of the order of -2% per year are found at 12 km in the midlatitudes.

REFERENCES

Ackerman, M., and Lippens, C., 1984: Material from the El Chichón volcano above Spain on 3 May 1982 — one month after the eruption. *Planet. and Space Sci.* **32** (1), 17-25.
Angell, J.K., 1988: Impact of El Niño on the delineation of tropospheric cooling due to volcanic eruptions. *J. Geophys. Res.* **93** (D4), 3697-3704.
Angell, J.K., 1993: Comparison of stratospheric warming following Agung, El Chichón and Pinatubo volcanic eruptions. *Geophys. Res. Lett.* **20** (8), 715-718.
Ansmann, A., Wandinger, U., and Weitkamp, C., 1993: One-year observations of Mount Pinatubo aerosol with an advanced Raman lidar over Germany at 53.5°N. *Geophys. Res. Lett.* **20** (8), 711-714.
Avdyushin, S.I., Tulinov, G.F., Ivanov, M.S., Kuzmenko, B.N., Mezhnev, I.R., Manchecorne, A., and Chanin, M.-L., 1993: 1. Spatial and temporal

evolution of the optical thickness of the Pinatubo aerosol cloud in the Northern Hemisphere from a network of shipborne and stationary lidars. *Geophys. Res. Lett.* **20** (18), 1963-1966.
Bradley, R.S., 1988: The explosive volcanic eruption signal in Northern Hemisphere continental temperature record. *Climatic Change* **12** (3), 221-243.
Barth, C.A., Sanders, R.W., Thomas, R.J., Thomas, G.E., Jakosky, B.M., and West, R.A., 1983: Formation of the El Chichón aerosol cloud. *Geophys. Res. Lett.* **10** (N11), 993-996.
Bluth, G.J.S., Doiron, S.D., Schmetzler, C.C., et al., 1992: Global tracking of SO_2 clouds from the June 1991 Mount Pinatubo eruptions. *Geophys. Res. Lett.* **19** (2), 151-154.
Bojkov, R.D., 1987: The 1983 and 1985 anomalies in the ozone distribution in perspective. *Mon. Wea. Rev.* **115**, 2187-2201.
Bojkov, R.D., Bishop, L., Hill, W.J., Reinsel, G.C., and Tiao, G.C., 1990: A statistical trend analysis of revised total ozone data over the Northern Hemisphere. *J. Geophys.* **95**, 9785-9807.
Boville, B.A., Holton, J.R., and Mote, Ph.W., 1991: Simulation of the Pinatubo aerosol cloud in general circulation model. *Geophys. Res. Lett.* **18** (12), 2281-2284.
Cadle, R.D., 1980: Some effects of the emissions of explosive volcanoes on the stratosphere. *J. Geophys. Res.* **C85** (8), 4495-4498.
Cadle, R.D., and Heidt, I.E., 1982: A comparison of constituents of Mount St. Helens eruption clouds with those of some other volcanoes. In Proc. Symp. *Atmospheric Effects and Potential Climate Impact of the 1980 Eruptions of Mount St. Helens,* A. Deepak (Ed.), NASA Conf. Publ. 2240, 155-159.
Cheng, R.J., 1982: The mechanisms of the particles generation and electrification during Mount St. Helens volcanic eruption. In Proc. Symp. *Atmospheric Effects and Potential Climatic Impact of the 1980 Eruptions of Mount St. Helens*, A. Deepak (Ed.), NASA Conf. Publ. 2240, 211-218.
Chung, V.S., Gallant, A., Fanaki, F., and Millan, M., 1981: On the observations of Mount St. Helens volcanic emissions. *Atmosphere-Ocean* **19** (2), 172-178.
Danielsen, E.F., 1982: Mount St. Helens plume dispersion based on trajectory analyses. In Proc. Symp. *Atmospheric Effects and Potential Climatic Impact of the 1980 Eruptions of Mount St. Helens,* A. Deepak (Ed.), NASA Conf. Publ. 2240, 141-154.
Davis, B.J., Johnson, I.R., Griffin, D.T., Phillips, W.R., Stevens, R.K., and Maughan, D., 1981: Quantitative analysis of Mt. St. Helens ash by X-ray diffraction and X-ray fluoresence and spectrometry. *J. Appl. Meteorol.* **20** (8), 922-933.

Deepak, A. (Ed.), 1982: *Atmospheric Effects and Potential Climatic Impact of the 1980 Eruptions of Mount St. Helens,* Proc. Symp, NASA Conf. Publ. 2240, 303 pp.
DeFoor, T.E., Robinson, E., and Ryan, S., 1992: Early lidar observations of the June 1991 Pinatubo eruption plume at Mauna Loa observatory, Hawaii. *Geophys. Res. Lett.* **19** (2), 187-190.
Deshler, T., Hofmann, D.J., Johnson, B.J., and Rozier, W.R., 1992: Balloon-borne measurements of the Pinatubo aerosol size distribution and volatility at Laramie, Wyoming during the summer of 1991. *Geophys. Res. Lett.* **19** (2), 199-202.
Deshler, T., Johnson, B.J., and Rozier, W.R., 1993: Balloon-borne measurements of Pinatubo aerosol during 1991 and 1992 at 41° N: Vertical profiles, size distribution, and volatility. *Geophys. Res. Lett.* **20** (14), 1435-1438.
Dutton, E.G., 1990: Comments on major volcanic eruptions and climate: A critical evaluation. *J. Climate* **3**, 587-588.
Dutton, E.G., and DeLuisi, J., 1983a: Spectral extinction of direct solar radiation by El Chichón cloud during December 1982. *Geophys. Res. Lett.* **10** (N11), 1013-1016.
Dutton, E.G., and DeLuisi, J., 1983b: Optical thickness features of the El Chichón stratospheric debris cloud. *5th Conf. on Atmospheric Radiation, Oct. 31-Nov. 4 1983, Baltimore, MD.,* Boston: American Meteorological Society, 361-363.
Dutton, E.G., and DeLuisi, J., 1987: Aerosol optical depth and ratios of diffuse-sky to total solar irradiance measured from aircraft following the eruption of El Chichón. *Rep. ARL-12, NOAA Environ.Res.Lab., Boulder, Colorado.*
Dutton, E.G., and Christy, J.R., 1992: Solar radiative forcing at selected locations and evidence for global lower tropospheric cooling following the eruptions of El Chichon and Pinatubo. *Geophys. Res. Lett.* **19** (23), 2211-2214.
Farlow, N.H., Snetsinger, K.G., Oberbeck, V.R., Ferry, G.V., Polkowski, G., and Hayes, D.M., 1982: Time variations of aerosols in the stratosphere following Mount St. Helens eruptions. In Proc. Symp *Atmospheric Effects and Potential Climatic Impact of the 1980 Eruptions of Mount St. Helens,* A. Deepak (Ed.), NASA Conf. Publ. 2240, 55-64.
Ferrare, R.A., Melfi, S.K., Whiteman, D.N., and Evans, K.D., 1992: Raman lidar measurements of Pinatubo aerosols over Southeastern Kansas during November-December 1991. *Geophys. Res. Lett.* **19** (15), 1599-1602.
Franceschini, M., Pitari, G., and Visconti, G., 1991: A study of the global distribution of sulphate aerosols with a 2D model including microphysics. *Il Nuovo Cimento* **14C** (4), 401-416.

Gadian, A.M., and Davies, D.R., 1981: A mathematical climate study of the effects of dust from the Mount St. Helens volcano eruption. *Weather* **36** (5), 136-138.

Galindo, I., 1992: Extinction of short wave solar radiation due to El Chichón stratospheric aerosol. *Atmosfera* **5**, 259-268.

Galindo, I., Kondratyev, K. Ya., and Zenteno, G., 1996: Determination of the atmospheric optical depth due to the El Chichón stratospheric aerosol cloud in the polluted atmosphere of Mexico City. *Atmósfera* **9**, 23-32.

Gandrud, B.W., Kritz, M.A., and Lazrus, A.L., 1983: Balloon and aircraft measurements of stratospheric sulfate mixing ratio following the El Chichon eruption. *Geophys. Res. Lett.* **10** (N11), 1037-1040.

Gardner, C.S., Sechrist, C.F. Jr., and Shelton, J.D., 1980: Lidar observations of the Mount St. Helens dust layers over Urbana, Ill. *Appl. Optics.* **19** (18), 292-303.

Gobbi, G.P., Congeduti, F., and Adriani, A., 1992: Early stratospheric effects of the Pinatubo eruption. *Geophys. Res. Lett.* **19** (10), 1997-1000.

Goldman, A., Murcray, F.J., Rinsland, C.P., et al., 1992: Mt. Pinatubo SO_2 column measurements from Mauna Loa. *Geophys. Res. Lett.* **19** (2), 195-196.

Groisman, P.Ya., 1992: Possible climate regional consequences of the Pinatubo eruption: An empirical approach. *Geophys. Res. Lett.* **19** (15), 1605-1605.

Hamill, P., Turco, R.P., and Toon, O.B., 1988: On the growth of nitric and sulfuric acid aerosol particles under stratospheric conditions. *J. Atmos. Chem.* **7**, 287-315.

Hansen, J.E., Lacis, A., Ruedy, R., and Sato, M., 1992: Potential climate impact of Mount Pinatubo aerosols. *Geophys. Res. Lett.* **19** (2), 215-218.

Hay, J.E., and Darby, R., 1984: El Chichón — Influence on aerosol optical depth and direct, diffuse and total solar irradiances at Vancouver, B.C., *Atmosphere-Ocean* **22** (3), 354-368.

Hirose, K., Dokiya, Y., Suzuki, Y., and Sigimura, Y., 1982: Global circulation of volcanic debris of Mt. St. Helens: Evidence from the changes of chemical constituents in the surface air. *J. Meteorol. Soc. Jap.* **60** (5), 1194-1202.

Hobbs, P.V., Hegg, D.A., and Radke, L.F., 1983: Resuspension of volcanic ash from Mount St. Helens. *J. Geophys. Res.* **C88** (6), 3919-3921.

Hoff, R.M., 1992: Differential SO_2 column measurements of the Mt. Pinatubo volcanic plume. *Geophys. Res. Lett.* **19** (2), 175-178.

Hoffer, J.M., Gomez, F., and Muela, P., 1982: Eruption of El Chichón volcano, Chiapas, México, 28 March to 7 April 1982. *Science.* **219** N4583, 1307-1308.

Hofmann, D.J., 1987: Perturbations to the global atmosphere associated with the El Chichón volcanic eruption of 1982. *Revs. of Geophys.* **25** (4), 743-759.

Hofmann, D.J., 1988: Aerosols from past and present volcanic eruptions. In *Aerosols and Climate*, P.V. Hobbs and M.P. McCormick (Eds.), Hampton, Virginia., A. Deepak Publ., pp. 195-214.
Hofmann, D.J., and Rosen, J.M., 1981a: Stratospheric aerosol and condensation nuclei enhancement following the eruption of Alaid in April 1981. *Geophys. Res. Lett.* **8** (12), 1231-1234.
Hofmann, D.J., and Rosen, J.M., 1981b: On the background stratosphere aerosol. *J. Atmos. Sci.* **38**, pp. 168-181.
Hofmann, D.J., and Rosen, J.M., 1982: Balloon-borne observations of stratospheric aerosol and condensation nuclei during the year following the Mt. St. Helens eruption. *J. Geophys. Res.* **87**, 11039-11061.
Hofmann, D.J., and Rosen, J.M., 1983a: Measurement of the sulfuric acid weight percent in the stratospheric aerosol from the El Chichón eruption. *Pap 18th. Gen. Assembly, IUGG, Hamburg, Aug. 1983*, 1-12.
Hofmann, D.J., and Rosen, J.M., 1983b: Balloon-borne particle counter observations of the El Chichón aerosol layers in the 0.01-1.8 μm radius range. *Dept. of Phys. and Astronom., Univ. of Wyoming, Rept. N AP-77, December 27, 11.*
Hofmann, D.J., and Rosen, J.M., 1983c: Stratospheric sulfuric acid fraction and mass estimate from the 1982 volcanic eruption of El Chichón. *Geophys. Res. Lett.* **10** (4), 313-316.
Hofmann, D.J., and Rosen, J.M., 1983d: Sulfuric acid droplet formation and growth in the stratosphere after the 1982 eruption of El Chichón. *Science* **222** (4621), 325-326.
Hofmann, D.J., and Rosen, J.M., 1984: On the temporal variation of stratospheric aerosol size and mass during the first 18 months following the 1982 eruptions of El Chichón. *J. Geophys. Res.* **D89** (3), 4883-4890.
Hofmann, D.J., and Rosen, J.M., 1987: On the prolonged life time of the El Chichón sulfuric acid aerosol cloud. *J. Geophys. Res.* **92**, 9825-9830.
Hofmann, D.J., Rosen, J.M., Reiter, R., and Jäger, H., 1983: Lidar- and balloon-borne particle counter comparisons following recent volcanic eruptions. *J. Geophys. Res.* **C88** (6), 3777-3782.
Inn, E.C.Y., Vedder, J.P., Concon, E.P., and O'Hara, D., 1982: Precursor gases of aerosols in the Mount St. Helens eruption plumes at stratospheric altitudes. In *Proc. Symp. Atmospheric Effects and Potential Climatic Impact of the 1980 Eruptions of Mount St. Helens*, A. Deepak (Ed.), NASA Conf. Publ. 2240, 109-116.
Jäger, H., 1992: The Pinatubo eruption cloud observed by lidar in Garmisch-Partenkirchen. *Geophys. Res. Lett.* **19** (2), 191-194.
Johnston, P.V., McKenzie, R.I., Reys, J.G., and Matthews, W.A., 1992: Observations of depleted stratospheric NO_2 following the Pinatubo volcanic eruption. *Geophys. Res. Lett.* **19** (2), 211-214.

Kent, G.S., 1982: SAGE measurements of Mount St. Helens volcanic aerosols. In Proc. Symp. *Atmospheric Effects and Potential Climatic Impact of the 1980 Eruptions of Mount St. Helens,* A. Deepak (Ed.), NASA Conf. Publ. 2240, 109-116.

Kent, G.S., and Yue, G.K., 1991: The modeling of CO_2 lidar backscatter from stratospheric aerosol. *J. Geophys. Res.* **96** (D3), 5279-5292.

Kerr, J.B., Evans, W.F.J., and Nateer, C.I., 1982: Measurements of SO_2 in the Mount St. Helens debris. In Proc. Symp. *Atmospheric Effects and Potential Climatic Impact of the 1980 Eruptions of Mount St. Helens,* A. Deepak (Ed.), NASA Conf. Publ. 2240, 219-224.

Kinne, S., Toon, O., and Prather, M., 1992: Buffering of stratospheric circulation by changing amounts of tropical ozone: A Pinatubo case study. *Geophys. Res. Lett.* **19**, 1927-1930.

Kolomeev, M.P., Nikonov, S.A., Sorokovikova, O.S., and Khmelevtsov, S.S., 1993: Modelling of the Northern Hemisphere climate response to the Pinatubo volcanic eruption. *Meteorol. and Hydrol.* **4**, 15-19 (in Russian).

Kondratyev, K. Ya., 1988: *Climate Shocks: Natural and Anthropogenic.* Wiley, New York e.a., 296 pp.

Kondratyev, K. Ya., 1989: *Global Ozone Dynamics.* VINITI, Moscow, 212 pp. (in Russian).

Kondratyev, K.Ya., 1992: Numerical modelling of global climate change. *Izv. Russian Geographical Sre.* **124**, N.3, p. 232-240 (in Russian).

Kondratyev, K. Ya., 1993: Complex monitoring of the Pinatubo volcanic eruptions. *Studying the Earth from Space.* **1**, 111-122 (in Russian).

Kondratyev, K. Ya., and Moskalenko, N.I., 1984: Radiative heat exchange in the atmosphere disturbed by volcanic eruptions. *Dokl. Akad. Nauk SSSR* **274** (4), 799-801 (in Russian).

Kondratyev, K. Ya., and Cracknell, A.P., 1996: *Observing Global Climate Change.* Taylor and Francis, London e.a. (in press).

Kondratyev, K. Ya., Moskalenko, N.I., and Pozdnyakov, D.V., 1983: Atmospheric Aerosol. Leningrad, *Gidrometeo.* 224 pp.

Kondratyev, K. Ya., Grigoryev, A.A., Pokrovsky, O.M., Shalina, E., 1983a: Remote sensing of Aerosols from Satellites. Leningrad, *Gidrometeo.* 216 pp.

Kotra, J. P., Finnegan, D.J., Zuller, W.H., Hart, M.A., and Moyers, 1983: El Chichon: Composition of plume gases and particles. *Science* **222,** N4627, 1018-1021.

Kuhn, P.M., Haughney, L.C., and Innis, R.C., 1981: Long-wave stratospheric transmission of Mount St. Helens ejects. *Opt. Lett.* **6** (1), 24-26.

Labitzke, K., and McCormick, M.P., 1992: Stratospheric temperature increases due to Pinatubo aerosols. *Gephys. Res. Lett.* **19**, 207-210.

Laulainen, N. S., 1982: Ash loading and insolation at Hanford, Washington, during and after the eruption of Mt. St. Helens. In Proc. Symp. *Atmospheric Effects and Potential Climatic Impact of the 1980 Eruptions of Mount St. Helens*, A. Deepak (Ed.), NASA Conf. Publ. 2240, 225-240.

Laver, J.D., 1982: Distribution of Mount St. Helens dust inferred from satellites and meteorological data. In Proc. Symp. *Atmospheric Effects and Potential Climatic Impact of the 1980 Eruptions of Mount St. Helens*, A. Deepak (Ed.), NASA Conf. Publ. 2240, 131-140.

Lerfald, G., 1982: Mount St. Helens dust veil observed at Boulder, Colorado, by optical techniques. In Proc. Symp. *Atmospheric Effects and Potential Climatic Impact of the 1980 Eruptions of Mount St. Helens*, A. Deepak (Ed.), NASA Conf. Publ. 2240, 241-250.

Lerzberg, E.A., Otterson, D.A., Roberts, W.K., and Papathakos, L.C., 1982: Aircraft sampling of the sulfate layer near the tropopause following the eruption of Mount St. Helens. In Proc. Symp. *Atmospheric Effects and Potential Climatic Impact of the 1980 Eruptions of Mount St. Helens*, A. Deepak (Ed.), NASA Conf. Publ. 2240, 251-260.

Lockwood, G.W., 1982: Spectrally resolved measurements of the El Chichon cloud from Flagstaff, Arizona. *EOS* **63,** 897.

Mankin, W.G., Coffey, M.T., and Goldman, A., 1992: Airborne observations of SO_2, HCl and O_3 in the stratospheric plume of the Pinatubo volcano in July 1991. *Geophys. Res. Lett.* **19** (2), 155-158.

Mass, C.F., and Portman, D.A., 1989: Major volcanic eruptions and climate: A critical evaluation. *J. Climate* **2**, 566-573.

Mathews, L.A., Roquemore, G.R., Amand, P.St., and Gibson, J.P., 1982: An incursion of dust in the southwestern United States from April 1980 eruptions of Mount St. Helens. In Proc. Symp. *Atmospheric Effects and Potential Climatic Impact of the 1980 Eruptions of Mount St. Helens*, A. Deepak (Ed.), NASA Conf. Publ. 2240, 261-268.

McCormick, M.P., 1981: Monitoring Mount St. Helens. *Nature* **290** (5802), 88-89.

McCormick, M.P., 1982: Ground-based and aircraft measurements of Mount St. Helens stratospheric effluents. In Proc. Symp. *Atmospheric Effects and Potential Climatic Impact of the 1980 Eruptions of Mount St. Helens*, A. Deepak (Ed.), NASA Conf. Publ. 2240, 125-130.

McCormick, M.P., 1983: Global distribution of stratospheric aerosols by satellite measurements. *AIAA Journal* **21** (4), 633-635.

McCormick, M.P., 1992: Initial assessment of the stratospheric and climatic impact of the 1991 Mount Pinatubo eruption: Prologue. *Geophys. Res. Lett.* **19** (2), 149-150.

McCormick, M.P., and Veiga, R.E., 1992: SAGE II measurements of early Pinatubo aerosols. *Geophys. Res. Lett.* **19** (2), 155-158.

McPeters, R.D., 1993: The atmospheric SO_2 budget for Pinatubo derived from NOAA-11 SBUV/2 spectral data. *Geophys. Res. Lett.* **20** (18), 1971-1974.
Michalsky, J.J., Kleckner, E.W., and Stokes, G.M., 1982: Mount St. Helens related aerosol properties from solar extinction measurements. In Proc. Symp. *Atmospheric Effects and Potential Climatic Impact of the 1980 Eruptions of Mount St. Helens,* A. Deepak (Ed.), NASA Conf. Publ. 2240, 269-274.
Michelangeli, O.V., Allen, M., and Yung, Y.L., 1989: El Chichón volcanic aerosols: Impact of radiative, thermal, and chemical perturbations. *J. Geophys. Res.* **94** (D15), 18429-18443.
Mojena, E., and García, O., 1984: Propagación sobre Cuba de la nube de ceniza de las erupticiones del volcán Chichón, marzo-abril, 1982. *Geofis. Int.* **23** (2), 454-459.
Mouginis-Mark, P., Rowland, S., Francis, P., Friedman, T., Garbeil, M., Gradie, J., Self, S., Wilson, L., Crisp, J., Glaze, L., Jones, K., Kahle, A., Pieri, D., Zebker, H., Krueger, A., Walter, L., Wood, Ch., Rose, W., Adams, J., and Wolff, R., 1991: Analysis of active volcanoes from the Earth Observing System. *Remote Sens. Environ.* **36**, 1-12.
Nardi, B., Chanin, M.-L., Hauchecorne, A., et al., 1993: 2 Morphology and dynamics of the Pinatubo aerosol layer in the Northern Hemisphere as detected from a ship-borne lidar. *Geophys. Res. Lett.* **20** (18), 1967-1970.
Newell, R.E., and Deepak, A. (Eds.), 1982: *Mount St. Helens Eruptions of 1980. Atmospheric Effects and Potential Climatic Impact.* A Workshop Report. NASA, Washington, D.C., 119 pp.
Patterson, E.M., Pollard, C.O., and Galindo, I., 1983: Optical properties of the ash from El Chichón volcano. *Geophys. Res. Lett.* **10** (N4), 317-320.
Pitari, G., 1992: On the possible perturbation of stratospheric dynamics due to Pinatubo aerosols. *Il Nuovo Cimento* **15C** (4), 485-489.
Pollack, J.B., Wittenborn, F.C., O'Brien, K., and Flynn, B., 1991: A determination of the infrared optical depth of the El Chichón volcanic cloud. *J. Geophys. Res.* **96** (D2), 3115-3122.
Post, M.J., Grund, C.J., Langford, A.O., and Profitt, M.H., 1992: Observations of Pinatubo ejecta over Boulder, Colorado by lidars of three different wavelengths. *Geophys. Res. Lett.* **19** (2), 195-198.
Rao, C.R.N., and Takasima, T., 1986: Solar radiation anomalies caused by the El Chichón volcanic cloud: Measurements and model comparisons. *Quart. J. Roy. Met. Soc.* **112**, 1111-1126.
Robock, A., 1983: The dust cloud of the century. *Nature* **301**, 373-374.
Robock, A., and Mao, 1992: Winter warming from large volcanic eruptions. *Geophys. Res. Lett.* **19** (24), 2405-2408.
Rose, W.I., and Hoffman, M.F., 1982: The May 1980 eruption of Mount St. Helens: The nature of the eruption with an atmospheric perspective. In

Proc. Symp. *Atmospheric Effects and Potential Climatic Impact of the 1980 Eruptions of Mount St. Helens,* A. Deepak (Ed.), NASA Conf. Publ. 2240, 1-14.
Rosen, J.M., and Hofmann, D.J., 1982: Dustsonde measurements of the Mount St. Helens volcanic dust cloud over Wyoming. In Proc. Symp. *Atmospheric Effects and Potential Climatic Impact of the 1980 Eruptions of Mount St. Helens,* A. Deepak (Ed.), NASA Conf. Publ. 2240, 65-82.
Rosen, J.M., Kjome, N.T., Fast, H., Khattatov, V.U., and Rudakov, V.V., 1992: Penetration of Mt. Pinatubo aerosols into the North Polar vortex. *Geophys. Res. Lett.* **19** (17), 1751-1754.
Ryzner, E., Weber, M.R., and Hallaron, T.S., 1981: Effects of the Mount St. Helens volcanic cloud on turbidity at Ann Arbor, Michigan. *J. Appl. Meteorol.* **20** (11), 1290-1294.
Sedlacek, W.A., Heiken, G.H., Mroz, E.J., Cladney, E.S., Perrin, D.R., Leifer, R., Fisenne, I., Hinchliffe, L., and Chuan, R.L., 1982: Physical and chemical characteristics of Mount St. Helens airborne debris. In Proc. Symp. *Atmospheric Effects and Potential Climatic Impact of the 1980 Eruptions of Mount St. Helens,* A. Deepak (Ed.), NASA Conf. Publ. 2240, 83-108.
Sheridan, P.J., Schnell, R.C., Hofmann, D.J., and Deshler, T., 1992: Electron microscopic studies of Mt. Pinatubo aerosol layers over Laramie, Wyoming during summer 1991. *Geophys. Res. Lett.* **19** (2), 183-186.
Stolarski, R.S., Bojkov, R.D., Bishop, L., Zerefos, C.S., Stahelin, J., and Zawondy, J.M., 1992: Measured trends in stratospheric ozone. *Science* **256**, 342-349.
Stolarski, R.S., 1993: Ozone loss and northern middle latitudes: Observation and theory. AGU 1993 Ann. Meeting, *EOS Suppl.,* p. 178.
Stowe, L.L., Carey, R.M., and Pellegrino, P.P., 1992: Monitoring the Mt. Pinatubo aerosol layer with NOAA-11 AVHRR data. *Geophys. Res. Lett.* **19** (2), 159-162.
Turco, R.P., Hamill, P., Toon, O.B., Whitten, R.C., and Kiang, C.S., 1979: A one dimensional model describing aerosol formation and evolution in the stratosphere, I: Physical processes and mathematical analogs. *J. Atmos. Ser.* **36** (N4), 699-717.
Turco, R.P., Toon, O.B., Whitten, R.C., Keese, R.C., and Hamill, P., 1982a: Simulation studies of the physical and chemical processes occurring in the stratospheric clouds of the Mount St. Helens eruptions of May and June 1980. In Proc. Symp. on *Atmospheric Effects and Potential Climatic Impact of the 1980 Eruptions of Mt. St. Helens,* NASA Conf. Publication 2240, NASA, 161-190.
Turco, R.P., Whitten, R.C., and Toon, O.B., 1982b: Stratospheric aerosol observations and theory. *Rev. Geophys. Space Phys.* **20** (2), 233-279.

Turco, R.P., Toon, O.B., Whitten, R.C., Hamill, P., and Keese, R.C., 1983: The 1980 eruptions of Mount St. Helens: Physical and chemical processes in the stratospheric clouds. *J. Geophys. Res.* **C88** (9), 5299-5320.

Valero, P.P.J., and Pilewskie, P., 1992: Latitudinal survey of spectral optical depths of the Pinatubo volcanic cloud-derived particles sizes, columnar mass loadings, and effects on planetary albedo. *Geophys. Res. Lett.* **19** (2), 163-166.

Vedder, J.F., Condon, E.P., Inn, E.C.Y., Tabor, K.D., and Kritz, M.A., 1983: Measurements of stratospheric SO_2 after the El Chichon eruption. *Geophys. Res. Lett.* **10** (N11), 1013-1016.

Wendler, G., and Kodama, Y., 1986: Effect of the El Chichón volcanic cloud on the surface radiative regime in Central Alaska. *J. Climatol. and Appl. Meteorol.* **25**, 1687-1694.

Wilson, J.C., Blancksear, E.D., and Hyun, J.H., 1983: Changes in the sub- 2.5 diameter aerosol observed at 20 km altitude after the eruption of El Chichon. *Geophys. Res. Lett.* **10** (N11), 1029-1032.

Winker, D.M., and Osborn, M.T., 1992: Airborne lidar observations of the Pinatubo volcanic plume. *Geophys. Res. Lett.* **19** (2), 167-170.

Wittenborn, F.C., O'Brien, K., Crean, H.W., Pollack, J.B., and Bilskik, K.H., 1983: Spectroscopic measurements of the 8- to 13- micrometer transmission of the upper atmosphere following the El Chichón eruptions. *Geophys. Res. Lett.* **10** (N11), 1009-1012.

World Meteorological Organization, 1991: *Scientific Assessment of Ozone Depletion*: 1991 WMO Global Ozone Research and Monitoring Project. Report No. 25.

Yue, G.K., McCormick, M.P., and Chion, E.W., 1991: Stratospheric aerosol optical depth observed by the Stratospheric Aerosol and Gas Experiment II: Decay of the El Chichón and Ruíz volcanic perturbations. *J. Geophys. Res.* **96** (D3), 5209-5219.

Zerefos, C.S., Bais, A.F., Ziemas, I.C., and Bojkov, R.D., 1992: On the relative importance of the QBO and ENSO in the revised Dobson total ozone records. *J. Geophys. Res.* **97**, 10135-10144.

CHAPTER 4

PROPERTIES OF AEROSOLS — RADIATIVE EFFECTS

The physical properties of aerosols, such as size, shape, refractive index and concentration in the atmosphere, control the aerosol interaction with radiation according to a set of derived properties, which are known as optical properties. Three of them are most frequently referred to because of their fundamental importance. These are the aerosol optical thickness—a measure of the size and number of particles present in a given column of air; the single scattering albedo—tghe fraction of light intercepted and scattered by a single particle; and the asymmetry parameter—an integrated measure denoting the portion of light scattered forward in the direction of the original propagation and the portion scattered backward towards the light source.

In order to determine theoretically the effects of volcanic particles on the radiation budget, several aerosol properties must be known. First, the quantity of aerosols, as a function of size; second, their composition, at least insofar as it is reflected in the bulk optical properties of the material, i.e., refractive index and emissivity. The imaginary part of the refractive index, which characterizes the absorption properties of the particle, is of particular interest. The wavelength dependence of the optical properties must be determined to accurately assess the radiative effects of aerosols. This is obtained from the information about size and bulk optical characteristics. For spheres, this is accomplished by means of Mie theory.

To make reliable estimates of the effect of stratospheric aerosols on the radiation budget of the Earth-atmosphere system, and to provide *a priori* information about aerosols for the interpretation of remote optical sounding, it is necessary to have reliable complex data on the optical properties of aerosols. Various aspects of this important issue are described by Ackerman, 1988; DeLuisi and Herman, 1977; DeLuisi et al., 1975, 1982, 1983; Elterman, 1976; Fegley and Ellis, 1975; Galindo, 1965, 1975, 1978, 1984, 1992; Galindo and Bravo, 1975; Galindo et al., 1996; Gardner et al., 1980; Jennings, 1993;

Kondratyev, 1969, 1976a, 1976b, 1981, 1983, 1991, 1992; Kondratyev and Pozdnyakov, 1980, 1981; Kondratyev et al., 1973, 1981a, 1981b, 1983a, 1983b; Patterson, 1975; Patterson et al., 1983; Pollack et al., 1976a, 1976b).

4.1 *In Situ* Measurements

Grams and Rosen (1978) made a brief review of the information on the concentration and size distribution of aerosols, the shape of particles, and the complex refractive index. They gave the results of calculating the phase functions for m = 1.500 + 005i for different values of the size parameter x = $2\pi r/\lambda$ (r is the particle's radius, λ is the wavelength). Comparison of the phase functions for the monodisperse and polydisperse aerosol illustrates the smoothing out of the angular fine structure of phase functions due to the polydispersiveness. Calculations reveal a very strong decrease of the scattering cross section at angles exceeding 15° with the increasing n_2, the imaginary part of m (i.e., the absorption index). More complete information may be found in Jennings (1993), Kondratyev (1991) and Kondratyev et al. (1983a, 1983b).

Although the effect of the stratospheric aerosol layer on the shortwave radiation transfer depends weakly on variability of the absorption index, the reverse situation is observed for the longwave radiation. The role of the aerosol size distribution (at least in the range 0.1–2 μm), is also very great for its variability determines the cooling or warming. Consideration of the effect of particles' nonsphericity on the phase function is rather important.

Clarke et al. (1983) performed laboratory measurements of the aerosol absorption coefficient at wavelengths 450, 550, 560, and 800 nm using samples of a gas plate covered with silicon rubber, with subsequent application of the integrating plate technique (the absorption coefficient b_a is determined from the transmission of light by a sample as compared to that by a pure plate). Results of measurements for the samples taken by the U-2 aircraft on routes about 1000 km long (to accumulate a sufficient mass of aerosol) before, during, and after the eruptions of Mount St. Helens and El Chichón showed the presence of an absorbing component in aerosols in all cases.

Typical values of b_a at 550 nm vary from 10^{-9} to 10^{-8} m^{-1}, except for cases when the samples were taken in the most dense parts of the eruptive clouds, immediately following the eruptions. Despite a large variability of their absorption coefficient, post-eruption values exceed those for early 1979. Data on the May 5, 1982 flight near the U.S.-Mexican border at a height of 21 km gave 8 μg/m^3 for the aerosol mass concentration, about half the mass falling on particles smaller than 5 μm in diameter.

The scattering coefficient was estimated at $1.2 \cdot 10^{-5}$ m^{-1}, and the single-scattering albedo varied within 0.875 to 0.958 (apparently, ash is characterized by a value greater than 0.988). The imaginary part of the complex refraction index was about 0.0034. Estimates for the period following the eruption of El Chichón have led to the conclusion about a volcanically induced increase of the global albedo (if the single-scattering albedo for the aerosol layer exceeds 0.94), and a decrease of tropospheric temperature (if the single-scattering albedo exceeds 0.98).

Rao and Bradley (1983) discussed observations of the total (0.29 to 2.9 μm) direct solar radiation and global radiation at Corvallis, Oregon, in late November and December 1982, when the stratospheric dust cloud from the eruption of El Chichón reached midlatitudes, manifested itself through the anomalous coloration of skies at dusk on December 30. Data of observations (instantaneous and mean annual values) in clear skies and without local pollution at near-noontime were considered.

The results obtained show a considerable effect of volcanic aerosol on the shortwave radiation field near the surface. For example, the atmospheric transmission coefficient for the global (direct) solar radiation decreased by about 13% (27%). The scattered-to-global radiation flux ratio increased by a factor 2.25. Results of observations of direct solar radiation at several adjacent locations showed its decrease varying between 18% and 25%. Similar transformation of the incoming shortwave radiation was observed after the Agung eruption in 1963.

Shortly after the May 18, 1980 Mount St. Helens eruption, aircraft lidar soundings were made in the eastern United States and the results were compared with data of ground-based lidar soundings. Results considered by McCormick (1982) made it possible to analyze the global distribution of eruptive aerosols. During the first flight (May 21), when an Nd laser was used (1.06 μm), a value of 107 was registered for the backscattering ratio (compared to the Rayleigh backscattering signal) from a 1-km aerosol layer at a 13.6-km height. Similar aerosol layers at altitudes of 12 to 14 km, characterized by large horizontal inhomogeneity, were observed on May 16 and May 27.

Apparently, the composition of the initial eruptive cloud in the lower stratosphere was mainly silicate. The ratios of backscattering signals at two wavelengths (ruby and Nd lasers) and depolarization values for the lower stratospheric aerosol (12 to 18 km) were reduced to background levels 1 month after the eruption. These reductions point to a rapid conversion of particles into spheres with a usual refraction index. At heights above 20 km the aerosol moved slowly westward, whereas the bulk of eruption material moved eastward with the speed and direction of motion, depending on height. The westward-moving aerosol was discovered by lidar stations in Japan and later by European and American stations. The circumglobal motion of the eruptive cloud took 60 days.

Data of aircraft lidar sounding in the United States in early September showed a smoothed spatial distribution of aerosols manifested in the formation of a vertically extended layer (14 to 21 km) with a maximum near 18 to 19 km and a moving layer at altitudes 21 to 22 km. In the case of ruby lidar, maximum values of the backscattering ratio were 1.3 to 1.5. Apparently by September the formation on the backscattering signal had almost completely been determined by sulfuric acid aerosols. Total production of sulfate aerosols due to gas-to-particle conversion was estimated at $5 \cdot 10^5$ tons, which corresponds to a doubled mass of background aerosols and is much below the aerosol production that followed other eruptions: $3 \cdot 10^7$ tons (Agung, 1963) and $5 \cdot 10^7$ tons (Fuego, 1974).

From observations of the outgoing shortwave and longwave radiance (OSR and OLR) using the Nimbus-7 AVHRR, (Schwedfeger et al. (1983) retrieved the spectral albedo of system A (channel 0.58 to 0.68 µm) and total OLR flux (channel 11.5 to 12.5 µm) under clear-sky conditions for February to September 1982 (a total of 44 days) in the Hawaiian region, where ground-based observations of atmospheric transparency were made at Mauna Loa Observatory. Data for the channel 0.58 to 0.68 µm were also used to retrieve the aerosol optical thickness of the atmosphere (τ) at a wavelength of 0.5 µm.

Comparison with results of ground-based observations has shown that errors in retrieving τ do not exceed ±0.05. The linear regression of A values gave the estimate of the derivative $\partial A/\partial \tau = 7.8\%$ (with the correlation coefficient 0.615 and MSD 1.3%), which characterizes the effect of volcanic aerosols on albedo. In the case of the total albedo (obtained from data of channels 0.58 to 0.68 and 0.725 to 1.1 µm) and the use of a more precise technique for retrieval, $\partial A/\partial \tau = 5.7\%$. In the case of OLR the derivative $\partial F/\partial \tau = 5.1$ W/m^2, but the correlation coefficient, constituting only -0.16, points to a prevailing contribution by other factors of the OLR variability. For the ERB, $\partial R/\partial \tau = -9.3$ W/m^2. The results are consistent with calculations by Harshvardhan (1979), taking into account nonequivalent observation conditions and calculations.

DeLuisi et al. (1983) discussed results of lidar sounding (at 694-nm wavelength), as well as observations of the atmospheric spectral transparency and direct solar and global radiation carried out at Mauna Loa Observatory in 1982. These results showed the effect on the atmospheric radiative regime of stratospheric aerosol of unknown origin (beginning January 28) and of the eruptive stratospheric cloud from El Chichón (April 9). Since cloudiness on the days prior to those mentioned had prevented observations, it is possible that the considered disturbances could have appeared some days earlier.

The effect of El Chichón on the radiative characteristics of the atmosphere was the largest of all similar effects observed earlier at Mauna Loa. Lidar sounding revealed a fine vertical structure of the eruptive cloud with an initial maximum of the aerosol concentration above 22 km, where the concentration of the background aerosol is also maximum, and revealed the subsequent descent of aerosol particles. However, later the layer near 22 km became more intensive than that at 26 km, although the latter was observed more frequently. In 3 months the vertical mixing determined the formation of homogeneous sun photometer data–the background AOT at 425-nm wavelength was 0.02; on 14 May it was 0.49, reaching 0.7 by the end of that day. The characteristic feature of the stratospheric aerosols was only a small variation of the wavelength dependence of AOT in June and July 1982, which reflected the efficiency of the gas-to-particle conversion mechanism for aerosol formation, as well as the effect of the mechanisms of coagulation and sedimentation.

Retrieving the aerosol size distribution from the spectral transparency suggested that it may well be approximated as lognormal with a broad maximum in the range of radii 0.2 to 0.6 μm. Estimation of the total aerosol content in the air column gave 0.06 g/m^2, which determines the aerosol mass (in the latitudinal belt $15°$ S – $35°$ N) to be about 10^7 tons (this estimate does not take into account the aerosols at altitudes below 21 km).

Actinometric observations registered a substantial post-eruption reduction of direct solar and global radiation. A decrease of global radiation at near-noon constituted 5.6% (from observations on June 6, 1982, as compared to June 5, 1981), and in the case of direct radiation it reached 21.3%. A decrease of the daily sum of global radiation for June 6, 1982 (as compared to June 5, 1981), constituted 2490 kJ/m^2 (7.7 ± 1.5%). It is probable, however, that the estimates obtained should be considered maximum, since the thickest part of the eruptive cloud was spreading over Mauna Loa at the time.

The scale of the eruption of El Chichón can only be compared with the Agung eruption in March 1973 in Indonesia, but the effect of Agung was confined mainly to the SH. The AOT at Mauna Loa was only ~ 0.02, but in Aspendale, Australia, it reached 0.15. Previous eruptions had also been characterized by a smaller vertical extent of the eruptive cloud. Only the AOT estimates for the Krakatau eruption (1883) gave maximum AOT of about 0.6.

As has been mentioned, the post-El Chichón observations revealed a serious disturbing effect of the eruptive aerosols on results of the remote sounding of the atmosphere and surface (e.g., SST retrieval). This effect can be particularly substantial with the use of limb or occultation techniques for sounding, and it is governed by such aerosol characteristics as the optical thickness, the phase function, and the complex refraction index.

King (1982) processed data of the atmospheric spectral transparency measured at Mauna Loa in July. He obtained mean monthly optical thicknesses of the stratospheric aerosol layer in the visible ranging from 0.5 to 0.25. These were 10 times greater than the background values. Retrieving the aerosol size distribution averaged over the atmosphere (with the prescribed refraction index 1.45) revealed the presence of a narrow maximum in the interval of radii 0.3 to 0.4 μm.

With the density of aerosol matter 2 g/cm³, the aerosol mass in the air column will be 0.0579 g/cm² (June) and 0.0602 g/cm³ (July). These estimates were found taking into account the size distribution in the range of radii 0.1 < r < 4.0 µm, which was found by extrapolation of the values over the measured sizes. The size distribution can be approximated by a modified gamma distribution:

$$dN/d(\log r) = Cr^\alpha \exp(-br). \qquad (4.1)$$

Estimation of the parameters for June (July) gave the following results: C = 1.549•10¹⁵ (9.897•10¹⁹), α = 8.26 (13.65), b = 24.4 (39.3) µm⁻¹, and r = 0.338 (0.347) µm.

On the assumption that the stratospheric aerosol consists of a 75% sulfuric acid-water mixture. King (1982) calculated the spectral dependence of the aerosol optical thickness, the single scattering albedo, and the factor of the indicatrix asymmetry for the wavelength region 0.25 to 25.0 µm. The use of results for 3.7, 10.8 and 12.0 µm has shown that the SST retrieval from data of measurements at these wavelengths, without taking into account the eruptive aerosols, underestimates SST by 0.6 K at night and by 1.2 K during the day. Calculations for 0.7 µm gave 0.668 for the factor of indicatrix asymmetry, 1000 for the single-scattering albedo, and 73.3 for the ratio between the total extinction coefficient and backscattering. The angular distribution of the polarization degree is characterized by an unusually large number of scattering angles, to which a neutral polarization corresponds.

Analysis of data of regular observations of direct solar radiation at Mauna Loa beginning in 1958 has shown that the total atmospheric transparency is a very reliable indicator of the volcanically caused variations in the stratospheric aerosol content. This was illustrated by actinometric observations by DeLuisi et al. (1982) after the May 18, 1980, Mount St. Helens eruption. From data of June 3 observations, when the eruptive cloud was again over Mauna Loa, having made a circle around the globe, the values of the ratio between global radiation variations and those of direct solar radiation were 0.256, 0.239, and 0.227 for solar zenith distances 61.6°, 59.7°, and 57.8°, respectively, since the mean weighted solar zenith distance for the illuminated side of the Earth is 60°. These results can be considered representative from the viewpoint of characteristics of the global scale effect of the stratospheric aerosol layer. The respective post-Agung values were 0.2 (1963), 0.25 (1964), and 0.30 (1966).

Data on the ratio between spectral scattered and direct solar radiation made it possible to estimate the size distribution and absorbing properties of the aerosols. If $\Delta\tau$ is the change in the optical thickness between July 3 and 4, then with the Junge size distribution,

VOLCANIC ACTIVITY AND CLIMATE

$$dN/d(\log r) \sim r^{-\nu^*} \quad (4.2)$$

where N is the number density and r the radius of particles, we obtain

$$\Delta\tau(\lambda) \sim \lambda^{-\gamma}, \quad (4.3)$$

where $\nu^* = \gamma + 2$ and $\gamma = 0.47$, according to observations at Boulder, Colorado.

Estimates of the imaginary part of the complex refraction index ranged between 0 and 0.002. The post-Agung aerosol was also weakly absorbing. Analysis of the secular trend of transparency made at Mauna Loa revealed reductions of transparency after different eruptions: 2% Agung, 0.75% Fuego, 0.57% Soufrière, 0.55% Sierra Negra, and 0.58% Mount St. Helens. Processing of a small number of lidar soundings at Boulder gave a maximum optical thickness of 1.6, but the effect of cirrus clouds is not excluded here.

If the effect of the stratospheric aerosol layer on the shortwave radiation transfer depends weakly on the variability of the absorption coefficient, an opposite situation is observed with respect to the long-wave radiation. The role of the aerosol size distribution is also pronounced (at least, in the range of radii 0.1 to 2.0 µm), depending on whose variability the effect can appear of either warming or cooling of the climate. Although sulfuric acid droplets are mainly spherical, an account of the effect of nonspherical particles on the phase function is also essential.

Studies of the heat budget of a stratospheric particle are very important for estimations of climatic impact. In this connection, Fiocco et al. (1976) calculated the temperature of an aerosol particle governed by the absorbed solar radiation, radiative and conductive heat exchange with the atmosphere, and phase transformations. Previous calculations had shown that at altitudes higher than 60 km the temperature difference ΔT between air and particle may exceed 100 K in the daytime.

The new calculations are made (without taking account of phase transformation in the 0–60-km altitude range) for winter and summer conditions at 45° N, for different values of the surface-atmosphere system albedo, and different times of the day. It has been shown that at altitudes below 50 km maximum noon values $\Delta T < 1$ K, and below 10 km $\Delta T < 0.01$ K. In near-noon hours $\Delta T > 0$, i.e., aerosol particles heat the atmosphere. At night the situation is reversed. From the data on the conductive heat exchange between particle and gas, information has been obtained on the rates of heating (or cooling) of the air at different altitudes depending on the particle's radius.

The analysis of the mean-diurnal values reveals a strong variability of the heating rate depending on the altitude and particle's radius. Near the stratopause (50 km) the presence of aerosols almost always cools the atmosphere. In the winter, low tropospheric layers of aerosol, as a rule, cause cooling. The cooling effect may also be observed in summer in the regions with low surface albedo. The aerosol mean-diurnal heating of the stratosphere (15–25 km layer) is about 0.05-0.1°/day, and after volcanic eruptions it can reach 1°/day in the equatorial zone.

Determination of the climatic implications of volcanic eruptions requires, first of all, the analysis of specific features of the aerosol effect on the radiative regime of the atmosphere. This problem has been discussed in a number of monographs (Kondratyev, 1976b, 1991, 1992; Kondratyev et al., 1973), and will be illustrated by the calculations of the vertical profiles of spectral upwelling and downwelling shortwave radiation fluxes performed by Kershgens et al. (1976) in the 200–3580 nm wavelength region for cloudless atmosphere conditions, with the surface imitating a rough ocean surface (albedo A = 9.8%) or bright desert surface (A = 30%), or summer Arctic ice (A = 52%). Calculations have been made aimed at the analysis of the effect of the molecular and aerosol absorption, as well as the surface albedo on the radiative heating of the atmosphere.

The calculation technique is based on the phase function being represented by a number of Legendre polynomials and consideration of subsequent orders of scattering. The molecular and aerosol scattering up to the 50-km altitude is described by the Elterman model (1968) using extrapolation up to 70 km. At all wavelengths and all altitudes, the aerosol's single scattering is assumed to be 0.85, which is computed for an imaginary part of the complex refraction index of 0.02. Almost all the calculations are made for the summer subtropical model atmosphere. The spectral range under consideration is divided into 37 intervals from 10 to 50 nm wide, and the whole atmosphere into not less than 40 layers. The extra-atmospheric spectral distribution of the solar radiation has been assumed after Labs and Neckel (1968) based on solar constant of 1358 Wm^{-2}.

The use of different data on the functions of molecular absorption by water vapor, carbon dioxide and ozone, represented as sums of exponentials, has revealed considerable differences between the vertical profiles of radiative heating, which shows the necessity of further experimental and theoretical studies to specify the data on the absorption functions. In conditions of the cloudless troposphere, the water vapor and aerosols are the basic absorbers.

Calculations for different aerosol models have shown the effect of the vertical profile of the aerosol attenuation coefficient and the spectral dependence of the aerosol absorption coefficient on radiative heating. Above 40–50 km, absorption is primarily determined by the ozone effect, but at lower altitudes the role of multiple scattering and absorption by aerosols and gaseous components is substantial: they could lead to a 200% increase of the radiative heating near the tropopause.

VOLCANIC ACTIVITY AND CLIMATE

At a high surface albedo a substantial increase in the radiative heating of the atmosphere takes place, especially near the surface and at small solar zenith angles. The surface albedo effect is observed even at high latitudes in the troposphere and lower stratosphere.

An analysis of the contribution of dust clouds has shown that the presence of a dense dust cloud in the lower troposphere causes almost an eightfold increase of the radiative heating at the 4-km altitude (near a maximum of dust concentration) as well as a considerable increase of the planetary albedo over ocean (up to 20.1%). The presence of the aerosol layer in the stratosphere could lead to an almost 50% increase of the radiative heating at the 22-km altitude (depending on the solar zenith angle). In this case the planetary albedo varies little.

Such theoretical studies show the complexity of reactions of the radiative flux divergence within all layers of the atmosphere to changes of radiative transfer parameters, such as the ground albedo, the dust content and layering, and also of gaseous transmission functions and the accuracy of their fit with finite series of exponentials. Under clear-sky conditions the solar heating of the atmosphere is rather sensitive to changes of these parameters.

Atwater (1975) performed numerical modeling of the aerosol effect on the Earth's surface-atmosphere system albedo, with due regard to the variability of the size distribution and complex refractive index of aerosols. The aerosol size distribution is approximated by the modified-distribution suggested by Deirmendjian (1973) that characterizes four types of aerosols: stratospheric, marine, continental and urban. The optical properties of aerosols are calculated using Mie formulae.

For the four aerosol types named, calculations have been made of the dependence on the balance albedo (i.e., the case when the albedo does not change with an addition of aerosols) on the imaginary part of the complex refractive index and on the refractive index. The selection of a model of the aerosol size distribution has considerable effect on the balance albedo value. So, for instance, in the transition from urban to continental aerosols and from the stratospheric to marine aerosols (at a constant refractive index), the albedo decreases, and the aerosols may cause either heating or cooling if the surface albedo is close to the balance albedo.

Similar results may be found for variations of either the real or imaginary parts of the complex refractive index. The aerosol balance albedo satisfies the inequality:

$$\frac{1}{(2[a/r + 1])} \le (a_s)_o \le \frac{1}{(a/r + 1)} \qquad (4.4)$$

where a and r are the absorption and reflectivity of the aerosol, respectively.

Table 4.1.1 characterizes the discrepancies in the estimates obtained by various authors, associated, with different input parameters but not the different models used. Here $<\cos \theta>$ is the parameter of the phase function asymmetry, and ω_o is the single scattering albedo. All the results considered have been obtained in the two-flux approximation, and therefore, they hold as a rule, only for optically thin layers.

Table 4.1.1. Estimates of the balance albedo and the climatic effect of aerosols.

Values of input parameters				$(a_s)_o$	Climatic Effect
$<\cos \theta>$	ω_o	a/η	a		
		0	1	1	Cooling
		1 – 4	0.1 – 0.3	0.1 – 0.5	Any
		0 – 8	0.15	0.8 – 1.0	Any
0.64	0.90	0.99	0.10	0.34 – 0.72	Cooling
		2	0.05 – 0.1	0.17 – 0.33	Cooling
0.64			0.08 – 0.17	0.47	Cooling
			0.6 (snow)		

4.2 Remote Sounding of Stratospheric Aerosol Properties

Extensive data on stratospheric aerosols have been obtained by remote sounding through measurements of atmospheric spectral transparency (Deirmendjian, 1973; Dutton and DeLuisi, 1983; Newmann and Gerber, 1976; Pueschel et al., 1974; Shaw, 1979; Volz, 1975a, 1975b; Whitten et al., 1980; Wilson et al., 1978), lidar sounding (McCormick et al., 1978; Russell et al., 1976a, 1976b, 1979, 1981a), and satellite sounding (Cochran and Pyle, 1978; Dmitrieva-Arrago and Gorbunova, 1980; Efimova, 1979; Fomin, 1979; Galindo, 1992; Grigoryev and Lipatov, 1974; Jayaweara et al., 1976; Guschenko, 1974; Kabanov, 1980; Kiseleva, 1980; Kondratyev, 1976a, 1976b, 1991, 1992; Kondratyev et al., 1973, 1981a, 1981b, 1983a, 1983b; McCormick, 1982; McCormick et al., 1979, 1982; McCormick and Swissler, 1983; McCormick and Wang, 1987; Pollack et al., 1976a, 1976b; Russell et al., 1976a, 1976b, 1979, 1981a; Stayler, 1977).

Rather than dwell on the detailed results of ground measurements of atmospheric transparency, some of which are described by Kondratyev (1969), only the longest series of recent observations is considered here.

Mendonca et al. (1978a, 1978b) analyzed the results of estimated cloudless-sky atmospheric transparency from the data of pyrheliometric observations in the Mauna Loa observatory in Hawaii for 20 years (1958–78). Comparison of the running means for 6 and 12 months with a sequence of most significant volcanic eruptions indicated that it is very likely the eruptions were the main factor in spectral transparency changes. During the whole 20-year period the annual course had its maximum in winter (December-January) and a minimum in May. An extension (1957–83) of the same record of annual means has demonstrated that above Mauna Loa during this period four noticeable transmission decreases occurred: one during 1963–1964 (Agung), the second in 1966–67 (Awu), the third in 1975–76 (De Fuego) and the outstanding one in 1982–1983 (El Chichón). During the first 5 years of the record (i.e., before the period of recent volcanic activity) a biennial oscillation had been observed (Mendonca et al., 1978a, 1978b).

Aerosol optical depth (thickness) is defined by the product of the aerosol mass extinction and the aerosol mass density, integrated vertically to the top of the atmosphere. Optical depth is therefore a relative measure of the total vertical aerosol mass if the extinction coefficient is constant. When the spectral variation of the optical depth is known, the size distribution of the aerosol particles can be reasonably determined by inversion, and then actual total mass can be calculated (DeLuisi et al., 1983).

Aerosol optical depth is determined from the portion of the measured direct solar beam extinction that cannot be attributed to known gaseous atmospheric constituents. Several solar wavelengths can be chosen in a way that the only known significant attenuators are air molecules, ozone, and aerosols. Thus, the aerosol depth value is simply calculated from the familiar Beer's law.

Table 4.2.1, partially adopted from Mendonca et al. (1978a), lists changes in atmospheric transparency, ΔT, and optical depth $\Delta \tau$, from before the Agung (1958–1962) level, till after specific eruptions or the recovery period. After the Agung eruption the atmospheric transparency dropped 2% which at the time was considered a drastic decline, but in 1982 was spectacularly surpassed by the effect of El Chichón. Only in 1976–77 was the transparency level restored to pre-Agung value. The observational results under discussion do not reveal any regular secular trend of atmospheric transparency.

The problem of the secular trend of atmospheric transparency under the influence of volcanic eruptions and/or anthropogenic aerosols is still being discussed. In this connection, Hoyt (1979) analyzed the trend of atmospheric transparency using data from pyrheliometric observations of the Smithsonian Astro Physical Observatory for the 1923–1957 period. These data showed a surprisingly constant transparency, which can be explained mainly by volcanic eruptions maintaining the atmospheric turbidity.

Table 4.2.1. Percentage changes of the atmospheric transparency deduced from measurements at Mauna Loa Observatory. After Mendonca et al. (1978a).

From	Until	ΔT	$\Delta \tau$
Before Agung 1958-1962	Agung-Awu 1963-1968	-2.0	-0.032
Before Agung 1958-1962	Recovery period 1970-1973	-0.2	-0.004
Before Agung 1958-1962	Fuego 1974-1976	-0.8	-0.012
Before Agung 1958-1962	El Chichón 1982 (estimated)	-7.0	

Early observations in the SH on the Montezuma mountain (22°40' S, 68°56' W, 2,711 m a.s.l.) showed increase in the opacity of the atmosphere after the eruptions of the volcanoes Puyehue (December 1921), Reventador (1926), Paluweh (August 1928), and Quizapu (April 1932). In the NH there were strong effects from the eruptions of Paluweh (August 1928), Quizapu (April 1932), and, probably, Lamington (January 1951) and Spurr (July 1953). Similar volcanic effects can be seen in pyrheliometric observational data for Table Mountain (34°22' N, 117°41' W, 2,286 m a.s.l.), which confirms the global spread of volcanic aerosols. Reports describing a drop in the NH SAT approximately 1 year after the eruptions also suggest relation with changing opacity. However, after 1950 episodical decreases in the Montezuma transparency changes were apparently caused by local anthropogenic aerosol sources and the record is not very useful thereafter. A similar situation exists at the Table Mountain station, though in this case the effects of the Lamington and Spurr eruptions might still be separable.

The annual course of transparency is determined (to a significant extent) by changing the water content of the atmosphere. The exception is the data for Mauna Loa, where the annual course of the water content is small, and therefore a change in the aerosol content is the determining factor. The results suggest that the cooling observed in the NH after about 1940 could not have been ascribed to the volcanic activity effect, particularly since this activity was reduced somewhat after 1930 as compared to the 1881–1930 period.

The Agung eruption in 1963, gave rise to a very sharp decrease in direct solar radiation and an increase in scattered radiation. From actinometric measurements made in Aspendale, Australia, maximum additional attenuation (as compared to background) reached 24% during the first passage of the volcanic dust cloud and was followed by a 100% increase in the scattered radiation. In 2 years these values decreased down to 10% and 30%, respectively.

A strong perturbation of the atmospheric optical state after the eruption enabled DeLuisi and Herman (1977) to apply a previously suggested technique for retrieving the values of the imaginary part n_2 of the complex refraction index for the volcanic dust from the measured ratio of the scattered radiation to the direct or the global solar radiation. The analysis of 2-month-averaged values of

VOLCANIC ACTIVITY AND CLIMATE

the scattered to global solar radiation ratio, D/Q, showed that before the Agung eruption these values had reached an annual maximum in May-June and had not changed by more than 10% from year to year. Immediately after the eruption (July-August 1963) the ratio increased by about 100% as compared to the previous years. At the same time the direct radiation decreased 24% below the average for the preceding years, which corresponded to an increase in the aerosol optical thickness of 0.14.

Comparison of the measured values of the derivative $\Delta(D/Q)/\Delta\tau$ for different years with the results of calculations for an effective wavelength of 0.5 μm for different values of the complex refraction index $n_1 - in_2$, in the surface albedo, solar zenith angle, and aerosol size distribution has led to the conclusion that the dependence of R on the albedo is rather weak. Therefore, a mean value albedo R = 0.10 has been assumed.

The dependence of D/Q on the aerosol optical thickness τ is practically linear. Thus, one can expect that the constancy of the derivative $\Delta(D/Q)/\Delta\tau$ at fixed R and n_2 does not depend on absolute values of D/Q and τ. With an assumed Junge's size distribution, it turns out that the derivative under consideration remains practically constant at the Junge parameter $\alpha = 2-3$, which is close to the case of volcanic aerosols.

The results obtained by DeLuisi and Herman (1977) indicate that the main factor determining the variability of the derivative $\Delta(D/Q)/\Delta\tau$ is the imaginary part of the complex refractive index (the real part assumed to be 1.5). An estimate of n_2 from measurement data has led to the conclusion that the imaginary part of the complex refractive index does not exceed 0.01 and is most probably close to zero.

The calculation of the backscattering to the total scattering gives increasing values with time, which points to a shift of the aerosol size distribution toward finer particles. Great stability of the small-size fractions favors a long duration of the additional attenuation of solar radiation by volcanic aerosols: during 3 years the direct solar radiation in Aspendale decreased by about 5%.

Of course, the mean-global decrease should be much less. On the assumption that the aerosol is uniformly distributed in a layer 100 mb thick (altitudes of 15–20 km), $\tau = 0.14$, n_2 0.01, R – 0.1, and the solar zenith angle is 60%, the fraction of the solar radiation absorbed by aerosols is 0.056, which is equivalent to an absolute value of the absorbed radiation by aerosols of 38 W/m² and radiative heating of 0.04°/day. This value considered as an upper limit agrees with the calculations of Pollack et al. (1976a, 1976b) but is lower than that obtained by other authors.

The 1974 Fuego eruption in Guatemala, although considered to be about five times weaker than the Agung had the most powerful effect on the stratosphere between Agung and late 1979. Lidar sounding in Boulder, Colorado

(40° N, 105° W), can be used to trace the pattern of stratospheric aerosols during an eruption and subsequent years. Similar lidar sounding were also made in Menlo Park, California, and Hampton, Virginia. Using the above-mentioned lidar observations, Fernald (1978) analyzed the temporal variations of the backscatter coefficient for a 17–22-km layer, where an increase in the post-eruption aerosol content and the stratospheric optical thickness (tropopause to 30 km) are most evident. Using actinometric data Galindo and Bravo (1975) found an optical thickness of 0.027, which corresponds to an average of atmospheric transparency reduction in México City of about 2.6% for the period November-March 1975. Volz (1975a, 1975b, 1975c) has derived an atmospheric transparency reduction of about 2.3% from the measured values at Sacramento Peak in New Mexico.

In the March 1975 to August 1976 period, the aerosol backscattering coefficient decreased exponentially, which led to an e-decay time of 7.6 months for the aerosol content in the 17–22-km layer. A much slower decrease was observed later (until September 1977) with a decay time of 18.9 months (for the 15–30-km layer the time constant was 50.2 months), which indicates an approach to equilibrium. The aerosol optical thickness of the stratosphere decreased from about 0.1 to 0.01 during the entire period in question.

Following the 1974 Fuego eruption no other ejections into the stratosphere were observed for 5 years. This is consistent with the observed extremely low aerosol content in the stratosphere. This quiescent period was followed by a quite active one starting with the November 1979 eruption of Sierra Negra (0.8° S) on the Galapagos Islands, then Mount St. Helens (46° N) in May 1980, followed by Ulawun (5° S) October 1980, Alaid (51° N) April 1981, Pagan (18° N) May 1981, then the Nyamuragira eruption in December 1981, and finally the eruption that probably caused the largest impact to the stratosphere for the previous 70 years, the April 4, 1982 eruption of El Chichón (17° N) in México. The instrumental observations and studies of these eruptions have clearly shown that they made the primary contribution to stratospheric aerosol enhancements (McCormick, 1982).

Lidar sounding performed by Itabe et al. (1977, 1980) in Fukuoka (Japan) after the Fuego eruption in October 1974, revealed a 10-fold increase in the aerosol backscattering coefficient β_m, at the level of maximum concentration, and a subsequent rather complicated variability till the spring of 1975, after which the backscattering began a gradual monotonic decrease.

An earlier comparison of the observed decrease of β_m after the spring of 1975 with calculations based on a 2-D model developed by Gudiksen et al. (1968) has shown that calculations overestimate the rate of decrease of β_m. Such a difference could be explained by the neglect of the process of transformation of gas into particles. Itabe et al. (1980) explained that this difference is caused by the neglect of the condensation growth of the stratospheric aerosol particles formed after the Fuego eruption. They try to explain the decay rate of β_m observed at Fukuoka by considering that the Fuego cloud was divided into many incompressible air parcels of the size of several cubic kilometers in each of which are included gases (SO_2, H_2O, etc.) and particles. These air parcels are diffused and transported by large-scale atmospheric motions in accordance with 2-D model.

From this and other studies it is clear that in order to discuss precisely the global dispersion of volcanic aerosols it is necessary to monitor the physical and chemical processes of the transformation of the volcanic eruption products (particles and such gases as SO_2, H_2O, etc.) in a noncompressible air column of about several kilometers in height.

The model applied by Itabe et al. (1980) for studying the aerosol kinetics makes it possible to simulate a temporal change in the size distribution of aerosols contained in the air volume, the growth of aerosol particles being caused by condensation of H_2SO_4 vapors, for which three equations are used. The first equation characterizes the process of SO_2 oxidation to H_2SO_4, the second estimates the concentration of H_2SO_4 vapors, and the third determines the temporal variation of the aerosol size distribution caused by the processes of condensation and coagulation.

Itabe et al. (1980) have found a numerical solution for these equations by assuming that the original size distribution corresponds to a haze model H (after Deirmendjian, 1973) or to a lognormal distribution, with the 0.01–0.4-µm range of particle radii taken into account. In the case of the haze model H, one obtains an agreement with the observed course of β_m with assumed values of the coefficients K of the rate of SO_2 transformation into H_2SO_4 and the initial concentration of the sulfur dioxide. The most probable value of K is $2.5 \cdot 10^{-7}$ s^{-1} and the number density of the sulfur dioxide at the initial moment (April 5, 1975) may vary within $1.5–2.5 \cdot 10^{10}$ cm^{-3}.

With a pre-set initial lognormal size distribution the aerosol kinetics model does not achieve an agreement with an observed decrease in the aerosol backscattering coefficient. To more reliably describe the process of the global transport of stratospheric volcanic aerosols, simultaneous observation of both gases (SO_2, H_2O, etc.) and aerosol components is needed, as well as better knowledge on the formation of H_2SO_4 vapors, an adequate model of aerosol transport that includes sedimentation.

In Kingston, Jamaica, Kent and Philip (1980) carried out a ruby laser lidar sounding of the stratospheric aerosol after the Soufrière eruption (13° N, 61° W) on April 13, 1979. The volcanic activity continued until April 25, 1979. On April 17 the ejected products reached an altitude of 18.5 km. Analysis of the wind data has shown that direct transport of the volcanic dust to Jamaica was possible only at altitudes about 4 km or above 20 km.

In the period under consideration lidar soundings covered the range of altitudes 16–40 km and were calibrated on the assumption that scattering by the 30- to 40-km layer is purely Rayleigh. Typical background regularities for the vertical profile of the ratio R–1 of the observed backscattering coefficient of the Rayleigh scattering are characterized by a main maximum at an altitude of 18–22 km and (at times) a secondary maximum in the 26–29-km layer. The parameter R–1 may vary by as much as a factor of 2, such variations normally being observed over a period of several days, but occasionally within a single night.

As has been discussed earlier, an outstanding contribution to observations of global stratosphere aerosol dynamics has been made by satellite remote sensing (Kondratyev, 1991; Kondratyev et al., 1983a).

Based on processing the aerosol remote sounding data from the SAGE-1 satellite for the period March 1979 to February 1980, without strong volcanic eruptions, McCormick and Wang (1987) proposed a base model of the background stratospheric aerosol in the form of zonally averaged (monthly means and zonal means) vertical profiles of the coefficient of aerosol extinction at wavelength 1 μm in the tropics, and middle and high latitudes of both hemispheres. (The corresponding data have been tabulated and graphed.) The similarity of the seasonal mean vertical profiles of the coefficient of the aerosol extinction β in these latitudinal bands (75° S – 40° S; 40° S – 20° S; 20° S – 20° N; 20° N – 40° N; 40° N – 75° N) has made it possible to suggest the following polynomial approximation:

$$\lg \beta = a + bz + cz^2 + dz^3, \quad (4.5)$$

where z is the height with respect to the troposphere, in kilometers. The values of the coefficient for different seasons are given in Table 4.2.2

Table 4.2.2. The coefficients of polynomial approximation of the vertical aerosol profile. After McCormick and Wang (1987). [© Elsevier Science Ltd. Reprinted with permission of the publisher.]

Season 1979–1980	a	b	c	d
March – May	-3.60	-3.59×10^{-2}	6.30×10^{-3}	-3.17×10^{-4}
June – August	-3.78	-1.79×10^{-2}	-5.66×10^{-4}	-1.27×10^{-4}
September – November	-3.67	-3.26×10^{-2}	-2.99×10^{-4}	-1.10×10^{-4}
December – February	-3.50	-5.42×10^{-2}	4.01×10^{-4}	-1.21×10^{-4}
Mean Global	-3.64	-4.77×10^{-2}	-1.46×10^{-3}	-1.71×10^{-4}

During May 1–5, about 3 weeks after the first eruption, an abnormal intensification of backscattering was registered at altitudes near 16 km. This was assumed to have been caused by a volcanic dust cloud that had gone round the globe driven by strong westerly zonal winds. Observations of the stratospheric cloud produced by the Soufrière eruptions have also been reported by McCormick et al. (1982). Two separate stratospheric plumes, believed to be derived from different volcanic eruptions, were traced by the SAGE satellite; the stratospheric cloud near the volcano was also studied using an airborne lidar. The major plume was observed to move northeastward over the Atlantic Ocean, while the minor plume (probably the same as that seen later over Jamaica after it had circled the globe) was tracked eastward, over the Atlantic Ocean to West Africa. The movement of both plumes agreed well with that expected from the independently measured values for the upper atmospheric winds.

From the meteorological satellite Nimbus-7 and the specialized satellite SAGE, remote sensing of stratospheric aerosols was made by measurements of the atmospheric attenuation of the solar radiation at sunrise and sunset with respect to the spacecraft (McCormick et al., 1979). The polar sun-synchronous (noon-midnight) orbit of Nimbus-7 (the height of this practically circular orbit is 944 km) enables one to make occultation measurements only in the latitudinal bands 64°– 80° N during sunsets and 64°– 80° S during sunrises. At an orbital period of ~ 90 min, 15 sunrises and 15 sunsets can be observed during 24 hours. The SAGE orbit has an inclination angle of 55°, a height of apogee 600 km, a height of perigee 548 km, and an orbital period 96.8 min, and it is subjected to a rapid precession, which creates conditions for occultation geometry of sun viewing (in different seasons) through almost the entire globe (79° N – 79° S).

The stratospheric aerosol monitor (SAM-2) developed by McCormick et al. (1979) and installed on Nimbus-7 is a modified one-channel ($\lambda \approx 1$ µm) sun photometer tested during the scientific program Soyuz-Apollo. SAM-2 has a precise biaxial pointing control system and a system for "slow" (0.25 s^{-1}) scanning of the solar disc (vertically, up and down alternately) at a viewing angle of 0.01°, which gives a vertical resolution of about 0.5 km.

Measurements of the transmission of solar radiation by the atmosphere make it possible to retrieve the aerosol optical thickness (the aerosol attenuation coefficient) as a function of height, since the contribution of Rayleigh scattering can be calculated. With a pre-set aerosol size distribution and the complex refraction index, one can calculate the vertical profile of the particles' number density. The accuracy of retrieving the aerosol attenuation coefficient is about 10% for the atmospheric layers where the aerosol attenuation exceeded the molecular scattering by more than 50%. Hence the attenuation coefficient for the stratospheric background aerosol (in the absence of the volcanic activity effects) can be retrieved to an accuracy of 10%.

The SAGE solar photometer has four channels centered on 0.385, 0.45, 0.60, and 1.0 μm. The holographic diffraction grating serves as a dispersing element of the instrument. An appropriate arrangement of radiation sensors provides for measurements of the solar radiation on the above-mentioned wavelengths. The results of 4-channel measurements can be used to retrieve the vertical profiles of air density, concentration of ozone and nitrogen dioxide, and the coefficient of aerosol attenuation. The ozone profile can be retrieved through the altitude range from the tropopause to 45 km, i.e., including the atmospheric layer 35–45 km where the anthropogenic (due to chlorofluoromethanes) effect on ozone should be most evident.

During the 1960s an important item of the program of stratospheric aerosol remote sounding was the control subsatellite aerosol measurements in various locations: Sonderstrom (Greenland), White Sands (USA), Natal (Brazil), Poker Flat (Alaska), and Wallops Island (USA). Similar observations were also made later in various countries: Australia, Belgium, France, Federal Republic of Germany, Japan, Russia, and United Kingdom.

A complex of subsatellite observations made in the region of Sonderstrom on November 20–25, 1978 during Nimbus-3 flights can serve as an example (Russell et al. 1979; 1981b). These observations included aircraft lidar sounding of the 8–30-km layer, aerosol sampling, and aircraft nephelometric measurements at altitudes up to 15 km, as well as dustsonde launchings.

An analysis of the satellite measurements has shown that the aerosol number density values in high altitudes agree with models of the stratospheric background aerosol, and with lidar data. In some cases it was possible to retrieve the aerosol profiles down to the lower troposphere, indicating the feasibility of obtaining information on tropospheric aerosols with the occultation technique.

The first SAM-2 data have shown that the aerosol layer reaches higher altitudes in the Arctic in autumn than in the Antarctic in spring. Higher attenuation coefficients in the Antarctic than in the Arctic correspond to the aerosol layer maximum.

These results confirm the successful functioning of the instrumentation on both satellites, and further analysis of the results contributes substantially to our understanding of the sources and sinks of stratospheric aerosols, the conditions determining their formation and, in addition, the properties of mother-of-pearl and noctilucent clouds. A 7-channel model of SAGE (SAGE-2) has been developed to be launched on a free-flyer satellite from the shuttle to expand this evolving stratospheric and tropospheric aerosol climatology. In recent years SAM instrumentation has been further improved and very efficiently applied.

The Mount St. Helens NASA workshop report (Newell and Deepak, 1982) contains the results of a number of remote sensing measurements,

VOLCANIC ACTIVITY AND CLIMATE

including ground-based sunphotometers, lidars (both groundbased and aircraftborne), balloonborne sensors, and satellite instruments like SAM-2, SAGE, and AVHRR on NOAA-6. It was concluded that volcanic material reached at least 23 km from the May 18 eruption, with many lower maxima during the day. As mentioned earlier, layers from 10 to 14 km moved rapidly eastward circling the globe in only 16 days. Lidar backscattering ratios were as large as 100 (at $\lambda = 0.6943$ µm, the ruby laser wavelength) from this intense layer early after the eruption, whereas background values are typically 1.1 (1.1 means that aerosols are backscattering 10% as much as local gas molecules). The high values of 100 are representative of a very fresh volcanic cloud downwind of the volcano, and for a few days after the eruption. At NH lidar ground stations these values were about 1.35 at 17.5 km in November 1980. Distinct enhancements (layers) were produced at various altitudes immediately after the eruption with separate characteristics dispersed worldwide. By September the bulk of the material had lost its fine vertical structure and appeared as one broad layer centered at about 18 km. Extinction coefficients at 1.0 µm immediately after the eruption were measured by SAGE to be as high as 0.01 km^{-1}. Optical depths for the stratosphere in August in the NH showed large meridional and zonal variations over scales of 1000 km. Values range from backgrounds of 0.001 to 0.005 at 1.0 µm. Changes in stratospheric optical depths in July over Mauna Loa were less than 0.003 at 0.5 µm. SAGE-derived mass calculations using appropriate optical models gave a value of $0.32 \cdot 10^6$ t (or $0.32 \cdot 10^{12}$ g) for the global enhancement of the stratospheric aerosol produced. This value so derived is probably the best that can be obtained due to the global nature of the SAGE data. It is to be compared with a $0.25 \cdot 10^{12}$ background value derived by SAGE in the quiescent early 1979 period (McCormick et al., 1982).

The advanced remote sensing technology of today was also available to monitor the recent eruption of El Chichón. As has been discussed in Chapter 3, worldwide lidars and other sensors have shown the material produced from this eruption to be very much larger than that from Mount St. Helens. In a series of airborne lidar missions, and with the aid of SAM-2 polar measurements, McCormick and Swissler (1983) have concluded that the stratospheric aerosol mass produced from El Chichón was about $12 \cdot 10^6$ tons in late October-early November 1982, about 40 times that produced by Mount St. Helens in 1980.

Stratospheric aerosol optical depths at five discrete solar wavelengths were calculated from solar extinction measurements made onboard NASA Convair-990 aircraft during nine meridional flights between 55° N and 6° S in December 1982 by Dutton and DeLuisi (1983). Figure 4.2.1 illustrates the latitudinal distribution of aerosol optical depths at 500 nm for all nine flights. The average optical depth along the 120° W meridian appears to be about 0.12. Calculations indicate that this enhancement is sufficient to cause an 18% decrease in direct broadband solar irradiance at a representative zenith angle of

60°. The major latitudinal variations shown in Figure 4.2.1 are the decrease south of the equator, a maximum at 5° N, a minimum at 25° N, and slight increase through the midlatitudes.

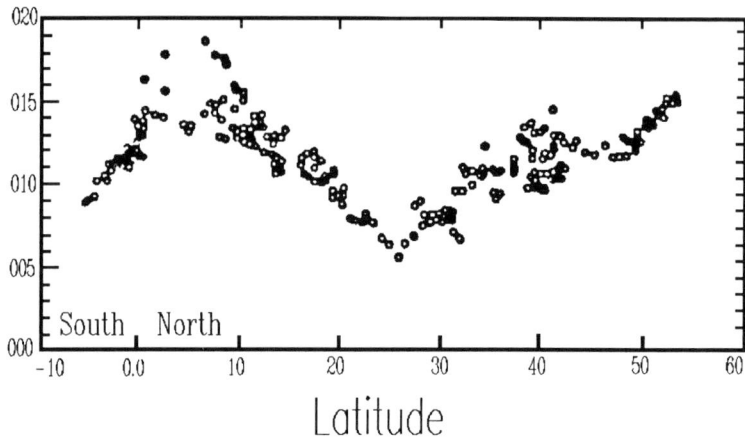

Figure 4.2.1. Latitudinal distribution of aerosol optical depths, at 0.500 μm measured during a total of nine CV-990 flights during December 1982. After Dutton and DeLuisi (1983). [Published by American Geophysical Union]

The El Chichón eruption produced two main layers, one centered at about 17 km, and the other at 26 km. The University of Wyoming dustsonde showed these layers to be of equal concentration, x 10 cm^{-3}, r ≥ 0.15 μm, with the upper layer having a mode radius of about 0.3 μm and the lower with a mode radius of about 0.1 μm (Rosen and Hofmann, 1982, 1983). The upper layer was found to be dominated by larger particles consisting of ≈ 80% H_2SO_4 to 20% H_2O (by weight) and the lower layers consisted of a 60%–65% acid aerosol. The upper more massive layer showed much more lidar backscattering and the lidar data showed this upper more massive layer to reside primarily between about 7° S and 36° N until November 1982 (McCormick and Swissler, 1983). Material at the altitude of the lower layer moved to the poles a few months after the eruption.

In Figure 4.2.2 are shown samples of lidar scattering ratio profiles taken aboard NASA Electra aircraft between 46° and 36° S roughly along the 80° W meridian from 19 to 31 October 1982 (McCormick and Swissler, 1983). The peak of the most dense layer was at 24–25 km and extended from 7° S to 36° N. The peak measured backscatter values (at λ = 694.3 nm) were about 25 compared with much greater backscatter of about 45 measured during July 1982. These results indicate than even 200 days after the eruption, the mass density over the places north of the equator was at least 4 to 5 times greater than over the SH.

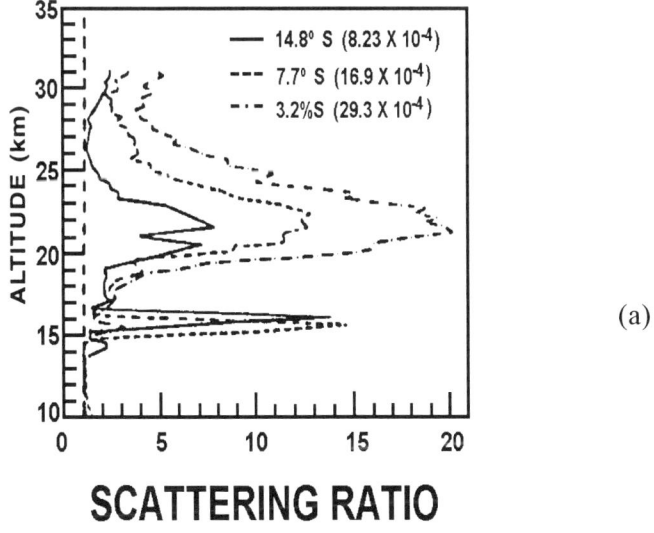

(a)

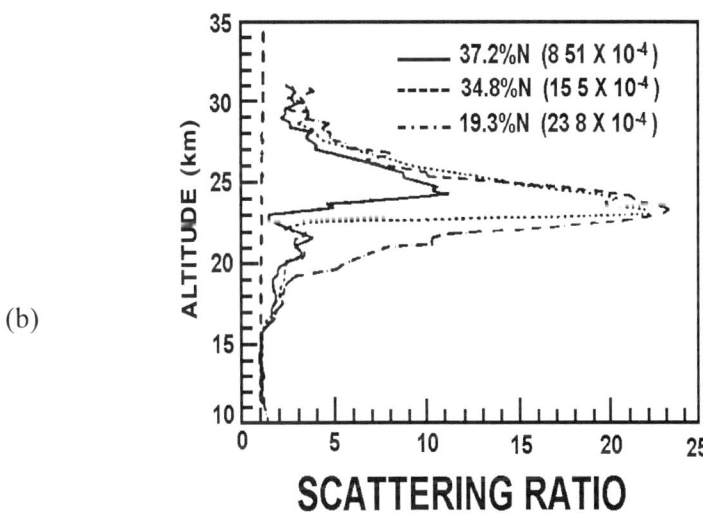

(b)

Figure 4.2.2. Representative vertical profiles of lidar scattering ratio at $\lambda = 694.3$ nm, for southern (a) and northern (b) latitudes along 80° W from flights between 19 and 31 October 1982. Integrated backscatter function, from the tropopause through the layer, is given in parentheses. The lower altitude ($\approx$ 16 km) enhancement over the southern latitudes is from a tropospheric cloud. From McCormick and Swissler (1983). [© by American Geophysical Union]

An excellent example of the dispersion of El Chichón-introduced aerosols over the northern midlatitudes during 1982 is given in Figure 4.2.3. It is drawn from 41 ruby laser ($\lambda = 694.3$ nm) profiles obtained at Garmisch-Partenkirchen (47° N) by Reiter et al. (1983). The very first observation of an aerosol layer, which they attributed to the El Chichón, was a layer of about 1 km thickness centered at 15.6-km altitude on May 3, 1982. This layer produced a backscattering ratio R = 3 (R is the ratio of observed to calculated molecular backscattering). As early as May 11, R = 6 was observed from the layer centered at 18 km. From its first appearance on May 3 over Central Europe this low altitude layer was observable until October. It had established itself between the tropopause and about 20 km. The highest value in this layer was observed on May 16 with R = 8 at 18 km.

On May 30 little peaks at 25 km indicated the arrival of another aerosol cloud, which was clearly observable on June 8 with R = 3 at 24 km. Until July 21 aerosol clouds at this higher altitude appeared sporadically above Central Europe; after this date a second layer was permanently registered above 20 km. The largest scattering ratio was seen on August 1 at 24 km (R= 14 with 600-m instrumental resolution, and R = 22 with 150-m resolution). From August onward, this layer produced higher scattering ratios than the low altitude layer.

During September the stratospheric circulation changes to the winter type western winds throughout the range of the lidar probing. The gap between the two layers was filled up slowly starting at the end of August and was no more observable after mid-October. After that a broad aerosol layer extended from the tropopause to about 30-km altitude (see Figure 4.2.3).

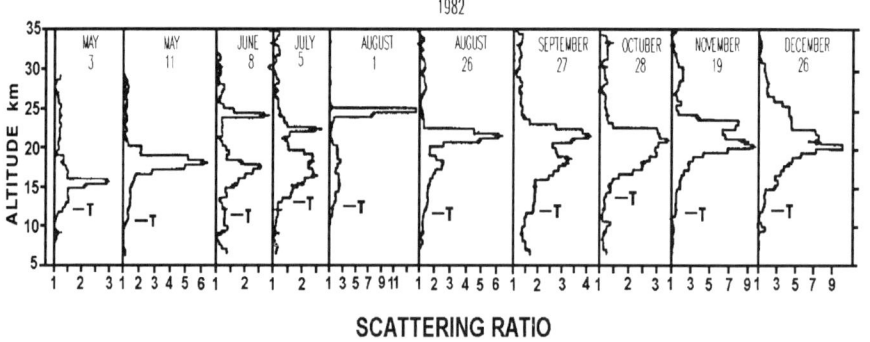

Figure 4.2.3. Selected lidar profiles showing the formation of aerosol layers following the April 28 eruption of El Chichón. The profiles are extinction corrected. T denotes the local tropopause. After Reiter et al. (1983). [© by American Geophysical Union]

Reiter et al. (1983) calculated that 9 months after the eruption of El Chichón the aerosol loading of the stratosphere above central Europe was 20 times more than that measured after Mount St. Helens of the same age. Optical depth data were shown to be record setting at Mauna Loa, downwind of the volcano, soon after the eruption. Initial values of 0.7 were recorded at 0.5 µm. These values remained at the 0.2 to 0.3 level for months after the eruption. Spinhirne (1983) aboard the NASA aircraft with the aforementioned lidar in October-November 1982 recorded values of about 0.14 at 0.5 µm between 8° S and 30° N and about 0.06 to 0.09 north and south of that latitude belt.

REFERENCES

Ackerman, T.P., 1988: Aerosols in climate modelling. In *Aerosols and Climate*, P.V. Hobbs and M.P. McCormick (Eds.), A. Deepak Publ. Hampton, Virginia, 335-348.

Atwater, M.A., 1975: Climatic effects of spherical polydispersions. In *Coll. Abstr. Second Conf. on Atmos. Radiation*, Arlington, Virginia, 29-31 October 1975, 131-134.

Clarke, D., Charlson, R.J., and Ogren, J.A., 1983: Stratospheric aerosol light absorption before and after El Chichón. *Geophys. Res. Lett.* **10** (11), 1017-1020.

Cochran, D.R., and Pyle, R.L., 1978: Volcanology via satellite. *Mon Wea. Rev.* **106**, 1373-1375.

Deirmendjian, D., 1973: On volcanic and other particulate turbidity anomalies. *Advances in Geophysics* **16**, 267-297.

DeLuisi, J.J., and Herman, B.M., 1977: Estimation of solar radiation absorption by volcanic stratospheric aerosols from Agung using surface-based observations. *J. Geophys. Res.* **82** (24), 3477-3480.

DeLuisi, J.J., Herman, B.M., Browning, S.R., and Sato, R.K., 1975: Theoretically determined multi-scattering effects of dust on Umkher observations. *J. Roy. Meteorol. Soc.* **101**, 325-331.

DeLuisi, J.J., Mendonca, B.G., and Hanson, K.J., 1982: Measurements of stratospheric aerosol over Mauna Loa, Hawaii, and Boulder, Colorado. In *Proc. Symp. Atmospheric Effects and Potential Climatic Impact of the 1980 Eruptions of Mount St. Helens*, NASA Conference Publication, **2240**, NASA 117-124.

DeLuisi, J.J., Dutton, E.G., Coulson, K.L., Defoor, T.E., and Mendonca, B.G., 1983: On some radiative features of the El Chichón volcanic stratospheric

dust cloud and a cloud of unknown origin at Mauna Loa. *J. Geophys. Res.* **88**, 6769-6772.
Dmitrieva-Arrago, L.R., and Gorbunova, T.N., 1980: The effect of stratospheric aerosols on the downwelling scattered shortwave radiation flux. *Trudy GGO* **410**, 96-102 (in Russian).
Dutton, E., and DeLuisi, J.J., 1983: Spectral extinction of direct solar radiation by the El Chichón cloud during December 1982. *Geophys. Res. Lett.* **10** (11), 1013-1016.
Efimova, N.A., 1979: Variations of direct solar radiation depending on volcanic activity. *Trudy GGI* **257**, 24-33 (in Russian).
Elterman, L., 1968: UV, Visible, and IR attenuation for altitudes to 50 km, 1968. Environmental Research Papers No. 285. Optical Physics Laboratory, U.S. Air Force Cambridge Research Laboratories, Office of Aero Space Research, p. 49.
Elterman, L., 1976: Aerosol measurements since 1973 for normal and volcanic stratospheres. *Appl. Opt.* **15**, 1113-1115.
Fegley, R.W., and Ellis, H.T., 1975: Optical effects of the 1974 stratospheric dust cloud. *Appl. Opt.* **14**, 1751-1753.
Fernald, F.G., 1978: Variability in the stratospheric aerosol loading associated with the 1974 eruption of volcano de Fuego. *Third Conf. Atmos. Radiation of the Amer. Met. Soc.*, Davis, California, Preprints, 235-237.
Fiocco, G., Grams, G., and Mugnai, A., 1976: Energy exchange and temperature of aerosols in the Earth's atmosphere (0.60 km). *J. Atmos. Sci.* **33** (2), 2415-2425.
Fomin, V.I., 1979: *Photogeological studies of volcanic belts*, Moscow, "Nedra" Publ. House, 214 pp. (in Russian).
Galindo, I., 1965: Turbidiometric estimations in Mexico City using the Volz sun photometer. *Pure Appl. Geophys.* **60**, 189-196.
Galindo, I., 1975: Physikalische und Mathematische Untersuchungen zur Atmosphärischen Trübung. *Arch. f. Met., Geophys. und Bioklim*, ser B **23**, 168-184.
Galindo, I., 1978: On the presence of saharian aerosol at the western part of the Atlantic Ocean. *Zeits. f. Meteorol*, Heft 6, B.**28**, 352-360.
Galindo, I., 1984: Anthropogenic aerosols and their regional scale climatic factors. In *Aerosols and Their Climatic Effects*, A. Deepak Publ. Hampton, Virginia, 245-259.
Galindo, I., 1992: Extinction of short-wave solar radiation due to El Chichón stratospheric aerosol. *Atmosfera* **5**, 259-268.
Galindo, I., and Bravo, J.L., 1975: On the presence of a volcanic stratospheric dust stratum over a polluted atmosphere: Mexico City. *Geofis. Int.* **15** (2), 157-167.

Galindo, I., Kondratyev, K.Ya., and Zenteno, G., 1996: Determination of the atmospheric optical depth due to the El Chichón stratospheric aerosol cloud in the polluted atmosphere of Mexico City. *Atmósfera* **9**, 23-32.

Gardner, C.S., Sechrist, C.F., and Shelton, J.D., 1980: Lidar observations of the Mount St. Helens dust layers over Urbana, Illinois. *Appl. Opt.* **19**, 2192-3031.

Grams, G.W., and Rosen, J.M., 1978: Instrumentation for in situ measurements of the optical properties of stratospheric aerosol particles. *Atmos. Technol.* **9**, 35-54.

Grigoryev, A.A., and Lipatov, V.B., 1974: *Dust storms studied from space*, Leningrad, Gidrometeoizdat, 31pp. (in Russian).

Guschenko, I.I., 1974: *Volcanic eruptions over the globe*. Catalogue, "Nauka" Publ. House, Mosocow **474** pp., (in Russian).

Harshvardhan, 1979: Perturbation of the zonal radiation balance by stratospheric aerosol layer. *J. Atmos. Sci.* **36** (7), 124-1285.

Hoyt, D.V., 1979: Atmospheric transmission from the Smithsonian Astrophysical Observatory, pyrheliometric measurements from 1923 to 1957. *J. Geophys. Res.* **C84**, 5015-5028.

Itabe, T., Fujiwara, M., and Hirono, M., 1977: Temporal variation of the stratospheric aerosol layer after the Fuego eruption observed by lidar in Fukuoka. *J. Met. Soc*, Japan, ser. II **55**, 606-612.

Itabe, T., Fujiwara, M., Hirono, M., and Igarachi, T., 1980: On the long-term decay rate of the post-Fuego stratospheric aerosols observed by lidar in Fukuoka. *J. Met. Soc.*, Japan, ser. II **58**, 127-136.

Jayaweara, K.O.L.F., Selfert, R., and Wendler, G., 1976: Satellite observations of the eruption of Tolbachic volcano. *EOS* **57**, 196-200.

Jennings, S.G. (Ed.), 1993: *Aerosol Effects on Climate*, The University of Arizona Press, Arizona, 305 pp.

Kabanov, A.S., 1980: Kinetics of the condenstion growth of stratospheric sulphate aerosol droplets in a homogeneous medium. *Izvestia AN SSSR, Fisika atmosfery i okeana* **16**, 172-177 (in Russian).

Kent, G.S., and Philip, M.T., 1980: Lidar observations of dust from the Soufrière volcanic eruptions of April 1979. *J. Atmos. Sci.* **37**, 1358-1362.

Kershgens, M., Raschke, E., and Reuter, U., 1976: The absorption of solar radiation in model atmospheres. *Beitr Phys. Atmos.* **49**, 81-91.

King, M.D., 1982: *Radiative characteristics of the Aerosols*. NASA Technical Memo 94859, Greenbelt, Maryland. *GSFC*, pp.3.3-3.13.

Kiseleva, M.S., 1980: On possible values of aerosol parameters in the stratosphere. *Izvestia AN SSSR, Fisika atmosfery i okeana* **16**, 536-539.

Kondratyev, K.Ya., 1969: *Radiation in the atmosphere*, New York, Academic Press, 912 pp.

Kondratyev, K.Ya., 1976a: Aerosol and Climate. *Trudy GGO*, **Issue 381**, 3-66 (in Russian).
Kondratyev, K.Ya., 1976b: *News in Climate Theory*, Leningrad, Gidrometeoizdat, 65 pp. (in Russian).
Kondratyev, K.Ya., 1981: Stratosphere and Climate. Uspekhi Nauki i Techniki, *Meteorologia i klimatologia* **6**, VINITI, Moscow (in Russian).
Kondratyev, K.Ya., 1983: *Earth Radiation Budget, Aerosol and Cloudiness.* Progress in Sci. and Technol., Meteorology and Climatology **10**, VINITI, Moscow, 316 pp. (in Russian).
Kondratyev, K.Ya. (Ed.), 1991: *Aerosols and Climate*. Gidrometeoizdat, Leningrad, 542 pp. (in Russian).
Kondratyev, K.Ya., 1992: *Global Climate*. NAUKA, St. Petersburg, 359 pp. (in Russian).
Kondratyev, K.Ya., and Pozdnyakov, D.V., 1980: Stratospheric Aerosol. Review. Ser. Meteorology, Obninsk, (7), pp. 39 (in Russian).
Kondratyev, K.Ya., and Pozdnyakov, D.V., 1981: *Atmospheric aerosol models*. Moscow, "Nauka" Publ. House, pp. 104 (in Russian).
Kondratyev, K.Ya., Vasilyev, O.B., Ivlev, L.S., Nikolsky, G.A., and Smokty, O.I., 1973: *Aerosol effect on radiation transfer: Probable climatic implications*. Leningrad State Univ. Publ. House, 266 pp., (in Russian).
Kondratyev, K.Ya., Moskalenko, N.I., and Terzi, V.P., 1981a: A closed modeling of the optical characteristics of the atmospheric aerosol. *Doklady AN SSSR* **253** (6), 1354-1356 (in Russian).
Kondratyev, K.Ya., Moskalenko, N.I., Terzi, V.P., and Skvortsova, S.Ya., 1981b: Modeling the optical characteristics of atmospheric aerosols. In *GARP, Aerosol and Climate*, Gidrometeozdat, Leningrad, pp. 130-153 (in Russian).
Kondratyev, K.Ya., Moskalenko, N.I., and Pozdnyakov, D.V., 1983: *Atmospheric Aerosols*. Leningrad, Gidrometeoizdat, 224 pp. (in Russian).
Kondratyev, K.Ya., Grigoryev, A.A., Pokrovsky, O.M., and Shalina, E.V., 1983a: *Remote Sounding of Atmospheric Aerosol from Space*, Gidrometeoizdat, Leningrad.
Labs, D., and Neckel, H., 1968: The Radiation of the Solar Photosphere from 2,000 Å to 100 μ. *Z. Astroph.* **69**, 1.
McCormick, M.P., 1982: Aircraft and ground-based lidar measurements of El Chichón stratospheric aerosols. Presented at the *7th Climate Diagnostics Workshop*, Boulder, Colorado.
McCormick, M.P., and Swissler, T.J., 1983: Stratospheric aerosol mass and latitudinal distribution of the El Chichón eruption cloud for October 1982. *Geophys. Res. Lett.* **10**, 877-880.
McCormick, M.P., and Wang, P.-H, 1987: Background stratospheric aerosol reference model. *Adv. Space Res.* **7** (9), 73-80.

McCormick, M.P., Swissler, T.J., Chu, W.P., and Fuller, W.H. Jr., 1978: Post-volcanic stratospheric aerosol decay as measured by lidar. *J. Atmos. Sci.* **35**, 1296-1303.

McCormick, M.P., Hamill, P., Pepin, T.J., Chu, W.P., Swissler, T.J., and McMaster, L.R., 1979: Satellite studies of the stratospheric aerosol. *Bull. Amer. Met. Soc.* **60**, 1038-1047.

McCormick, M.P., Kent, G.S., Yue, G.K., and Cunnold, D.M., 1982: Stratospheric aerosol effects from Soufriere volcano as measured by the SAGE satellite system. *Science* **216**, 1115-1118.

Mendonca, B.G., Hanson, K.J., and DeLuisi, J.J., 1978a: Secular trends in clear sky transmissions at Mauna Loa Observatory - Perturbations in stratospheric aerosols. *Third Conf. Atmos. Radiation of the Amer. Met. Soc.*, Davis, California, 28-30 June 1978, preprint, pp. 330-332.

Mendonca, B.G., Hanson, K.J., and DeLuisi, J.J., 1978b: Volcanically related secular trends in atmospheric transmission at Mauna Loa Observatory, Hawaii. *Science* **202**, 513-515.

Newell, R.E., and Deepak, A. (Eds.), 1982: *Mount St. Helens Eruptions of 1980: Atmospheric Effects and Potential Climatic Impact,* NASA SP-458, NASA, Washington, D.C.

Newmann, J., and Gerber, M., 1976: Stratospheric particles and solar radiation over northeast Australia after Mt. Agung's eruption in 1973. *Australian Met. Magazine* **24**, 1-16.

Patterson, E.M., 1975: Optical absorption coefficients of soil aerosol particles and volcanic ash between 1 and 16 µm. In *Coll. Abstr. Second Conf. on Atmos. Radiation*, Arlington, Virginia, 29-31 October 1975, 177-180.

Patterson, E.M., Pollard, C.O., and Galindo, I., 1983: Optical properties of the ash from "El Chichón" Volcano. *Geophys. Res. Lett.* **10**, 317-320.

Pollack, J.B., Toon, O.B., Sagan, C., Summers, A., Baldwin, B., and Van Camp, A., 1976a: Stratospheric aerosols and climatic changes. *Nature* **263**, 551-555.

Pollack, J.B., Toon, O.B., Sagan, C., Summers, A., Baldwin, B., and Van Camp, A., 1976b: Volcanic explosions and climatic change: A theoretical assessment. *J. Geophys. Res.* **82**, 1071-1083.

Pueschel, R.F., Garcia, C.J., and Hansen, R.T., 1974: Solar radiation: Effects of atmospheric water vapor and volcanic aerosols. *J. Appl. Meteorol.* **13**, 397-401.

Rao, C.R.N., and Bradley, W.A., 1983: Effects of the El Chichón volcanic dust cloud on insolation measurements at Corvallis, Oregon (USA). *Geophys. Res. Lett.* **10** (5), 389-391.

Reiter, R., Jäger, H., Carnuth, W., and Funk, W., 1983: The El Chichón Cloud over Central Europe, observed by lidar at Garmisch-Partenkirchen during 1982. *Geophys. Res. Lett.* **10**, 1001-1004.

Rosen, J.M., and Hofmann, D.J., 1982: Dustsonde measurements of the Mount St. Helens volcanic dust cloud over Wyoming. In Proc. Symp. *Atmospheric Effects and Potential Climatic Impact of the 1980 Eruptions of Mount St. Helens*, A. Deepak (Ed.), NASA Conf. Publ. 2240, 65-82.

Rosen, J.M., and Hofmann, D.J., 1983: Measurement of the sulfuric acid weight percentage in the stratospheric aerosol from the El Chichón eruption. Paper A21A-7, Spring 1983 AGU Meeting, Baltimore, Maryland, *EOS* **64**, p. 197.

Russell, P.B., Hake, R.D., and Viezee, W., 1976a: Lidar measurements of the post-Fuego stratospheric aerosol. Final Report, *Stanford* Research Institute, Menlo Park, California, April 1976, 48 pp.

Russell, P.B., Hake, R.D., and Viezee, W., 1976b: The post-Fuego stratospheric aerosol: Lidar measurements and radiative implications. *Proc. IAMAP Symp. on radiation in the Atmosphere*, Garmisch-Partenkirchen, 141-143.

Russell, P.B., Swissler, T.J., and McCormick, M.P., 1979: Methodology for error analysis and simulation of lidar aerosol measurements. *Appl. Optics* **18**, 3783-3797.

Russell, P.B., McCormick, M.P., Swissler, T.J., Chu, W.P., Livingston, J.M., and Pepin, T.J., 1981a: Satellite and correlative measurements of the stratospheric aerosol. I: An optical model for data conversions. *J. Atmos. Sci.* **38**, 1279-1294.

Russell, P.B., McCormick, M.P., Swissler, T.J., Chu, W.P., Livingston, J.M., Fuller, W.H., Rosen, J.M., Hofmann, D.J., McMaster, L.R., Woods, D.C., and Pepin, T.J., 1981b: Satellite and correlative measurements of the stratospheric aerosol. II: Comparison of measurement made by SAM II, dustsondes and an airborne lidar. *J. Atmos. Sci.* **38**, 1295-1312.

Schwedfeger, A., Stowe, L.L., and Gruber, A., 1983: Sensitivity of earth radiation budget parameters for El Chichón volcanic aerosol as estimated from NOAA-7 AVHRR data. *5th Conf. on Atmospheric Radiation, Oct. 31-Nov. 4. Baltimore, Maryland.* Boston: American Meteorological Society, 354-356.

Shaw, G.E., 1979: Aerosols at Mauna Loa: Optical properties. *J. Atmos. Sci.* **36**, 862-869.

Spinhirne, J.D., 1983: Latitudinal variation for spectral optical thickness of the El Chichón eruption cloud. Paper A21A-05, AGU Spring 1983 Meeting, Baltimore, Maryland, *EOS* **64**, pp. 197.

Stayler, W.F., 1977: Determination of stack plume properties from satellite imagery. *AIAA 25th Aerospace Sciences Meeting*, Los Angeles, California, 24-26 January 1977, 10 pp.

Volz, F., 1975a: New volcanic twilights from the Fuego eruption. *Science* **189** (48).

Volz, F., 1975b: Distribution of turbidity after the 1912 Katmai eruption in Alaska. *J. Geophys. Res.* **80**, 2643-2648.

Volz, F., 1975c: Burden of volcanic dust and nuclear debris after injection into the stratosphere at 40°-50° N. *J. Geophys. Res.* **80**, 2649-2653.

Whitten, R.C., Toon, O.B., and Turco, R.P., 1980: The stratospheric sulphate aerosol layer: Processes, models, observations, and simulations. *PAGEOPH* **118**, 86-127.

Wilson, L., Sparks, R.S.J., Huang, T.C., and Watkins, N.D., 1978: The control of volcanic column heights by eruption energetics and dynamics. *J. Geophys. Res.* **B84**, 1829-1836.

CHAPTER 5

ASSESSMENT OF THE EFFECT OF VOLCANIC ERUPTIONS ON CLIMATE

As already mentioned in the Introduction and Chapter 2, varied opinions can be found in the literature on the climate implications of volcanic eruptions (see also Kondratyev and Grassl, 1996; Lamb, 1970, 1972, 1977; Newell, 1984; Rampino and Self, 1982; Schneider, 1983, 1984; Self et al., 1981). It has been suggested, for instance, that volcanic activity was the cause of past ice ages. However, the volcanic theory of ice ages has suffered a decline in popularity for two reasons. First, the ice age record has greatly improved. It is now clear that there have been a large number of glacial advances during the past few million years, and these glacial advances have been highly periodic. The periods of the ice ages correspond well with the characteristic periods of variation in some of the Earth's orbital parameters, but there is no evidence that volcanic activity is highly periodic. Secondly, a detailed record of volcanic activity shows that there was no increase of activity at the beginning of the glacial advance.

5.1 Observational Data and Radiation Budget

Several attempts have been made to demonstrate that the volcanic effect is distinguishable, by itself, in the record of surface air temperature (Bryson and Goodman, 1980, 1982; Loginov, 1984; Loginov et al., 1983a, 1983b; Mass and Schneider, 1977; Oliver, 1976; Robock, 1978, 1984; Robock and Mao, 1995; Taylor et al., 1980; Vinnikov and Groisman, 1981; Yamamoto, 1981), and they have all found some indication of a correlation, even though they used data sets that are not consistent. In particular, the reliability of the data varies greatly from one period to another and from one area to another, and none of these chronologies provides estimates of the sulfur contribution to the stratosphere.

VOLCANIC ACTIVITY AND CLIMATE

Other recent studies by Ellsaesser (1977, 1983), Gilliland (1982), Gilliland and Schneider (1984), Hansen et al. (1977, 1978, 1981, 1992), Portman and Gutzler (1996), Rind et al. (1990, 1992) who took account of volcanic effects, El Niño and also CO_2 and solar activity, were able to account for or "explain" 90 per cent of the variability in the past 100-year record.

Examination of the observational data revealed that, as a rule, during the first or second year after powerful eruptions the surface air cooled by 0.5 to 1.0 K, but in some cases such an event was not observed. Figure 5.1.1 shows the results obtained by Oliver (1976).

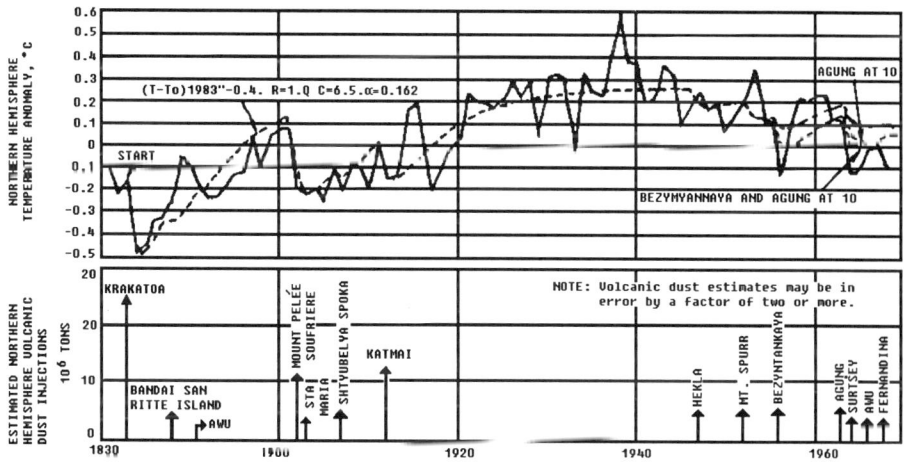

Figure 5.1.1 NH Surface temperature anomalies and volcanic eruptions. After Oliver (1976).

It should be added, however, that some experts have looked for the volcanic effect on temperature unsuccessfully (Landsberg and Albert, 1974; Mason, 1976). Furthermore, the data usually used to measure the amplitude of the forcing from individual eruptions prior to 1963 are based on subjective observations of the volcanic cloud the "DVI" as summarized by Lamb (1970). He adjusted the DVI according to the temperature cooling that he thought had occurred after each eruption on the NH temperature subsequent to 1750. Oliver (1976) achieved agreement only after he modified Lamb's estimates of relative strengths of eruptions that occurred during the past 100 years. The data on volcanic activity used by Gilliland and Schneider (1984) came from a very different source, namely, the record of acidity in an ice core from Greenland (Hammer, 1977). The acidity was assumed to be related to sulfate from the stratosphere that was deposited with the annual snowfall, but it is likely that this method would exaggerate the effects of local high latitude eruptions, even ones that did not reach the stratosphere. Vinnikov and Groisman (1981) in their diagnostic study of the temperature record were unable to detect any CO_2 signal when they took the DVI into account (implying that it was not statistically significant). Instead, they argued that the clear-day actinometric measurements were a more suitable indicator of stratospheric aerosol. When they used this indicator, a relation was found.

Lough and Pritts (1987) performed an analysis of the SAT variability for the period 1602–1900 from dendroclimatic data at 77 locations in the United States territory and southwest Canada, to reveal the effects of volcanic eruptions on climate. The results obtained show that, though over most of the U.S. territory the post-eruption temperature decreases had taken place, in some states in the western USA a warming was observed. This warming reflected the inhomogeneity of the spatial distribution of the climatic system's response to volcanic eruptions, with a clearly expressed annual course of cooling and warming in individual regions. So, for example, the warming in the western USA is most substantial in the summer. The effect of eruptions was the strongest in low latitudes. These conclusions have been confirmed by independent dendroclimatic data for regions outside the one considered.

In the works of Rampino and Self (1982) and Rampino et al. (1988) the authors reviewed the data on the effect of volcanic eruptions on climate and reached the conclusion about a small decrease of SAT after strong explosive volcanic eruptions and much more substantial heating of the stratosphere (up to 3–4 °C). The results of analysis of ice core acidity are a reliable indicator of the chronology and scale of eruptions. All these eruptions had been, however, much weaker than those revealed from paleodata, which could have led to the onset of a "volcanic winter."

The first large volcanic eruption monitored by the global radiosonde network was that of Agung in 1963. In the tropical lower stratosphere temperature increases of more than 5 °C were observed within a month or so. The region of positive temperature anomalies propagated polewards somewhat like the radioactive trace substances from tropical nuclear tests that entered the stratosphere. Examination of the various factors that control tropical tropospheric free air temperature showed that the most important was equatorial Pacific sea surface temperature. The free air temperature was characterized by the difference in geopotential height at pressures of 700 and 300 hPa (3 and 10 km) for a group of stations selected to represent a zonal mean (a technique initiated by Angell and Korshover, 1975). When a regression analysis was performed between the sea temperature and air temperature two seasons ahead, the residuals, corresponding to the unexplained variations of the air temperature, matched closely with the atmospheric transmission at Mauna Loa (Newell,

1970, 1984). The maximum tropospheric temperature difference from the long-term mean corresponded to a cooling of about 0.4 °C and did not occur until over a year after the eruption, whereas the stratospheric warming occurred almost immediately. More details of this type of study have been given by Angell and Korshover (1975, 1983a, 1983b).

Ellsaesser (1977, 1983) questioned the conclusions drawn by Angell and Korshover about the probable effect of the 1963 Agung eruption on the change in the mean surface temperature that had been summarized by them as follows. Figures 5.1.2 – 5.1.7 show these variations.

- In 1963 the surface temperature dropped by 0.6 °C in the northern extra-tropical latitudes, 0.2 °C in the tropics, and 0.5 °C in the southern-tropical latitudes.
- The temperature of the 850–300 hPa layer in the southern extra-tropical latitudes dropped by 0.7 °C, and of the 300–100 hPa layer it increased by 0.7 °C. In this connection an assumption has been made about probable changes from the previously observed climate cooling toward warming in the early 1970s though not only due to the Agung eruption.

The doubts concerning the justification for the above-mentioned conclusions were based on the following considerations:

(i) Since the stratospheric dust cloud resulting from the Agung eruption was tenfold denser in the SH than in the NH, it is difficult to understand why the eruption caused surface temperature changes in the extra-tropical latitudes of the NH which were one and a half times greater and lasted twice as long as those in the SH.

(ii) Warming of the 300–100 hPa layer extratropical latitudes by 0.7 °C has occurred simultaneously with the similar warming in 1962–1963 in the NH, but the latter cannot be ascribed to the eruption effect since the dust cloud only reached the southern boundary of the USA in September 1963 (probably, both these phenomena have been driven by the same still unknown mechanism, which has nothing to do with this volcanic eruption).

(iii) If the eruption was the cause of temperature variations, then it is not clear why the surface temperature in the extra-tropical latitudes of the NH did not drop before 1963–1964, in contrast to the cooling of the 850–300 hPa layer and warming of the 300–100 hPa layer.

(iv) Also, it is not clear in what way the large volcanic dust particles with lifetimes in the stratosphere of not more than 30 days could have produced the increase in the mean annual temperature of the upper troposphere, which continued until 1964.

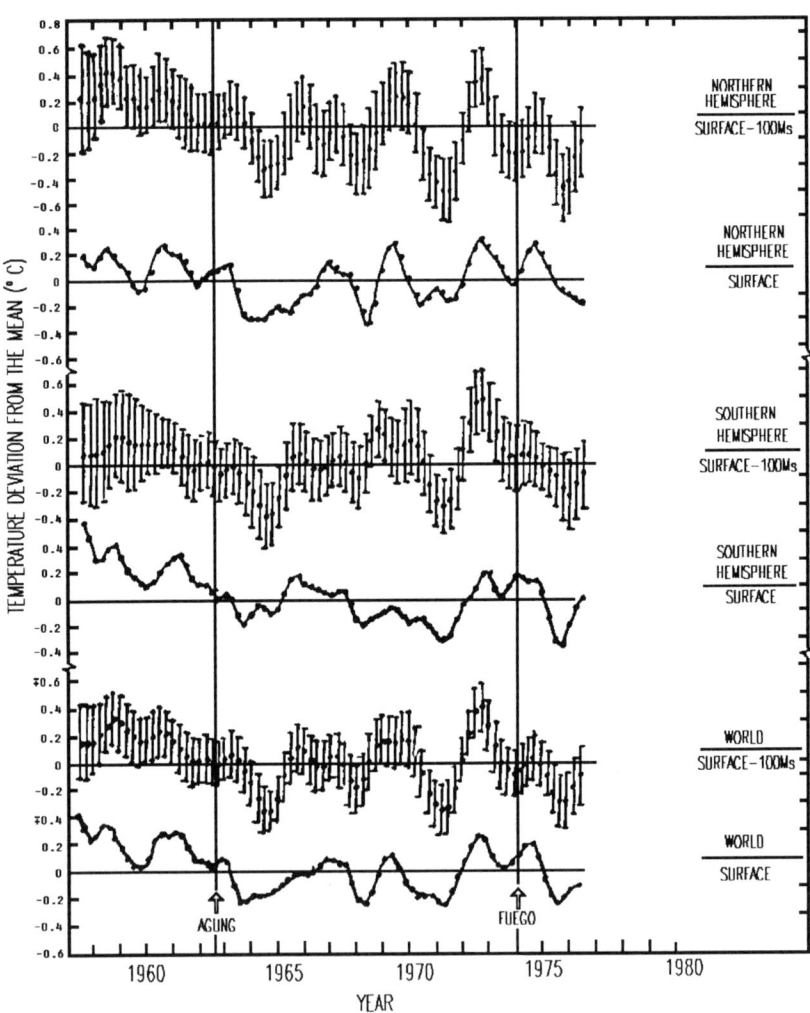

Figure 5.1.2. Deviations of mean seasonal surface temperature and atmospheric temperature averaged over geographical regions. After Angell and Korshover (1977).

VOLCANIC ACTIVITY AND CLIMATE

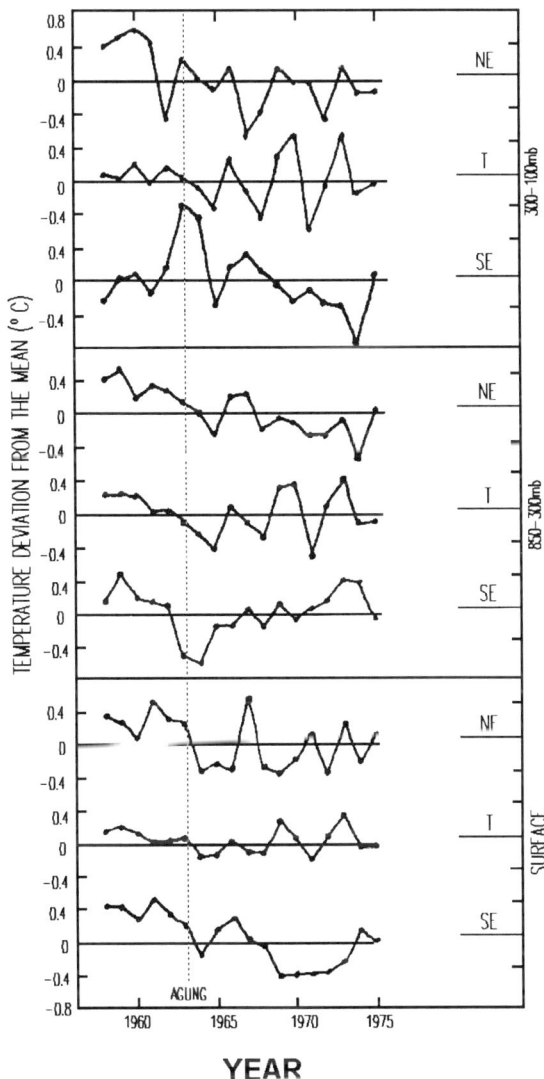

Figure 5.1.3. Nonsmoothed mean annual deviations of temperature for the extratropical latitudes of the NH and SH and in the tropics. The vertical dashed line marks the moment of the Agung eruption. After Angell and Korshover (1977).

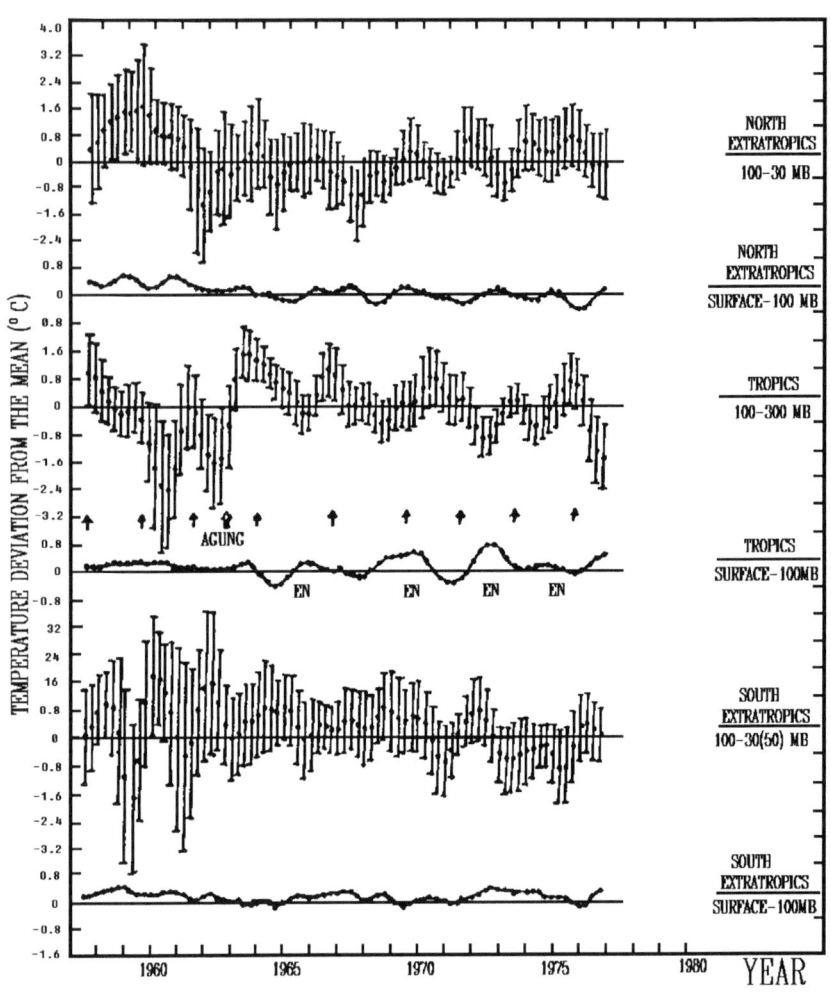

Figure 5.1.4. Same as Figure 5.1.2, but for different atmospheric layers EN, and El Niño. After Angell and Korshover (1977, 1978).

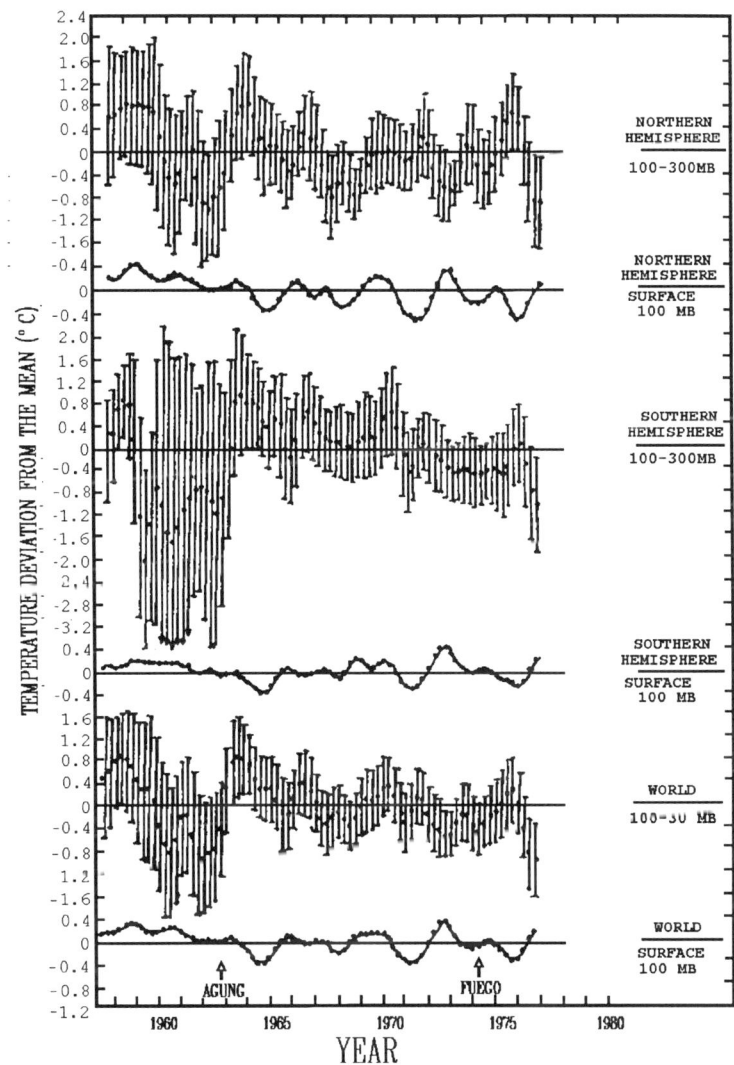

Figure 5.1.5. Same as in Figure 5.1.4, but for both hemispheres and the globe. After Angell and Korshover (1977, 1978).

In view of these controversies, Taylor et al. (1980) performed an analysis of the temperature data for the 1815-1963 period using the technique of superimposed epochs that attempts to select small signals by subsequency of discrete events, in data series with noise characteristics of natural temperature variations. The data discussed had been obtained from the same 42 stations located in various parts of the globe used earlier in a similar study by Mass and Schneider (1977).

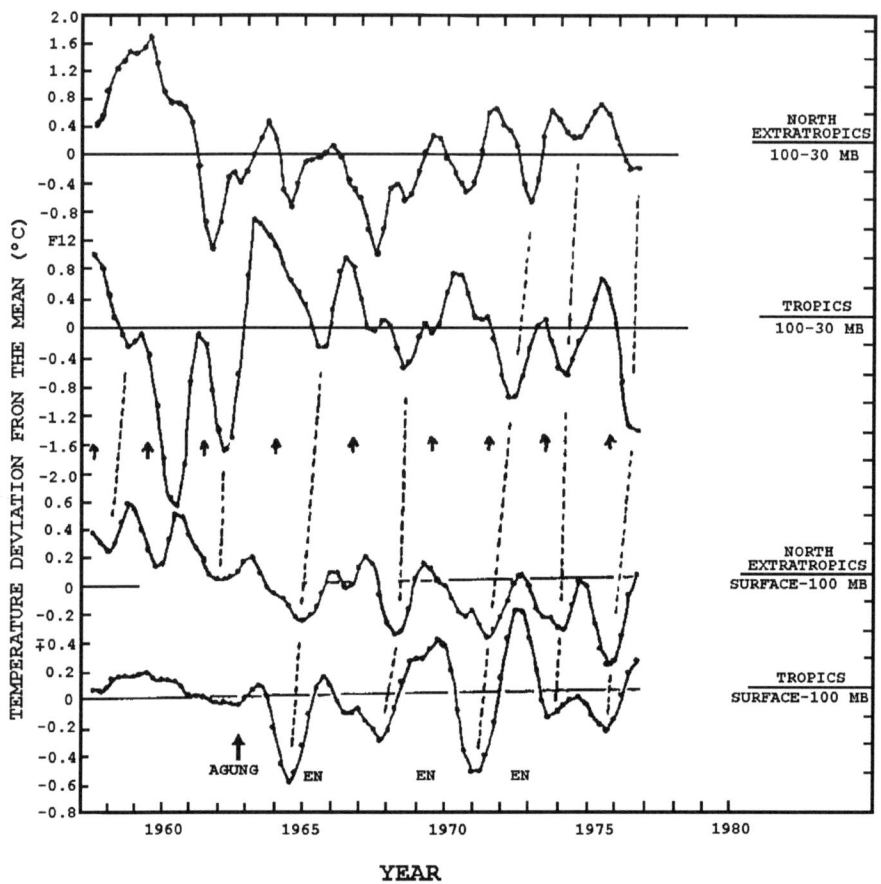

Figure 5.1.6. Smoothed deviations of mean seasonal temperature values from seasonal averages, which reflect the upward and northward spreading of temperature waves due to positive SST anomalies in the eastern tropical Pacific, EN, and El Niño. After Angell and Korshover (1977, 1978).

Only temperature data of the most powerful eruptions have been analyzed, after Lamb (1970). These correspond to a (DVI) of not less than 100 (DVI characterizes an additional attenuation of the solar radiation caused by the loading of the atmosphere with the volcanic dust). During the period under consideration, 25 similar eruptions took place in 18 years. Separately the 13 most powerful eruptions with DVI > 500 were the following (DVI in parentheses): Tambora, Indonesia, 1846 (1000); Cosiguina, Nicaragua, 1875

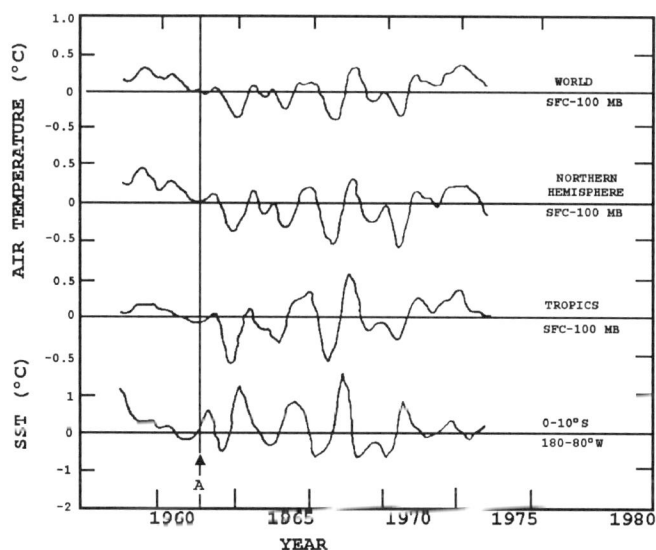

Figure 5.1.7. Variations in SST and surface air temperature in the eastern Pacific (0 –10° S; 180–80° W). A, Agung eruption. After Angell and Korshover (1983a).

(400); Amargura, Fiji, 1815 (3000); Krakatau, Indonesia, 1883 (1000); Agung, Indonesia, 1963 (800). Application of the technique of superimposed epochs revealed a weak but statistically significant "volcanic" signal in the long series of temperature observations. Attributable to powerful volcanic eruptions, the "global" temperature drop amounted to about 0.5 K and usually continued during the first and second year after the eruption. The time of temperature return to the normal level was about 2–5 years. The statistical significance level for the volcanic signal (an identical drop of temperature due to random variations) varied within 0.5 °C to 4 °C.

As an example of the superimposed epoch analyses of temperature sets for eruption years is shown in Figure 5.1.8 (Mass and Schneider, 1977). The composition is made from records of eruptions of large magnitude (DVI > 100) with at least 5 years' separation from any other major event. A sharp temperature fall of about 0.3 °C in the first year, and somewhat less in the second year after the eruption followed by a monotonic rise is obvious. The deviations of the first 2 years after the eruptions are different from the mean at the 99% significance level. Thus, Mass and Schneider (1977) concluded that a definite, albeit weak, volcanic signal is identifiable in the stations' temperature records.

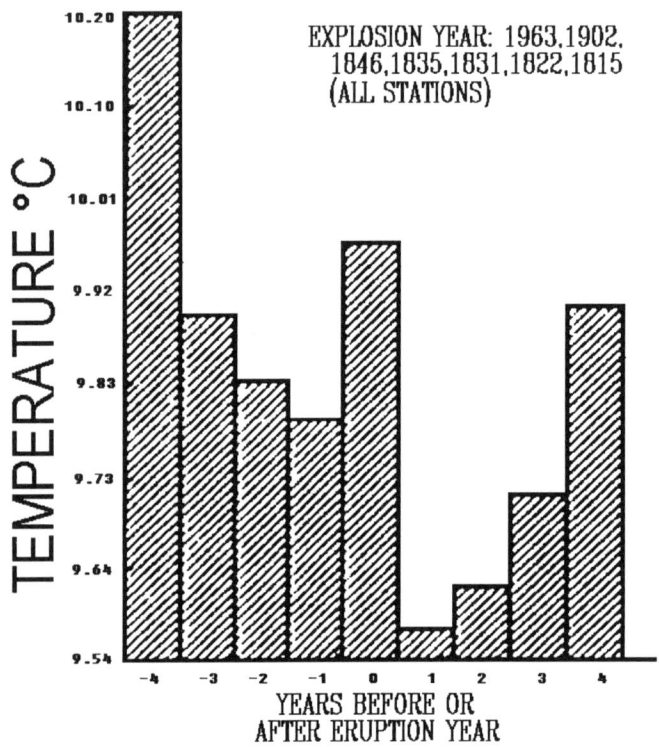

Figure 5.1.8. Superimposed epoch diagram of mean temperature at 42 stations during the years before and after several large eruptions. A statistically significant decrease of about 0.3 °C occurs during the year following an eruption. After Mass and Schneider (1977).

The effect of eruptions shows itself more strongly in high latitudes, but in this case it is combined with a much greater variability of the temperature field. Therefore, the statistical significance of a high-latitude signal is practically lower, and the duration of the signal is shorter than in low latitudes. Data on the annual change of temperature show that, as a rule, in summer the signal turns out to be stronger and lower. However, the reliability of this conclusion is limited because of the absence of information about the spatial and temporal distribution of the ejected products as well as the lack of a sufficiently homogeneous and representative series of temperature observations. Therefore, an important objective of further studies is a repeated analysis of data with the use of more homogeneous (although shorter) series as well as observational results for the last decades, which are more representative both in space and in time. Therefore, an objective analysis aimed at constructing global temperature fields may be useful.

Analyzing the data of Taylor et al. (1980), Ellsaesser (1983) emphasized the controversy in the interpretation of observational data from the viewpoint of the climatic effect of volcanic activity. Consideration of the chronology of eruptions has led to the conclusion about nonrepresentativeness (i.e., fragmentary character) of observations before 1960.

After the 1963 Agung eruption, Newell (1970) detected a 6 K warming of the stratospheric layer from 50 to 80 hPa (20 to 18 km) over Australia, but from data by McInturff et al. (1971) the temperature increase near the 50-hPa level in the tropics did not exceed 2 K. This showed the great difficulty of identifying a volcanically caused disturbance against variations due to other factors. Later studies by Angell and Korshover (1983a, 1983b) pointed to a 3 K warming of the layer from 30 to 100 hPa (24 to 16 km) in 1963, with a subsequent 2 K temperature decrease but did not reveal the occurrence of any stratospheric warming at extratropical latitudes.

Using the NOAA-6 temperature sounding data and aerological sounding, Parker and Brownscombe (1983) analyzed changes in the stratospheric temperature field following the eruption of El Chichón (the vertical resolution of satellite data was only 15 km). Data from the four tropical aerological stations point out a temperature increase in the lower stratosphere (at 30- and 50-hPa levels) up to 6 K at 30 hPa in July 1982 and 5 K at 50 hPa in August. The latter date of maximum warming at the 50-hPa level suggests that the temperature disturbance propagated downward (as after the Agung eruption).

An analysis of satellite indirect sounding data on the annual change of mean monthly temperatures for the latitude band of $15° N - 15° S$ and for the entire globe in 1980, 1981, and 1982 at levels 80, 50, 15, and 2 hPa (the "sharpest" weighting function and clearest observations correspond to the 50-hPa level) showed that in 1982 a temperature increase took place in the tropics at 50 hPa (as compared to 1980–81), with a maximum of 2 K in July and August. Maximum warming at 15 hPa (28 km) was observed in May, and near 2 hPa and 5 hPa cooling took place in July and August. Temperature variations at 80 hPa were uncertain.

Variations in the mean global temperature whose day-to-day MSDs were only 0.03 K revealed a considerable warming in 1982 at 80, 50, and 15 hPa, but probably at the two lower levels (80 and 50 hPa) it had begun before the eruptions. The latter circumstance, as well as a symmetry of warming with respect to the equator, suggests the supposition that it was partly determined by the QBO. This assumption is favored by the view that the eruptive cloud covered only the latitudinal band of $60° N - 20° S$, and by the probability of the downward-spreading disturbance connected with the quasibiennial oscillation.

However, since the amplitude of the observed warming exceeded that corresponding to the quasibiennial oscillation, there is no doubt that a volcanically caused stratospheric temperature increase was also manifested.

Although there are causes for the possible substantial warming of the stratosphere following large volcanic eruptions, even the analysis of data on the stratospheric temperature field after the Agung eruption (8° S, 115° E) in March 1963 was not sufficiently convincing from the viewpoint of filtering out all the factors, except the eruption. Therefore, Quiroz (1983) undertook a filtering out of the El Chichón signal, taking into account the following factors:

(i) the annual, semiannual change, and QBO;
(ii) sudden warming of the high-latitude stratosphere followed by cooling in the low latitudes; and
(iii) SST anomalies, especially those connected with the El Niño event.

Since the concentration of the eruptive stratospheric aerosols reaches its maximum several months after the eruption, emphasis was placed on variations in the temperature field in mid and late 1982. As for the contribution by the annual and semiannual change, it can be easily excluded by extracting climatological monthly means from the observed temperature values. It is much more difficult with other factors. Analysis of observational data of the mean zonal temperature at 15° N at 30 hPa showed that between April and June 1982 (after the eruption) a 2 °C temperature increase took place, but it is difficult to identify this warming as having been caused by the eruption, because it had begun in January, long before the eruption.

It seems that this warming might be produced by the December 1981 Nyamuragira (Zaire) volcano eruption, which is considered to be a major factor in the appearance of the "mysterious" volcanic cloud during the months preceding the El Chichón eruption.

Previous studies have led to the conclusion that about 6 months after a strong SST increase in the eastern equatorial Pacific (El Niño), a warming by 1 °C of the tropical troposphere takes place, followed by a cooling of the lower stratosphere. Since the 1982 SST increase started only in the spring and could be pronounced only in the fall and winter, no doubt the effect of the summer SST anomaly was negligible, but it must later be taken into account. Thus of major importance is an accounting of the QBO.

Based on an analysis of the spatial and temporal variability of QBO for the previous years, Quiroz (1983) showed that an altitude-dependent phase shift is typical of QBO at 30 hPa. Since the trend of stratospheric warming is manifested simultaneously, it becomes possible to consider a volcanically induced warming. The exclusion of the QBO contribution gave values of the eruptive residual variability varying within 1.0 to 3.0 °C in the 0–30° N band (with an error of 0.5 °C to 1.0 °C), whereas the total observed temperature anomaly at 30 hPa reached 6 °C at several locations, including Singapore (1° N).

Within the considered latitudinal belt a slightly increased warming took place south of 15° N, although the volcano's latitude is 17° N. This was probably connected to the maximum outgoing longwave radiation near the Equator, the absorption of which by the eruptive aerosols is a major energy source for a stratospheric warming. Bearing in mind substantial errors in estimates, it is in fact possible that a maximum warming took place in the latitudinal belt of 10–15° N. Although the volcanic signal is pronounced at other latitudes (e.g., at the 10-hPa level), the reliability of the respective data of observations must be thoroughly analyzed.

An interesting analysis of the effect of volcanic eruptions on the temperature field of the tropical troposphere was performed by Parker (1984). Having considered data of aerological soundings for the latitudinal belt of 20° S – 20° N for the period of 1950–1984 with seasonal averaging and the filtering out of the contribution by the Southern Oscillation to the formation of the temperature field, Parker concluded that the mean temperature of the levels 850 to 500 and 500 hPa during the year following the March 1963 Agung eruption was higher than before the eruption. A maximum temperature increase, reaching only 1 °C was observed 2 years after the eruption. No cooling was observed after the April 1982 El Chichón eruption (until the end of 1982). The cooling in the NH in the winters of 1982–1984 did not differ from the cooling in 1961 and 1971, neither of which was caused by an eruption. The results of observational data analysis that revealed the impact of eruptions are also given in the monograph by Loginov (1984) and in Loginov et al. (1983a, 1983b).

Definite correlations between climate variability and volcanic eruptions were found from an analysis of paleoclimatic data. It was shown, for example, in studies by Stothers and Rampino (1983a; 1983b) that between 1500 BC and the present time there had been a high correlation between volcanic eruptions and such characteristics as dry fogs in western Europe and acid rains in Greenland. Volcanic eruptions in the Mediterranean Sea and in Iceland are a reason for at least five of the nine layers of increased acidity detected in Greenland ice cores for this period.

A complete analysis of the present-day global temperature field variations between 1958 and 1982 was made by Angell (1986, 1989) and Angell and Korshover (1983a, 1983b). The analysis revealed a global-scale cooling of about 0.5 °C between 1958 and 1970 and a subsequent weak warming. From spring 1981 to spring 1982 a substantial cooling was, however, registered in the NH. Apparently, after the 1963 Agung eruption a 0.3 °C decrease of the surface air temperature in the NH took place.

Taking into account existing data on aerosols, calculations have been made of the variations of solar radiation absorbed by the surface-atmosphere system, and the outgoing longwave radiation (Kondratyev and Moskalenko, 1984). To calculate radiative fluxes, the doubling technique was used with due regard to multiple scattering in the case of shortwave radiation. The longwave (thermal) emission was calculated with the inclusion of molecular and aerosol absorption. Calculations of the surface temperature variations revealed the important role of the longwave radiation contribution causing a substantial decrease in surface cooling.

The calculations are in general agreement with cooling near the surface and warming of the stratosphere observed after powerful eruptions. The main contribution to the stratospheric heating is produced by the absorption of the upwelling longwave flux and not by solar radiation. Variations in volcanic activity from the late 19th century to the early 1940s show some agreement with the temperature changes. The sparse volcanic activity during 1940–1960 are inadequate to explain the observed global cooling. Apparently, also the cooling during the "Little Ice Age" cannot be ascribed to volcanic activity. Estimation of the climatic effects due to shuttle-generated aerosols shows that they are negligible. Similar results are found for the effect of supersonic stratospheric aviation (SSA). Even assuming a very optimistic increase of SSA, the T_s variations should not exceed 0.1 K. Probable trends in the volcanic activity and in anthropogenic aerosol sources indicate that these factors are likely to be smaller than the predicted increasing greenhouse effect of carbon dioxide and other radiative gases such as O_3.

Because simulation modeling of volcanic climatic impact has been mostly concentrated on assessment of a long-term impact of stratospheric sulfur acid aerosol resulting from gas-to-particle volcanic SO_2 conversion, Gerstell et al. (1995) have considered the radiative and dynamical perturbations associated with the short-lived but more strongly absorbing sulfur dioxide and ash clouds. Using radiative transfer model calculations as well as observational data from satellites, aircraft, and ground-based instrumentation, Gerstell et al. (1995) have estimated the amplitudes of the stratospheric radiative heating rate perturbations produced by each of these components during the first few weeks after the El Chichón eruption, before the sulfur acid aerosol cloud fully developed.

One week after the April 4, 1982 eruption, net radiative heating rate perturbations exceeding 20 K per day were found at altitudes near 26 km, which were mainly due to absorption of solar radiation by the silicate ash. The sulfur dioxide gas and sulfuric acid aerosols each produced net heating perturbations that never exceeded 3 K per day. Observations indicate, however, that in spite of strong heating due to ash absorption stratospheric temperatures never increased by more than a few degrees Kelvin. Gerstell et al. (1995) show that the

VOLCANIC ACTIVITY AND CLIMATE

radiative heating was largely balanced by upwelling and adiabatic cooling. The results considered demonstrate a necessity to take into account the radiative forcing by the ash and the sulfur dioxide gas, in particular in connection with its influence on the vertical and horizontal dispersal of the volcanic plume.

Calculations for the post-Agung stratospheric conditions revealed that, for this case, a stratospheric temperature increase was caused by the increased absorption of upwelling longwave radiation. At later stages, when the small-size fraction of silicate and H_2SO_4 aerosols is dominant, the surface temperature decreases. Since the cooling effect is the longest and most substantial, the whole post-eruption period is characterized by cooling of the Earth's surface, the extent of which is determined primarily by the effect of the H_2SO_4 aerosol (Kondratyev and Moskalenko, 1984).

After an individual large-scale volcanic eruption, the optical thickness of the stratosphere in the visible range can reach 0.3 in some regions, 0.2 in many regions and may remain so for several months. This is exactly the case after the El Chichón eruption. An optical thickness of about 0.1 is quite typical for post-eruption periods, and usually lasts for at least a year. Modeling such an increase of optical thickness gives a decrease of mean global temperature of the Earth's surface by about 1 K. An analysis of possible climatic implications of volcanic eruptions (injection of water vapor to the stratosphere, photochemical processes in the ozone layer, etc.) leads to the conclusion that the radiative effects of stratospheric aerosols play the main role. Note, however, that possible change of tropospheric cloud characteristics due to aerosols is an important feedback neglected in the calculations which may seriously affect estimations of ΔT_s.

Earlier calculations have shown that the main influence of stratospheric aerosols on climate arises from the growing reflection of solar radiation that depends substantially on surface albedo: the contribution of aerosols to the decrease of atmospheric transparency for longwave radiation being comparatively small. Harshvardhan and Cess (1975; 1976) and Harshvardhan (1979) undertook new calculations of the sensitivity of the Earth's surface-atmosphere system albedo to the properties of stratospheric aerosols at varying values of surface albedo and solar zenith angle.

In the absence of the stratospheric aerosol layer, the system's albedo can be written as:

$$A_p = A_c c + A_s (1-c) \tag{5.1}$$

where A_s is the albedo for clear sky and A_c is the albedo in presence of clouds; c is the cloud amount. The expression for A_p may be used on a global basis or for individual latitudinal belts.

The system's albedo is calculated for 10°-latitude belts with due regard to observational data. Calculations yielded mean-monthly mean-zonal A_p profiles, as well as a profile averaged over a year, taking account of monthly values of insolation (as weighting factors). Mean-annual values of the system albedo obtained in such a way closely correspond to those of minimum albedo found by Vonder Haar and Ellis (1975) from satellite observations. (Substantial discrepancies occur only near the equator and in high latitudes due to cloud effects not excluded completely in the processing of satellite observations.)

The technique mentioned has been used to calculate variations of mean-global albedo caused by stratospheric aerosols for a cloud amount taken from climatological data, as well as variations of mean-zonal albedos for individual months. A good agreement was obtained between calculated (0.308) and observed (0.306) values of mean-annual albedo for the NH.

Consideration of the effect of stratospheric aerosols on the outgoing longwave radiation allowed estimation of changes in the radiation budget of the Earth's surface-atmosphere system. Sensitivity of the system albedo A_p for a hemisphere to stratospheric aerosols is determined by the derivative $dA/d\tau$, where τ is the optical thickness of the aerosol layer at $\lambda = 0.55$ μm. Since for the stratospheric aerosol layer $\tau \ll 1$ the albedo change is directly proportional to optical thickness change.

Harshvardhan (1979) had assumed the aerosol to consist of droplets of a 75% solution of sulfuric acid, its size distribution being described by modified γ-distribution, and phase function by the Henyey-Greenstein function. In this case the extinction coefficient at $\lambda = 0.55$ μm is identical to the scattering coefficient $1.1 \cdot 10^{-4}$ km^{-1}, the single-scattering albedo is 1, and the asymmetry of the phase function is 0.73. The effect of stratospheric aerosols has been calculated using the approximation of an optically thin layer. An infrared radiative transfer model is used to estimate the increased greenhouse effect attributed to the aerosol layer. The infrared heating tends to compensate for the albedo effect in altering the radiation balance.

The results show that the dominant influence of an optically thin stratospheric aerosol layer appears as an increase in reflected solar radiation over the globe, except for the winter polar regions where there is no solar insolation. Analysis of mean monthly values of $\Delta A/\Delta\tau$ reveals substantial dependence on solar zenith angle shown by a latitudinal-temporal variability within 0.08–1.0 (in percent at $\Delta\tau = 0.01$). In high latitudes the effect of surface albedo is also great.

The results of numerical modeling confirm the inadequacy of the assumption (used in energy balance climate models) of the equivalent climatic effects of solar constant changes and increases in stratospheric aerosol concentrations, since even in the presence of a globally homogeneous aerosol layer its effect on the distribution of solar radiation attenuation turns out to be inhomogeneous.

For latitudes below 50° variations of the radiation budget are sufficiently homogeneous and small, due to the compensating contribution of the enhanced greenhouse effect in the atmosphere. Calculations of the sensitivity of the greenhouse effect (expressed in Wm^{-2}/τ_{vis}) have shown that it is maximum in the tropics (13.9 at clear sky and 7.7 at overcast) and it is minimum in the winter sub-Arctic (5.3 and 4.0, respectively). Thus, an increase of optical thickness by 0.1 in the clear-sky tropics leads to enhancement of the greenhouse effect by 1.4 W/m^2.

An increase in the temperature of the aerosol layer due to absorption of the upwelling longwave radiation flux makes a substantial contribution to variation of the greenhouse effect. The strongest perturbations occur in spring and fall in the polar regions, when the difference in the equator-to-pole radiation budget increases by 6.5–7.0 W/m^2 for an aerosol optical thickness of 0.1 at λ = 0.55 μm. In the polar night, the radiation budget of the system grows due to enhancement of the greenhouse effect. Being caused by stratospheric aerosols, variations of the radiation budget are small as compared to values of the budget that vary from -175 W/m^2 to 11 W/m^2, and at $\Delta\tau$ = 0.1 do not exceed the errors of estimation of mean-monthly values (3–14 W/m^2). However, even such small changes of the radiation budget can substantially affect climate.

Calculations with due regard to spectral dependence of albedo (determined by Rayleigh scattering, in particular) have shown that selection of the effective wavelength equal to 0.53 or 0.55 μm usually leads to an almost maximum value of $dA_p/d\tau_{vis}$ (A_p is the spectrally averaged albedo of the system). Also of great importance is consideration of the dependence of the albedo on the solar zenith angle. Calculations of $dA_p/d\tau_{vis}$ performed by Cess et al. (1981) with due regard to the above-mentioned factors have led to the conclusion about the unacceptability of "one-wavelength" approximation and the necessity to bear in mind the dependence on the solar zenith angle. A mean-global value of $dA_p/d\tau_{vis}$ for average cloud conditions was found to be 0.092, i.e., it agreed well with the data of Pollack et al. (1976a, 1976b). An important conclusion resulting from the albedo sensitivity computations is the necessity of including solar zenith angle effects in considering albedo enhancement.

Assessment of post-eruption variations in the ERB is a key aspect of the climatic impact of volcanic eruptions. Using the two-stream approximation and a three-layer model of the surface-troposphere-stratosphere system, Lenoble (1986), as shown in Table 5.1.2, estimated the effect of the stratospheric aerosol layer consisting of a 75% water solution of sulfuric acid droplets or background aerosol particles on the system's radiation budget and its components, as well as on the surface temperature (the surface is considered to be an isotropic reflector).

Table 5.1.2. The Size Distribution and Optical Properties of Stratospheric Aerosols*. After Lenoble (1986). [© by Elsevier Science Ltd. Reprinted with permission of the publisher.]

Parameters Aerosol model	α	δ	σ	$\omega_s$eff	$b_s$eff	$c_s=\tau_{IR}\text{eff}/\tau_s\text{eff}$
Background aerosol	1.0	18	1.0	0.998	0.189	0.0383
Volcanic aerosol	1.0	16	0.5	0.994	0.169	0.0217

* $\omega_s$eff, $b_s$eff are effective values of the single scattering albedo and backscattering; $\tau_s$eff, τ_{IR}eff are effective values of the optical thicknesses for the shortwave and longwave radiation, respectively; τ_s=0.03.

The distribution of the particles number density (n) by radius (r) is prescribed by a modified gamma function:

$$n(r) = A^{\alpha} \exp(-br^{\gamma}), \tag{5.2}$$

where A, α, b and γ are parameters. A comparison of the applied two-stream approximation has shown that the errors of an approximate calculation of the diffuse transmission do not exceed 10%, with the single scattering albedo $\omega \leq$ 0.6, optical thickness $\tau \leq$ 0.5, and normally incident solar rays. Here the system's albedo is calculated accurately, and the error in calculations of albedo variations, due to changed properties of the stratospheric aerosol layer, does not exceed 10%.

The calculations of the sensitivity of mean monthly albedos of the system to the choice of the aerosol model almost always revealed, in both cases, an increase of albedo by several tenths of a percent. The Table 5.1.3 data illustrate variations in the system's radiation budget decreases since the contribution by backscattering prevails.

An examination of the mean global surface temperature decrease (τ_p = 0.03) gave -0.045 °C (volcanic aerosol) and -1.1 °C (background aerosol). An account of the albedo feedback resulted in an intensified cooling down to -0.8 °C and -2.15 °C, respectively. A weaker effect of volcanic aerosols on climate would exist if there were a lower single-scattering albedo (e.g., a stronger absorption).

Gilliland (1982), and Gilliland and Schneider (1984) developed a new seasonal energy balance climate model in which 10 units are considered that characterize average conditions for the hemispheres: the atmosphere over the land, the atmosphere over the ocean, land surface, mixed layer and deep layers of the ocean, the SAT being a predictant. All principal input parameters—depth of the mixed layer, cloud amount, albedo, insolation, hemispherical energy

VOLCANIC ACTIVITY AND CLIMATE

Table 5.1.3. Variations in the System's Radiation Budget (RB), ΔB (W/m^2). After Lenoble (1986). [© by Elsevier Science Ltd. Reprinted with permission of the publisher.]

Aerosol model	RB variations due to:			
	Backscattering	Shortwave Radiation Absorption	Greenhouse effect	Total
Background aerosol	-28.5	0.0923	1.56	-26.0
Volcanic aerosol	-26.5	2.28	0.887	-23.2

exchange—vary during a year and are, whenever possible, prescribed from observational data or assessed from physical considerations. The values of unreliably estimated parameters, for example, the coefficients characterizing the interaction between the units have been chosen under condition of minimum MSD from observed seasonal mean SAT. The time step is 1 day, the interval of averaging 1 month. The spatially weighted hemispherical mean SAT over the land in the Northern (5°–85° N) and Southern (0°–45° S) Hemispheres were taken as the principal characteristics of the observed climate, which provides for the most reliable comparison with observational data (though in this case, too, the limitation of the observational data should be borne in mind).

Based on the use of this model, Gilliland and Schneider (1984) obtained estimates of the contribution to climate change of three most widely discussed external factors of climate: volcanic eruption, increasing CO_2 concentrations, and solar constant variations. An analysis of the numerical modeling results has made it possible to reveal the existence of independent volcanic impacts on climate in both hemispheres, whereas the manifestations of the effects of CO_2 and extra-atmospheric insolation (the solar constant sinusoidal cycle was taken with a period of 76 years) are of global scale. An estimation of the mean global SAT increase due to a CO_2 doubling gave 1.6 ± 0.3 °C; its statistical significance remains, however, uncertain because of the use of numerous adjusting parameters. With an account of this estimate, the increase of CO_2 from 314 ppm in 1959 to more than 340 ppm in 1983 (i.e., by 8%) should have caused an equilibrium warming contributing 0.33 ± 0.17 °C.

With the time period for the content of anthropogenic sulfate aerosol to double in the atmosphere assumed to be 10 years (this agrees with some observational data), then in 3 to 4 decades, the content of the background stratospheric aerosol will reach the level observed after strong volcanic eruptions: Agung, El Chichón, Pinatubo. Naturally, in these conditions the CO_2-induced warming will be substantially compensated. This testifies to an extreme urgency of monitoring the dynamics of stratospheric aerosol and detecting CO_2 climatic signal. One of the most reliable ways of detecting the CO_2 signal consists in improving the reliability of the estimates of other factors of climate change, whose contributions must be filtered out for the CO_2 signal to be detected.

Apparently, any climate signal, i.e., that caused by either volcanic eruptions or increasing CO_2 concentration, can be detected only with observational data averaged over vast regions. The most rational will be an averaging with weight, taking into account the results of numerical modeling from the viewpoint of revealing the geographical regions and periods of the year with the most substantial climate changes. Belle and Abdullah (1985) considered an example of such averaging of the SAT observational data for 1958–1981 as the background ones and for a later period as a source of information about the signal of the El Chichón eruption, with an account of numerical modeling results. According to these results a maximum post-eruption SAT decrease should have been observed in 1983, reaching 0.3–0.5 °C in the tropics and 0.7–0.2 °C in polar regions. The processing of data for 1983 revealed, however, a warming, whose signals exceed 2σ, which should, apparently, be explained by the effect of the almost simultaneously onset of an ENSO strong event.

From the data of a global network of 63 aerological stations for the period winter 1957–58 to summer 1983, Angell and Korshover (1983a) analyzed the variability of tropospheric temperature after the 1963 Agung and 1982 El Chichón eruptions. Seven climatic zones in both hemispheres, and the globe on the whole were considered. For extratropical latitudes, information has been obtained on SAT and the temperature of the layer 850–300 hPa (1.5–9 km), and for the tropics, calculations have been made of the temperature of the layer 300–100 hPa (9–16 km). The representativity of the data for the SH needs further studies.

The results obtained by Angell and Korshover (1983a) show that during the first year after the eruption of Agung, a decrease of average SAT for the NH by 0.34 °C was observed, and the El Chichón eruption was followed by a temperature increase by 0.37 °C (at the level of 0.5% statistical significance, in accordance with the t-test). These changes in the hemispherical mean temperature are caused, largely, by the contribution of the midlatitude band, where the respective variations reached -0.36 °C for Agung and 1.27 °C for El Chichón. Apparently, the trend of temperature increase is connected with the effect of an abnormally strong positive SST anomaly in the eastern equatorial Pacific associated with the ENSO event of 1982–83. The connection between the SST anomaly and SAT changes in the tropics is direct, and in midlatitudes indirect.

VOLCANIC ACTIVITY AND CLIMATE

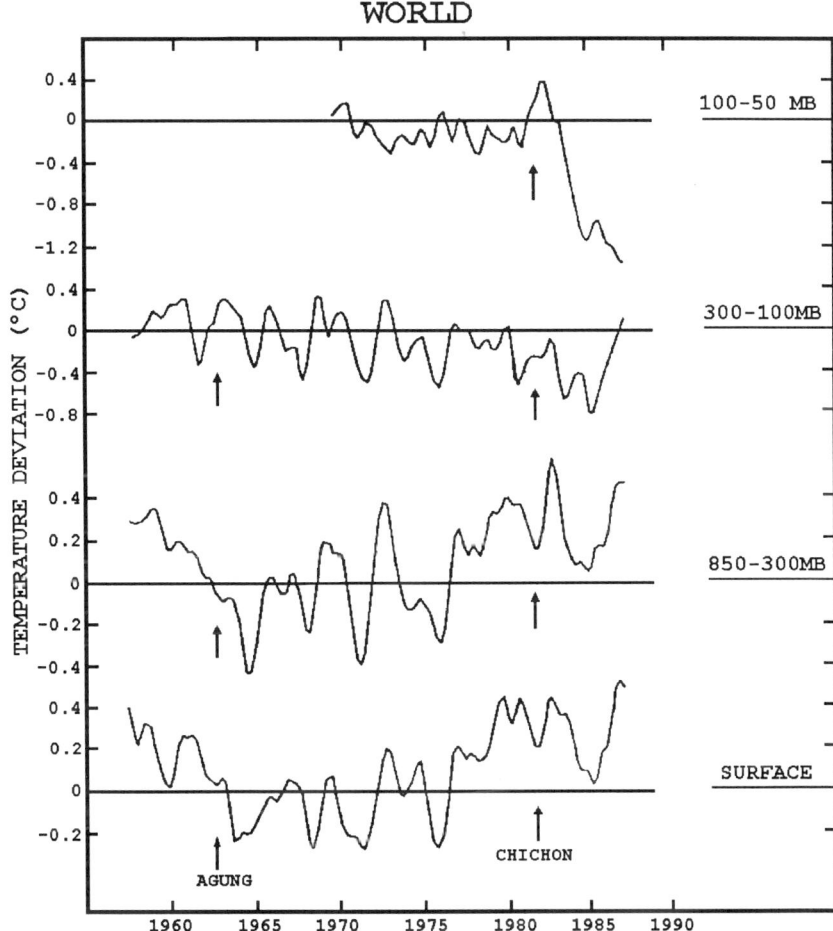

Figure 5.1.9. Mean global temperature anomalies at the surface and in the free atmosphere. After Angell (1989).

If we use the results obtained to filter out the El Niño signal, we find the same cooling of the atmosphere, about 0.5° C, by amplitude, took place after the El Chichón eruption, as after the Agung eruption but more shortlived. Application of such correction procedure to observational data after six strong volcanic eruptions in the tropics for the last century—Krakatau, Soufrière, Pelée, Santa María, Agung, El Chichón—has led to the conclusion that a decrease of the mean surface temperature of the NH continents is about 0.3 °C, whereas after the Katmai eruption in Alaska it was only 0.1 °C. As for the cooling of the whole Earth's surface, in this case the data on the VS in the past

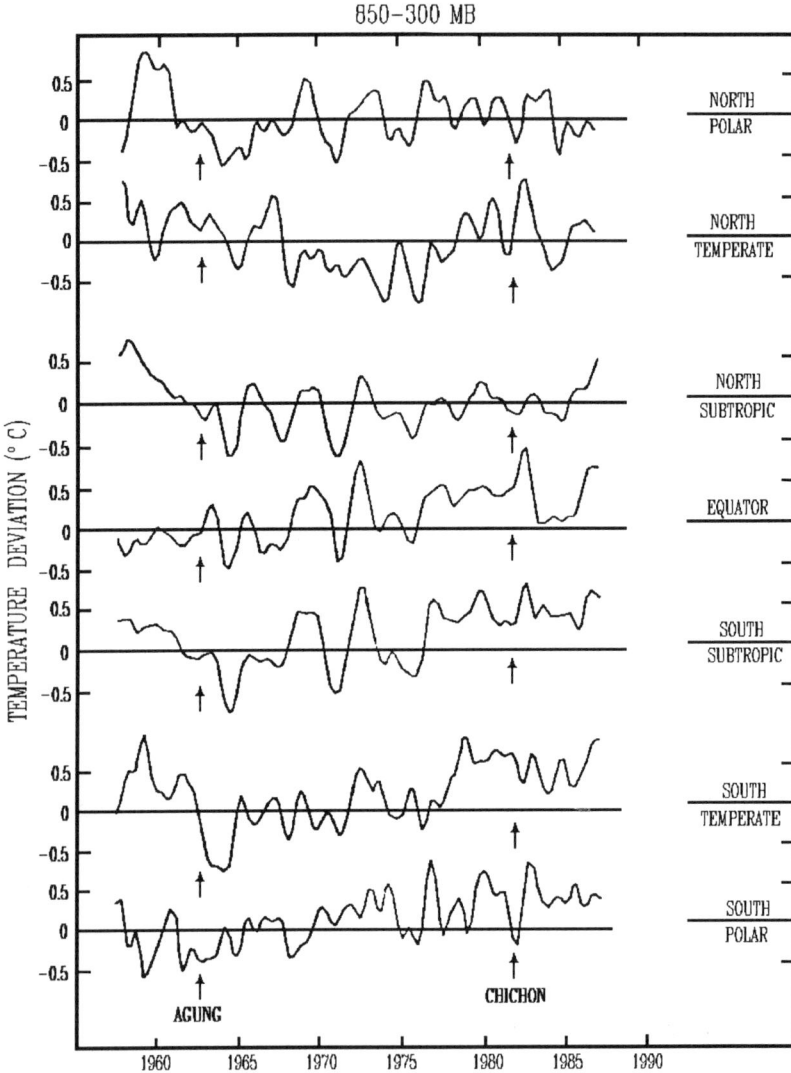

Figure 5.1.10. Temperature anomalies for the 850- to 300-hPa layer for various latitude belts. After Angell (1989).

are uncertain and controversial. Angell (1989) undertook a new effort to analyze the available observational data to detect the volcanic signal, as shown in Figures 5.1.9 through 5.1.13.

VOLCANIC ACTIVITY AND CLIMATE

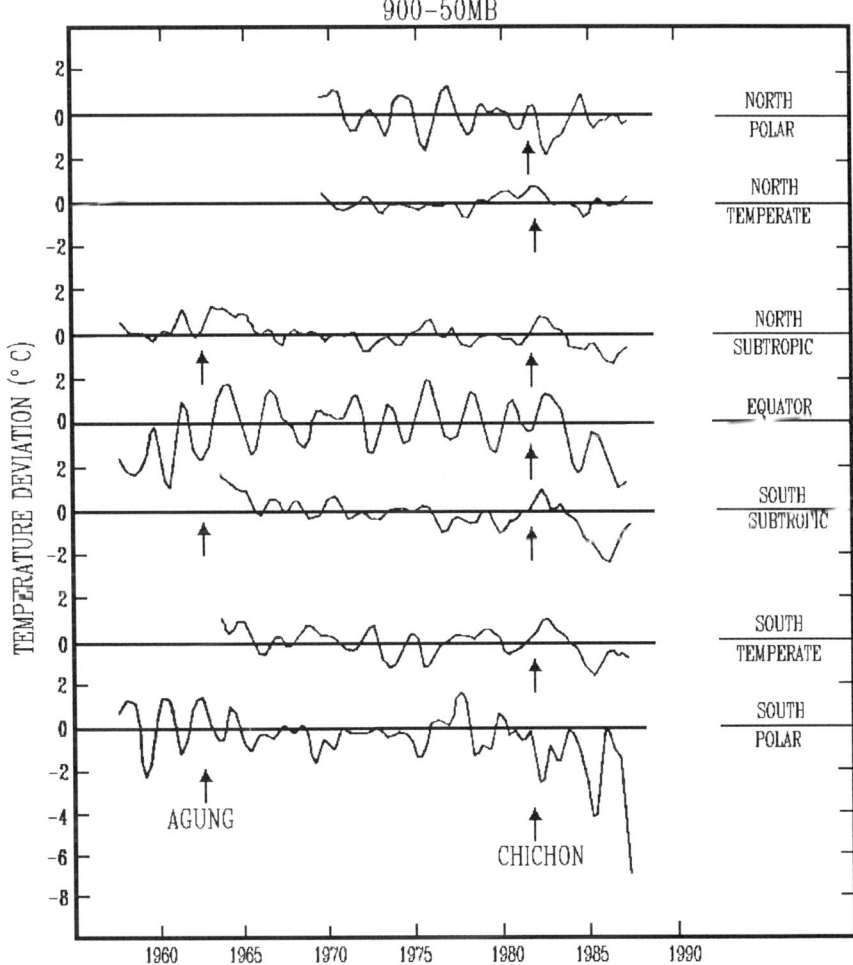

Figure 5.1.11. Temperature anomalies for the 100- to 50-hPa layer for various latitude belts. After Angell (1989).

So far, the most convincing evidence for the post-El Chichón tropospheric cooling, in accordance with theoretical estimates, is a decrease of zonal mean SAT in the subtropics of the NH (10–30° N). In contrast to this, a sharp warming took place in the region 0–10° S, 180–80° W. Since the SST had returned to the norm by late 1983, one could expect a manifestation of the El Chichón-induced temperature decrease in 1984. The results considered suggest a conclusion about the impossibility to detect the VS without taking other factors of climate formation into account (see also Chapter 1).

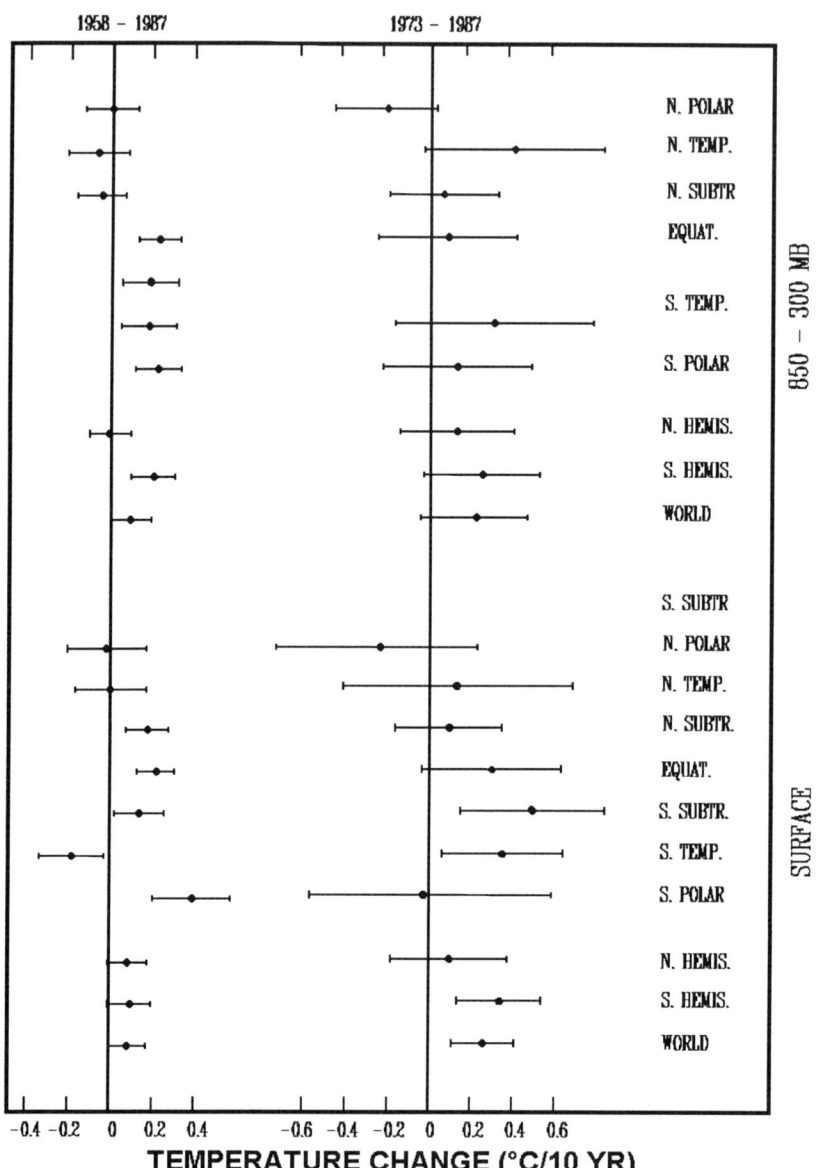

Figure 5.1.12. Decadal temperature changes at the surface and in the 850- to 300-hPa layer (°C/10 years) based on 1958–1987 and 1973–1987 observational data. After Angell (1989).

VOLCANIC ACTIVITY AND CLIMATE

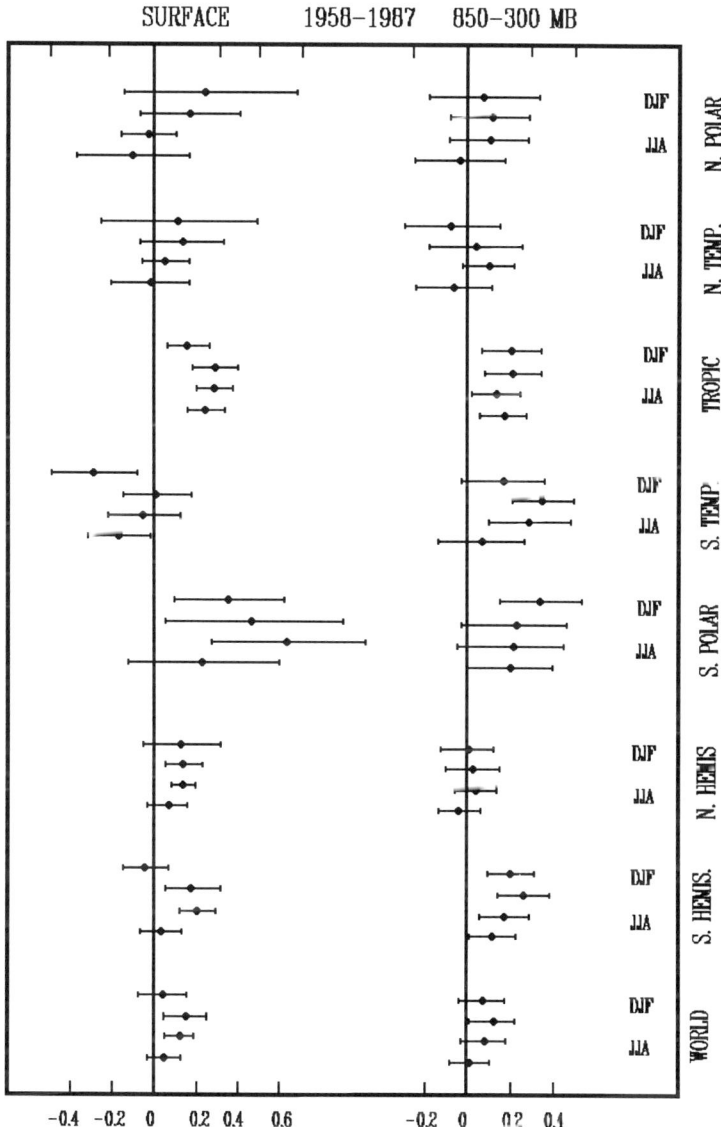

Figure 5.1.13. Temperature changes at the surface and in the 850- to 300-hPa layer for the time period 1958–1987 in various seasons. After Angell (1989).

To estimate approximately the effect of an increasing concentration of stratospheric aerosols on the atmospheric temperature regime in low latitudes where the Agung effect was maximum, Hansen et al. (1977, 1978, 1981, 1992) and Lacis et al. (1992) used a one-dimensional model of radiative convective equilibrium. The case of a "normal" atmosphere at a stratospheric aerosol optical thickness of 0.005 was taken as a control background level. It was assumed that 20%, 60%, and 80% of the post-eruption aerosols were located, respectively, in the layers 20–6, 19–22, and 16–19 km, and that the aerosols were sulfuric acid. The heat capacity of the surface layer was equivalent to a 70-m layer of ocean water.

Comparison of calculated temperature changes with those observed in Port Hedland, Australia, at 60 and 100 hPa, showed reasonable agreement, especially noting that part of the observed temperature increase could have been caused by the QBO. Calculated tropospheric mean temperatures in the 30° N – 30° S latitudinal belt, starting from the time of eruption until early 1966 also agreed with observational data, a temperature lowering of about several tenths of a degree being observed, with a time constant of about a year. The significant increase of temperature at ~ 20 km in the stratosphere was shown in Fig. 5.1.11.

Table 5.1.4 shows calculations for the cases of the sulfate aerosol that weakly absorbs solar radiation, and the silicate aerosol that strongly absorbs it.

Table 5.1.4. Calculated Temperature Changes in the Stratosphere and Troposphere Caused by Sulfate and Silicate Aerosols. After Hansen et al. (1977, 1978). [©American Association for the Advancement of Science. Reprinted with permission of the publisher.]

Number of days	30	60	120	180	360	540	720	1000
Sulfates								
T_{55}	0.63	2.15	4.77	5.34	3.75	1.90	0.84	0.02
T_{tr}	-.01	-.03	-.12	-.23	-.48	-.54	-.51	-.44
Silicates								
T_{55}	3.8	8.1	12.8	13.1	10.5	7.4	5.2	2.8
T_{tr}	-.01	0.03	0.08	-.13	-.22	-.23	-.19	-.13

T_{55} – Stratospheric temperature at 55 LPa
T_{tr} – Mean tropospheric temperature

The considerable difference between the results show the need for adequate treatment of aerosol optical properties. The properties of "real" aerosols are probably intermediate to those of sulfates and silicates.

The results show that the sign and phase shift of temperature changes in the troposphere and stratosphere calculated with a one-dimensional climate model, agree well with those observed after the Agung eruption. They also indicate the usefulness of an approximate model to evaluate the climatic effect of changing atmospheric composition.

The problem of the reliable identification of volcanic climatic signal still attracts serious attention. The most complete analysis of volcanic climatic impact (with regard to surface air temperature variations) has been recently conducted by Robock and Mao (1995) who examined the regional and seasonal patterns of SAT effects of large volcanic eruptions using climate records of the past 140 years. This study was focused on the short-term (a few years) effect of historic major volcanic eruptions, even though a long-term effect is possible, due to long-time scales of oceanic response for a few decades after the eruptions as shown earlier (see, for instance, Robock, 1978, and Rind et al., 1990, 1992).

Robock and Mao (1995) have persuasively demonstrated that the major explosive volcanic eruptions produced clear temperature effects in different seasons and location. An important aspect of the data processing has been filtering out of the low-frequency variations and El Niño/Southern Oscillation signal. In general, volcanic effects are cooling but include a winter warming signal in high-latitude NH continents. For 2 years following great volcanic eruptions, the surface cools significantly by 0.1–0.2 °C in the global mean, in each hemisphere, and in the summer on the latitude bands 0°–30° S and by 0.3 °C in the summer in the latitude band 30°–60° N. By contrast, in the first winter after major tropical eruptions and in the second winter after major high latitude eruptions, North America and Eurasia warm by several degrees, while northern Africa and southwestern Asia cool by more than 0.5° C.

In cases of simultaneous ENSO events, the warming produced by the ENSO masked the volcanic cooling during the first year after the eruption. The timescale of the ENSO response is only 1 year while the volcanic response timescale is 2 years, so the cooling in the second year is evident whether the ENSO signal is removed or not.

The results mentioned, both the global cooling and NH continental winter warming, agree with general circulation model calculations. Robock and Mao (1995) have emphasized that several issues involving volcanoes and climate remain to be investigated, including the possible cause-and-effect relationship with ENSO and the possible long-term effects from the period of volcanism, such as the warming of the 1920s and 1930s during a period of virtually no volcanoes. More efforts are required to separate ENSO and volcanic signals more reliably, in particular over North America. It is still not quite clear how much of the winter warming over North America is volcanic and how much is due to ENSO.

Portman and Gutzler (1996) have applied the superimposed epoch method to search for the volcanic and ENSO signals in U.S. surface climate records. For this purpose anomalies for monthly mean SAT and total precipitation (TP) taken from the U.S. Historical Climatology Network were composed (averaged) over years of major explosive volcanic eruptions, ENSO warm events and ENSO cold events since the year 1900. It was assumed that volcanic eruptions and ENSO events occur independently of each other. After assessing for significance of all composite anomalies with regard to several statistical and physical criteria the composite ENSO-related anomalies were subtracted from anomalies of temperature and precipitation associated with the volcanic eruption.

After filtering out ENSO-related anomalies it has been found out that volcanic signals are strongly suggested east of the Continental Divide, for example, where positive monthly temperature anomalies exceeding 1 °C occur during the first fall and winter after eruption. As Portman and Gutzler (1996) found, negative temperature anomalies occur west of the Continental Divide during the first winter and spring after eruptions and in the southern United States during the summer of the first post-eruption calendar year. Positive monthly precipitation anomalies exceeding 15 mm in magnitude are found in the southeastern United States during the first winter and spring after eruptions. Precipitations anomalies that are smaller in magnitude and yet significant, such as positive anomalies in the northwestern United States, are found during the summer of the first post-eruption calendar year.

Table 5.1.5 data represent a partial summary of some recent studies of an impact of volcanic eruptions on SAT changes in the U.S.A. An important difference between various sets of data is the removal of ENSO signals. The data under consideration demonstrate clearly the importance of filtering out ENSO signals.

Since in the study by Portman and Gutzler (1996) the composite variations in SAT anomalies that are associated with the largest volcanic eruptions closely resemble those composite variations associated with ENSO cold events, removal of signals associated with the coincident ENSO events (all are warm) was found to strength rather than weaken apparent volcanic signals over many grid regions.

Portman and Gutzler (1996) have emphasized difficulties to characterize the volcanic aerosol forcing due to numerous variable parameters, such as eruption latitude, time of year, plume height, and amount of sulfur produced. Large variations of such parameters from one eruption to the next, plus the small number of major volcanic eruptions during modern times, seriously complicate an empirical analysis of observational data and illustrate the importance of the relevant climate modeling in the form of sensitivity calculations.

VOLCANIC ACTIVITY AND CLIMATE

Table 5.1.5 Approaches and results of some recent investigations of volcanic eruptions and USA temperature variations. After Portman and Gutzler (1996).

	Groisman (1985–1992)	Lough and Fritts (1987)	Robock and Mao (1992, 1995)	Portman and Gutzler (1996)
Was any ENSO signal removed?	No	No	Yes	Yes
Shortest averaging interval	Season	Season	Month	Month
Earliest-latest eruption years	1815–1982	1631–1888	1875–1991	1902–1982
Number of eruption years	9–10	26	12	10
Summary of major results	Cool in summer, U.S. East Coast.	Warm (especially in winter), western U.S. Cool (especially, central in summer) and eastern U.S.	Warm in first winter after tropical eruption (below 30" lat.) and in first or second winter after other eruption, continental U.S.	Warm in first fall and winter, eastern U.S. Cool in first winter and spring, western U.S. Cool in summer, southern U.S.

5.2 Numerical Modeling of the Climatic Impact of Stratospheric Aerosols

An effective way to examine the "observed" impact of volcanic eruptions is through the use of sophisticated climate models, making use of observational data. Several model calculations of the effects of volcanic dust on climate have been performed during the recent years. These included use of one-dimensional radiative convective models (Hansen et al., 1978; Kondratyev and Moskalenko, 1984; Pollack and Ackerman, 1983; Pollack et al., 1976a, 1976b, 1983, 1993), zero-dimensional empirical models (Oliver, 1976), zero-dimensional energy balance models (Harshvardhan and Cess, 1976; Schneider and Mass, 1975), a one-dimensional energy balance model (Robock, 1978, 1984); a two-dimensional zonal model (MacCracken, 1976; MacCracken and Potter, 1975; MacCracken and Luther, 1983a, 1983b), and three-dimensional general circulation models (Hansen et al., 1992; Hunt, 1977, 1978; Pollack et al., 1993; Rind et al., 1992; Tanre et al., 1984). The models all concentrate on surface air temperature as the calculated climate variable, although Hansen et al. (1978) and Pollack et al. (1976a, 1976b) also looked at stratospheric temperatures, and Hunt (1977) investigated circulation effects. Other aspects of the climate system are undoubtedly affected by volcanic eruptions but have not been included in modeling studies.

In some of the models the ash and sulfate from the eruption are treated separately with different absorption properties. In addition, in the work by Bryson and Goodman (1980) ash is considered to fall out rapidly, with mean stratospheric residence time about 100 days while sulfate is considered to evolve with a gas-to-particle conversion time constant of 115 days and a mean residence time of 400 days.

Although different assumptions about the details of the volcanic forcing in the climate models, and different sets of surface temperature records were used in the studies, all of them agree on one point: large volcanic eruptions cause a reduction of hemispheric or global average surface temperature for a period of a year or two and may be an important cause of climate change on the interannual to 500-year time scale. More definitive quantitative results await future improvements in volcanic dust chronologies, temperature reconstructions, and climate models. Some of the most recent publications on models have been considered in the Introduction.

To a reasonable approximation, the lower 50 km of the atmosphere can be modeled as a two-layer system, consisting of the troposphere in convective equilibrium and the stratosphere in approximate radiative equilibrium. The general circulation of the troposphere is characterized by heat sources located at low levels in equatorial and subtropical latitudes. Heat sinks are located in the middle and upper atmosphere and in polar latitudes. Stratospheric circulation is generated by the latitudinal gradient of radiative heat flux divergence in the 20- to 50-km layer, which is equatorward in the summer hemisphere and poleward in the winter hemisphere. Also of great importance is the energy transfer from beneath. The general features, together with boundary layer conditions, determine the static stability and motions in the troposphere and stratosphere, and the difference in the relative contributions of vertical and horizontal motions to the transfer of heat, momentum and various pollutants. An important feature is the great difference between the characteristic lifetimes for trace gases in the troposphere (1–2 weeks) and in the stratosphere (1–2 years). Also, the role of minor gaseous constituents in chemical and photochemical processes differ considerably in the troposphere and stratosphere.

MacCracken (1976) applied these concepts in a two-dimensional zonal model of atmospheric general circulation ZAM-2 to estimate climate sensitivity to variations in the ozone layer, stratospheric aerosols and water vapor, as well as to variations in the solar constant. The first stage of the ZAM-2 test calculations was aimed at comparing calculated fields of meteorological parameters with the observed ones.

Calculations made with insolation held constant during the year and then with due regard to annual and diurnal changes showed that a persistent climatic regime sets in much faster in the first case. Using mean-annual insolation, calculated and observed temperature fields fit rather well, except for the lower stratosphere at polar latitude where the calculated temperatures are overestimated. Agreement between mixing ratio fields is also satisfactory. The calculated precipitation distribution underestimates the precipitation minimum in the subtropical belt. The fields of the total atmospheric water vapor content and cloud amount fit rather well.

Significant differences are found in the mean-annual fields of the zonal velocity component, but the results appear to be better when the annual change of insolation is taken into account. The calculated meridional circulation shows only one Hadley cell rather than the observed three cells, but the calculated meridional energy transfers agree well with those observed, being primarily advection in low latitudes and macroturbulent transfer in mid and high latitudes. In addition, ZAM-2 satisfactorily reproduces the various heat budget components for the surface and in the atmosphere.

On the whole, the version of the MacCracken (1976) model with fixed mean-annual insolation, appears to be useful for estimations of the sensitivity to various climatic factors. Calculations of the effect of aerosol variations in the 75- to 150-hPa layer were made for aerosol contents of zero and 0.43 μgcm^{-2} (background content), and for values exceeding this background content by factors of 2 and 4 The complex refractive index of particles is assumed to be 1.45 ± 0.005, which corresponds to weakly absorbing particles.

Analysis of calculated surface temperature shows that the effect of aerosols is stronger and less regular in high latitudes. Similar conclusions could be drawn from the analysis of observational data discussed previously. In the case of the fourfold increase of the aerosol content (about equivalent to a 1% decrease of the solar constant), the surface temperature is lower by about 1 K in low latitudes. There is a maximum decrease of several degrees near latitude 70° N, which gives rise to snow cover on land, but not far from the southern boundary of the snow. A warming effect (about 2 K) is observed in association with decreases in cloud amount and planetary albedo. The decreases were due to lower evaporation in midlatitudes. Precipitation also decreases here by 3.6%. As can be concluded, consideration of varying cloudiness may lead to some radical changes, e.g., a decrease rather rather than an increase, in the planetary albedo with the dust-loading of the stratosphere. Calculations for mean-global conditions showed that for a tripling of the stratospheric aerosol content (apparently the maximum possible effect of SSAs), the planetary albedo rises by 0.4%–0.6% due to the increased scattering by aerosols. The actual situation is, of course, more complicated due to the interaction between aerosols and cloudiness.

The effect on the hydrological cycle of increasing stratospheric aerosol content turned out to be unexpectedly strong. A decrease in the northward transport of energy is rather significant in those latitudes where the cooling takes place. An increase in dust loading of the stratosphere lowers evaporation from the surface and turbulent heat exchange, but raises the surface net long-wave radiation (the latter due to decreasing atmospheric emission resulting from reduced water content of the atmosphere). It is significant that total shortwave radiation absorbed by the surface remains practically constant (even with decreasing solar constant),which is explained by a decrease in humidity and cloudiness compensating aerosol attenuation.

The heat budget of the atmosphere is characterized by a decrease in latent heat and absorbed solar radiation. As far as the stratosphere is concerned, in the case of the fourfold increase of the aerosol content, its temperature rises by about 4 K. The albedo of the planet remains practically constant.

Comparison of estimates of decreasing mean-global surface temperature, for the case of the fourfold increase of the aerosol content (a 1% decrease of the solar constant) with similar earlier estimates, led to the conclusion that there is considerably lower sensitivity of climate to stratospheric aerosols in the zonal model. The problem of sensitivity requires further serious study.

With the mean-global content of ozone amount decrease by 11.7%, the MacCracken (1976) model predicted a mean surface temperature increase of 0.24 K (in the midlatitude of the NH the increase was twice as high), a stratospheric temperature decrease by 0.5–1.5 K depending on latitude and altitude, and the planetary albedo decrease somewhat due to changes in the hydrological cycle. One is aware that since 1975 considerable improvement in ozone photochemistry has taken place, and close to 11.7% ozone depletion is now found on site; the MacCracken (1976) findings are mentioned for completeness of the radiative part of the discussion only.

The calculations showed that the effect of increasing water vapor content in the stratosphere is not important. An increase of the mixing ratio above the 150 hPa level by 1 ppm led to a rise in the surface mean temperature by 0.094 K and a drop in the stratospheric temperature by 0.2 K. MacCracken (1976) emphasized that the results obtained were preliminary and further studies are needed, particularly those incorporating processes involved in the water cycle.

MacCracken and Potter (1975) performed simulations of climate changes on the assumption that either the aerosol content in the global stratosphere increased by a factor of 10 (up to 4.3 µg/cm^2), which is approximately the maximum dust loading of the stratosphere after the 1963 Agung eruption, or a solar constant decrease of 3%. The complex refractive index for aerosol particles was again assumed to be 1.45 ± 0.005. Since the perturbed simulation did not attain a stable regime, these results should be considered preliminary. In both cases, similar climate changes took place. The mean global surface temperature decreased from 291.5 K, in the test calculation for present conditions, to 286.9 K., i.e., by 4.6 K. The temperature fields are similar with a temperature decrease of 4 K in the equatorial zone and 15 K in polar regions. The air temperature in the troposphere is also lower, but temperature variations in the stratosphere are more complicated. When the stratosphere becomes dust-loaded, it is strongly heated (8 K), whereas when the solar constant decreases, the lower stratosphere warms slightly and the upper stratosphere cools.

The surface net longwave radiation increased (especially in polar regions) due to an atmospheric emission decrease, which is combined with a decrease in global solar radiation and evaporation (the latter decreases by more than 12%). There is an increase in the planetary albedo due to increasing cloud amount (by 2%) when the solar constant decreases, and also when the stratosphere becomes dust-loaded. For stratospheric dust loading, the boundary of the snow cover moves southward by 10 degrees of latitude (the zone of snowfalls drifts southward by 20 degrees of latitude, reaching 60° N). In the case of a solar constant decrease, there is snow accumulation in the mountains at 60° N, and snowfalls reach 50° N. These numerical modeling results indicate, as has been pointed out earlier, that stratospheric dust loading and a solar constant decrease are not equivalent from the point of view of their climatic impact, as some authors have suggested. This non-equivalency arises because the feedback relationships due to the hydrological cycle and variations in the heat budget components are more important than the albedo feedback. The most important interacting factors are as follows:

1. A decrease in incoming solar radiation leads to cooling in the troposphere, which gives more cloudiness, and in turn an increase in albedo and a decrease in absorbed solar radiation.
2. A decrease in solar radiation absorbed by the surface, together with atmospheric cooling, leads to a substantial decrease of evaporation and water content in the atmosphere.
3. The latter determines a decrease in atmospheric thermal emission and increase in net longwave radiation, especially in polar regions, where the water content is small.
4. In connection with a decrease in the meridional energy transfer in the form of latent heat, the meridional circulation intensifies and the role of diabatic heating with downward fluxes increases.

These results of MacCracken and Potter (1975) illustrate the need to allow for various feedback mechanisms in modeling studies of climate changes.

Pollack and Ackerman (1983) using a 1-D radiative-convective equilibrium model performed calculations of the vertical profiles of radiative fluxes and temperature during the first 6 months after the El Chichón eruption. On the assumption that there are water clouds in the atmosphere amounting to 50%, separate calculations were made for clear and cloudy atmospheres. The optical thickness of clouds is taken to be 7.5, which corresponds to a system albedo of 0.6. The volcanic aerosol is prescribed as a polydisperse ensemble of silicate nuclei (their share is an adjusting parameter) covered with the sulfuric acid shell. The optical thickness of the volcanic aerosol at 0.55 µm is taken as 0.3, and the lognormal size distribution for the modal radius 0.6 µm and n = 1.5 have been prescribed. The aerosol consists (by volume) of a 90% H_2SO_4 water mixture (concentration 75%) and 10% silicate ashes. The aerosol layer is located in the altitude range 16 to 31 km with a maximum concentration at 28 km.

Calculations showed that the formation of the eruptive cloud causes an increase of the surface-atmosphere system albedo at $\mu_o = \cos\theta_o = 2/\pi$ (θ_o is the solar zenith angle) by 0.027 (by about 10%); the albedo increases in proportion to the aerosol optical thickness, increasing from 0 to 0.5. It is, however, practically insensitive to the share of ash, f, (varying from 0 to 0.9) and to the modal radius of particles. Calculations of the volcanically caused reduction of the global solar radiation in clear-sky conditions at Mauna Loa (2.2 ± 0.4%) gave results consistent with observations.

A stratospheric temperature increase caused by the absorption of the upward longwave radiation flux by the eruptive aerosol was maximum at a height of 24 km (30-hPa level), reaching 3.6 K, which agrees well with the summer 1982 observation data. These calculations contain an adjusting element consisting in setting $\mu_o = 0.60$, which provides better agreement with the observed temperature profile in the tropics before the eruption. Maximum warming at 24-km height (but not 28 km, where the aerosol concentration is maximum) is connected with an additional contribution to the warming at the expense of the absorption of the downward longwave radiation emitted by the aerosol layer above. If f (the relative share of ash) > 0.2, a considerable contribution to the solar radiation absorption is made by volcanic ash. The warming of the eruptive aerosol layer is much greater over the ocean than over land, since the land surface temperature markedly decreases.

The results obtained by MacCracken and Potter (1975) underscore the need to take into account various feedbacks when studying the factors that determine climate changes. MacCracken and Luther (1983a; 1983b) performed calculations of the effect of El Chichón volcanic stratospheric aerosols on the atmospheric radiative regime. The results of the calculations were used in the numerical modeling of the climatic impacts of volcanic eruptions with a 2-D zonal dynamical-statistical climate model in which a latitude-dependent annual change of cloudiness was prescribed. Table 5.2.1 presents results of calculations of variations in direct (S), scattered (H), and global (G) solar radiation at the surface level, with the prescribed aerosol optical thickness 0.255 at 0.55 μm wavelength (this corresponds to conditions of July 1982 in Hawaii).

Table 5.2.1. Volcanically Induced Changes (%) in Shortwave Radiation Fluxes. MacCracken and Luther (1983a, 1983b).

Parameter	Sun Elevation	
	60°	30°
S	-19	-30
H		
Without aerosol	0.04	0.07
With aerosol	0.28	0.46
G	-1.2	-4.9

These results agree well with observations. In estimating the equilibrium effect on climate it is assumed that the aerosol optical thickness in the infrared constitutes 5.9% with respect to the visible.

A major result of numerical modeling of the equilibrium state consists not only of the conclusion about a gradual global-scale cooling during the year following the eruption but also a revealing of the meridional redistribution of rainfall. In the presence of an eruptive aerosol layer at altitudes of 23 to 29 km in the latitude band 50–35° N calculations showed a 1.0 K decrease of air temperature at 20° N, increasing to 1.4 K at 70° N. Averaged values were 0.9 K for the NH and 0.3 K for the SH.

Calculations revealed a southward shift of the ITCZ (Intertropical Convergence Zone) and an intensification of the Hadley cell in the NH, which favored the "pumping" of heat from the SH oceans, thereby determining an indirect climatic effect of eruptions in the SH. The shifting of the ITCZ intensifies rainfall in the subtropical SH. The intensity of the global moisture cycle weakens and is manifested through decrease in rainfall and evaporation.

A prescribed temporal change in the aerosol optical thickness substantiated by Robock (1984) made it possible to calculate the climate dynamics between April 1, 1982 and July 1, 1983. These calculations showed that between late 1982 and July 1983 a uniform 0.3°C cooling took place in the NH, whereas in the SH only a slight cooling was observed. A latitudinal intensification of cooling turned out to be much weaker than in the case of a stationary mean annual regime. A year after the eruption the mean global SST rose by only 0.15 K. Calculations suggested the recovery of the mean global temperature by mid-1983.

Since the estimates of the equilibrium climate suggested the possibility of detecting the effects of the El Chichón eruption for several years after the eruption, MacCracken and Luther (1983a, 1983b) estimated a time-dependent regime using a seasonal model with a 70-m oceanic layer. In this case a time-dependent meridional profile of the aerosol optical thickness is prescribed with a maximum of 0.3 in June. Calculations for the period between 1 April 1982 and 1 July 1983 revealed a gradual global-scale temperature decrease of about 0.2 K for the last half year of integration. The results do not permit one to estimate the possible duration of cooling and its subsequent intensification. Numerical modeling showed a decrease of rainfall from 5% to more than 10% north of the ITCZ, during the year following the eruption.

MacCracken and Luther (1983b) emphasized the necessity to continue numerical experiments to use the "volcanic experiment" in order to check the reliability of climate models applied to estimate the sensitivity of climate to a growing CO_2 concentration. Such experiments will provide a better understanding of the causes of climate changes during the next century.

Charlock (1983) calculated the vertical profile of the IR radiative warming of the stratosphere (at altitudes of 12 to 28 km) at the equator in July for conditions simulating a disturbed stratosphere after the El Chichón eruption. A 4-km eruptive aerosol layer (optical thickness 0.25 at 0.69 µm) is assumed to be located at altitudes either 18 or 25 km. In both cases (a cloudless atmosphere is considered) nearly equal relative warming of the aerosol layer takes place, caused by the absorption of the surface thermal emission in the atmospheric transparency window at 760 to 1240 cm^{-1} which is only partially compensated for by cooling due to the self-emission of the layer.

An absolute value of the radiative heat flux divergence in the lower aerosol layer is about twice as large as in the upper layer. The aerosol-induced warming is determined only by the contribution from the transparency window, since in the remaining spectral intervals the radiative heat flux divergence is negative. The total radiative temperature change in the lower layer is positive, while in the upper layer it is negative. An increase in the total downward thermal emission flux due to the volcanic stratospheric aerosol at the surface level constitutes only 0.1 to 0.3 W/m^2, and at 10-km height it varies within 0.7 to 1.2 W/m^2 (these calculations were made for latitudes 0° and 35° N). No doubt the effect of aerosols on the tropospheric climate is determined by their impact on the shortwave radiation transport. With clouds in the troposphere not only the value but also the sign of the aerosol impact on the stratospheric radiative regime can vary, since here the aerosol-absorbed upward thermal emission flux decreases. The change in the radiative heat flux divergence is in approximate proportion to cloud-top height above 4 km.

VOLCANIC ACTIVITY AND CLIMATE 315

Vupputuri and Blanchet (1984) undertook a numerical modeling of the effect of the El Chichón eruption on the temperature field and chemical composition of the atmosphere, using a 1-D interactive radiative-convective climate model, with an account of photochemical reactions including sources and sinks. Relative humidity was prescribed and considered a 1-layer cloudiness with a constant top height of 6.5 km and cloud amount 0.446 with surface albedo 0.2. Calculations of radiative flux divergence were made with the use of the delta-Eddington approximation. In the photochemical model, the reactions of oxygen, hydrogen and nitrogen cycles were taken into account (concentrations of H_2O, H_2, and CH_4 were taken from observations). The interaction of volcanic aerosol and photochemical processes is determined by the change in the rate of photodissociation of O_3 and NO_2. Numerical modeling was made for a mean global atmospheric model, with a constant optical thickness of the eruptive aerosol of about 0.1 (model 1), as well as with an account of varying optical thickness as it is shown in Figure 5.2.1 (model 2).

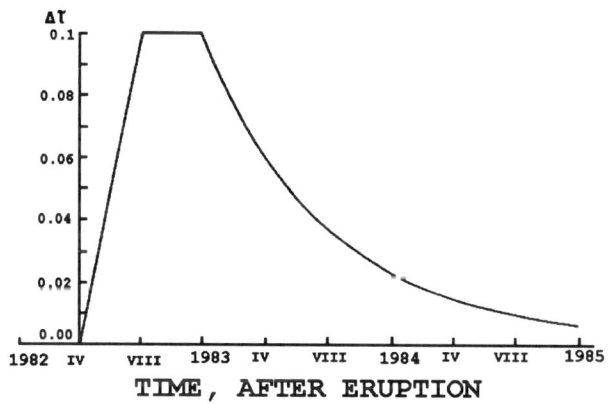

Figure 5.2.1. Considered time variation of optical thickness after the El Chichón eruption. After Vupputuri and Blanchet (1984).

Figure 5.2.2 shows results of calculations of the vertical temperature profile with and without account of the interaction of volcanic aerosol with the ozone (model 1). As is seen, the interactive calculation provides good agreement with data for the standard atmosphere, except for the tropospheric layer, which can be explained by the lack of account of the horizontal and vertical heat transport.

Figure 5.2.3 illustrates the results of calculations of the vertical profile of the ozone mixing ratio, which confirm the qualitative similarity of the profiles but a systematic underestimation of calculated quantities.

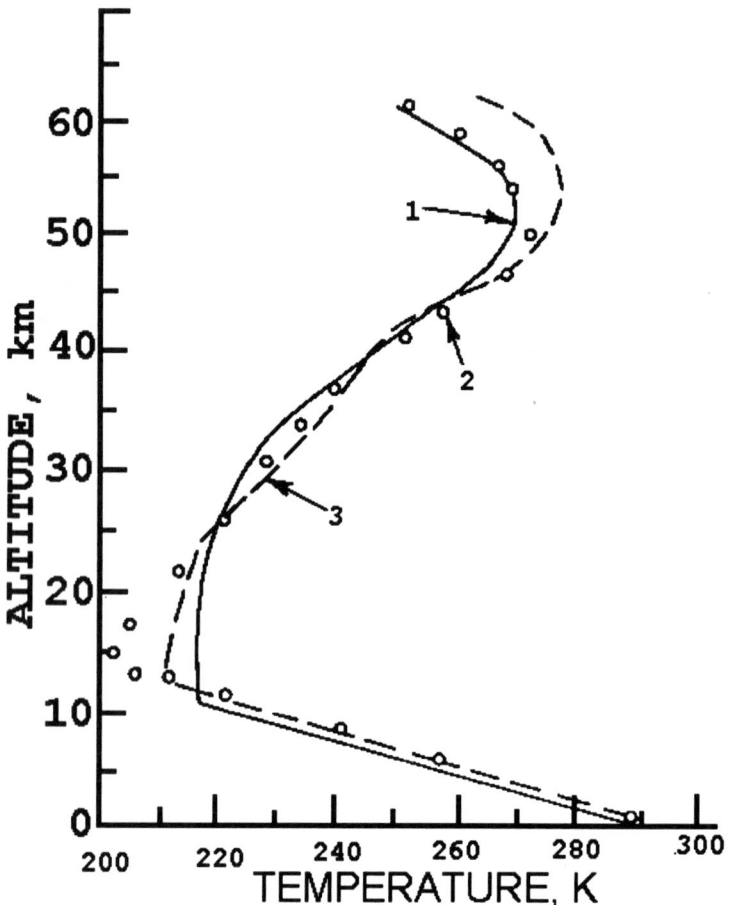

Figure 5.2.2. Computed vertical temperature profiles after the El Chichón eruption. After Vupputuri and Blanchet (1984).
1. Standard atmosphere, 2. Model 1. With ozone consideration,
3. Model 1. Without ozone consideration.

Figure 5.2.4 shows the computed vertical profile of temperature changes, which characterizes the effect of the optical properties of the volcanic aerosol and the role of the account of radiative flux divergence due to longwave radiation. These data reveal a strong effect of the optical properties on the stratospheric heating. An account of the thermal emission provides considerable warming of the stratosphere (due to the absorption of the upward radiation fluxes in the transparency window). If the stratospheric aerosol is of sulfuric acid, the stratospheric heating is almost totally determined by the absorption of the upward longwave radiation flux.

VOLCANIC ACTIVITY AND CLIMATE 317

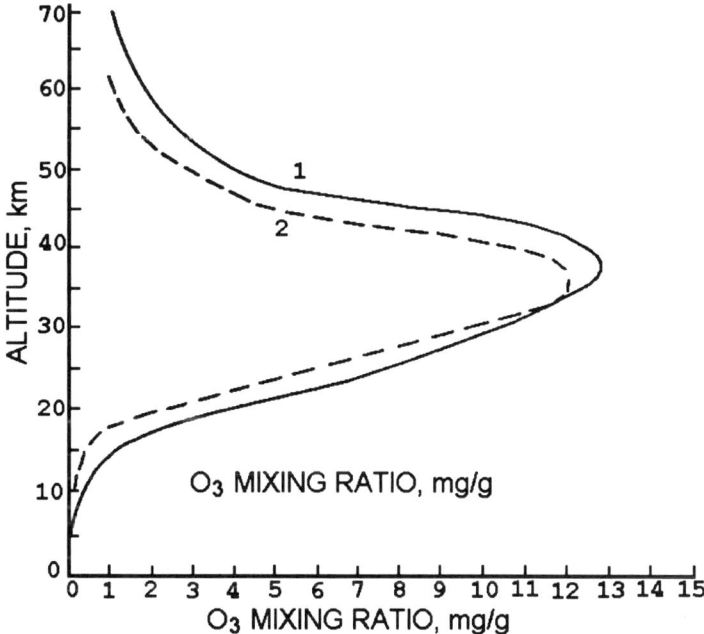

Figure 5.2.3. Computed vertical profile of the ozone mixing ratio. After Vupputuri and Blanchet (1984). 1. Standard atmosphere; 2. Model 1.

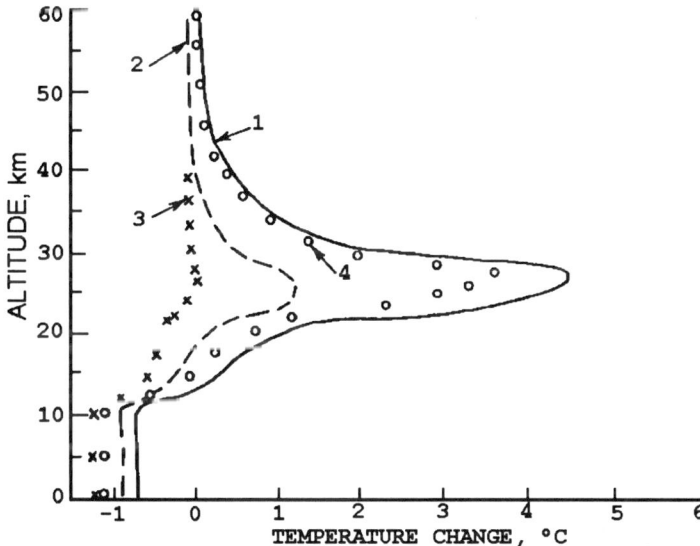

Figure 5.2.4. The effect of El Chichón eruption on vertical temperature structure. After Vupputuri and Blanchet (1984).

Data presented in Figure 5.2.5 show the effect of volcanic aerosol on the vertical profile of atmospheric ozone, and Figure 5.2.6 illustrates the role of the height of maximum aerosol concentration in the formation of the vertical temperature profile. Lowering of the level of maximum concentration brings forth a considerable reduction of stratospheric heating and tropospheric cooling.

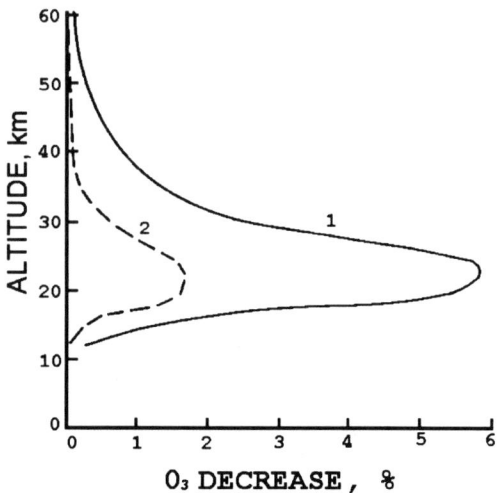

Figure 5.2.5. Computed effects of volcanic aerosol on the ozone vertical profile. After Vupputuri and Blanchet (1984). 1. Without aerosol; 2. With aerosol.

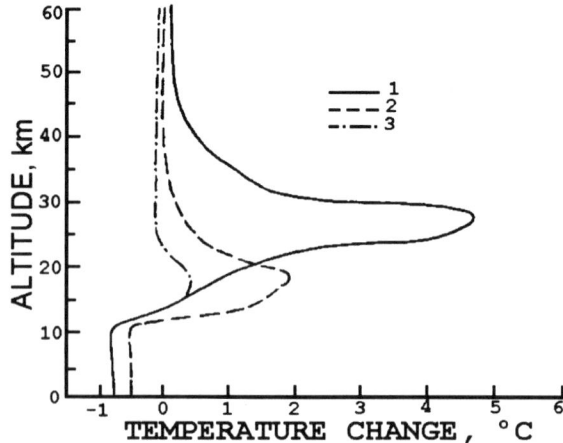

Figure 5.2.6. Computed vertical temperature profile as a function of height of maximum aerosol concentration. After Vupputuri and Blanchet (1984).

Figures 5.2.7 through 5.2.9 show the results of numerical modeling of the evolution of the post-eruption temperature field (with the aerosol content dynamics from data of Figure 5.2.1) and of the radiative regime and water vapor concentration, and Figure 5.2.9 shows the decrease of the ozone content. As follows from Figure 5.2.7, the response of the stratosphere to volcanic disturbance is much faster than that of the troposphere. At the level of maximum aerosol concentration (25 km), the stratospheric warming five months after the disturbance reaches about 4 °C, whereas maximum cooling near the surface (about 0.9 °C) occurs in about 10 months.

Changes in the water vapor concentration (Figure 5.2.9) follow the temperature variations. The same, but to a lesser extent, refers to the ozone content dynamics, shown in Figure 5.2.10— a maximum of stratospheric heating in about a month. From the data of satellite observations, the post-El Chichón drop of ozone constituted 5%.

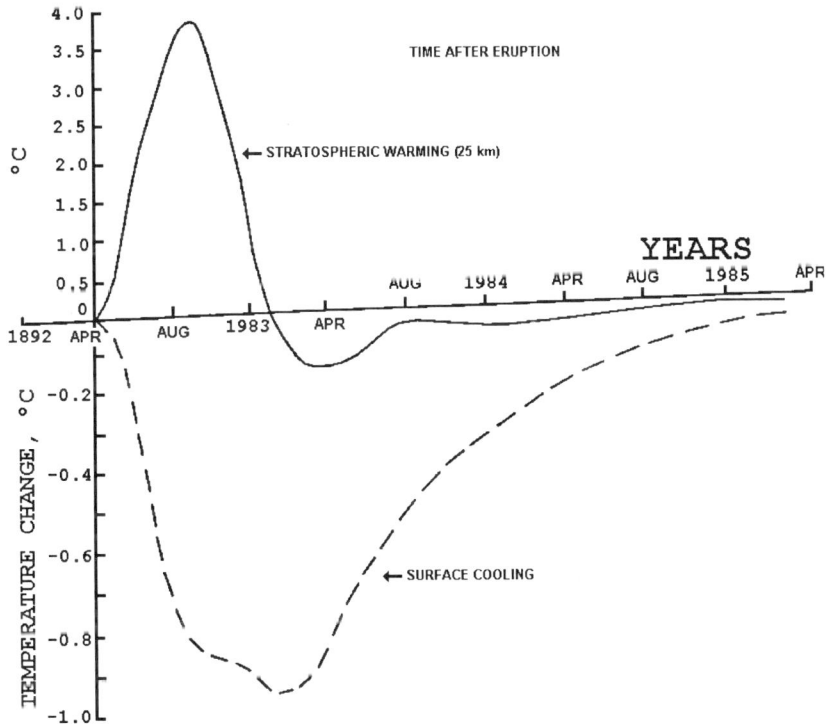

Figure 5.2.7. Variation of stratospheric and surface temperature changes as a function of time for the El Chichón stratospheric aerosol cloud model shown in Figure 5.2.1. After Vupputuri and Blanchet (1984).

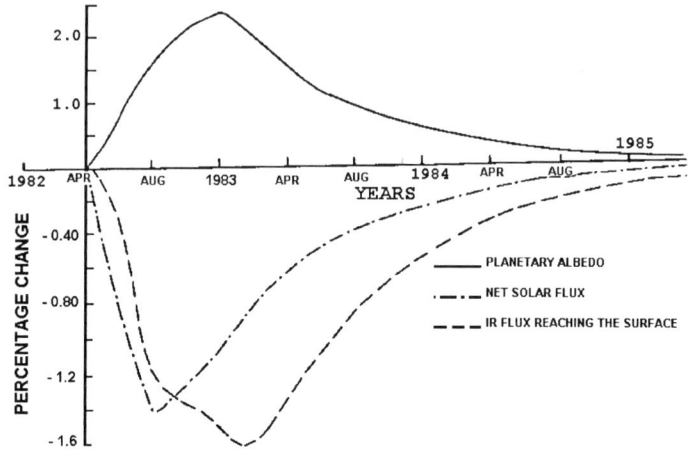

Figure 5.2.8. Variation of thermal IR and net solar flux changes at the surface and changes in the planetary albedo (in %) as a function of time for the assumed El Chichón stratospheric aerosol cloud model. After Vupputuri and Blanchet (1984).

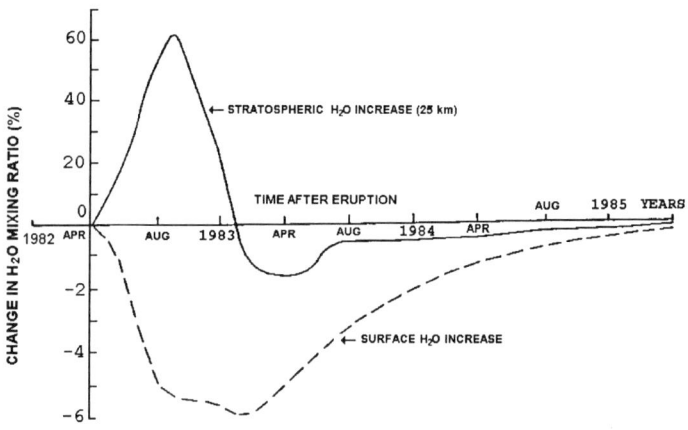

Figure 5.2.9. Variation of water vapor mixing ratio changes in the stratosphere and at the surface as a function of time for the assumed El Chichón stratospheric aerosol cloud model. After Vupputuri and Blanchet (1984).

VOLCANIC ACTIVITY AND CLIMATE

Figure 5.2.10. Variation of total ozone depletion (%) as a function of time for the assumed El Chichón stratospheric aerosol cloud model. After Vupputuri and Blanchet (1984).

MacCracken and Luther (1983a; 1983b) made calculations of the effect of the post-El Chichón stratospheric aerosol on the radiative regime of the atmosphere, the results of which were used in numerical modeling of the effect of eruptions on climate for the 2-D zonal dynamic-statistical climate model, with a prescribed latitude-dependent annual change of cloudiness. Table 5.2.2 shows the results of calculations of variations in the fluxes of direct solar (S) and scattered (H) as well as global (G) radiation at the level of the Earth's surface, with the given aerosol optical thickness of 0.255 at the wavelength 0.55 μm (this corresponds to conditions of July 1982 on the island of Hawaii).

The results agree well with the observational data. In assessing the equilibrium effect on climate, it is assumed that the aerosol optical thickness in the infrared region constitutes 5.9% with respect to the visible. These computed variations of stratospheric heating and surface cooling are similar to those reported by Hansen et al. (1978) for the Agung eruption. It is relevant to mention that Pollack and Ackerman (1983), also using a one-dimensional, radiative convective model, have calculated that El Chichón caused an increase in planetary albedo of 10%, a decrease in total solar radiation of 2%–3% (at the ground on cloudless days). These calculations are compatible with the observations considering their respective error bias.

Table 5.2.2. Volcanically Induced Changes in the Shortwave Radiation Fluxes. After MacCracken and Luther (1983a).

	Sun Elevation	
Parameter	60°	30°
ΔS	-19%	-30%
Without aerosol H/S:	0.04	0.07
With aerosol	0.28	0.46
ΔG	-1.2%	-4.9%

Assessments of the equilibrium climate have led to the conclusion of the possibility of detecting the consequences of the El Chichón eruption for several years after the event. MacCracken and Luther (1983a, 1983b) obtained estimates of the time-dependent regime using a seasonal model of a 70-m layer of the ocean. In this case, a time-dependent meridional profile of the aerosol optical thickness was prescribed, with a maximum of 0.3 in July. The calculations for the period April 1, 1982 to July 1, 1983 revealed a gradual global temperature decrease of about 0.2 K for the last 6 months of integration. The results obtained do not permit one to assess a possible duration of cooling and its subsequent intensification. Numerical modeling has led to the conclusion that a decrease of precipitation of 5% to more than 10% north of the ITCZ occurs during the first post-eruption year. MacCracken and Luther (1983a, 1983b) emphasized the necessity to continue numerical experiments in order to test the reliability of the climate models applied to assess the climatic sensitivity to increasing CO_2 concentrations. Such experiments will improve our understanding of the causes of climate change during the century to come.

Important aspects of the problem of global climate change are the interrelationships between greenhouse and aerosol contributions to climate change. Earlier unjustified exaggerated assessments of anthropogenically induced global climate warming due to the increase of greenhouse gas concentrations have been recently followed by more objective estimates with the consideration of such climatically significant factors as cloudiness, ERB variations, including its components SC, absorbed ASR, OLR, and changes in optically active gases as well as AOT of the atmosphere. In the context of the problem mentioned Ardanuy et al. (1992) have considered global relationships between ERB and its components, cloudiness, SC, volcanic aerosols, and SAT. Interannual variability and correlations between Nimbus-7 THIR/TOMS cloud amount, ERB WFOV (wide field of view) longwave, shortwave, and net

VOLCANIC ACTIVITY AND CLIMATE

radiation, and SAM II aerosol optical thickness, along with mean global SAT data have been assessed for the period 1979–1990.

Table 5.2.3 data illustrate observed variability of ERB and other variables. The areas of the highest OLR values are found over the southern tropical Atlantic, Pacific, and Indian oceans, as well as over Saudi Arabia.

Table 5.2.3. The Observed Range of Variation of RB and related Geophysical Variables on Annual, Global Scale. After Ardanuy et al. (1992). The bracketed values are relative quantities.

Parameter	Observed range of variation from mean
Outgoing longwave radiation, W/m^2	± 0.4 (± 0.2%)
Absorbed shortwave radiation, W/m^2	± 0.6 (+ 0.2%)
Radiation budget, W/m^2	± 0.5
Surface air temperature, °C	− 0.7 to + 1.3 (0.5 to +0.1%)
Total cloud amount, %	− 0.5 to + 0.3 (± 0.8%)
Aerosol optical thickness at 1 μm	± 0.002 to 0.06

In all cases, for this spatial resolution, maximum OLR values of 272 W/m^2 are recorded, corresponding to an effective broadband emitting temperature of -10 °C. The low OLR values due to the convection over the Amazon and Congo rainforests are evident, as is the signature of monsoon over Asia and Indonesia (-23 °C). The range of albedos is 19% to 38% in the tropics, while the albedo exceeds 80% in the polar regions. Northern Africa is an exception because of the Sahara desert high surface albedo. This is why radiation budget here drops up to -20 W/m^2.

Obviously, maximum ERB values are observed in the tropics (~80 W/m^2) while minimum values take place in the Arctic and Antarctic (~-110 W/m^2).

Ardanuy et al. (1992) have discussed that SC is apparently related to global average SAT values for the time period 1979–1990. The 0.40 °C range in observed global temperatures may be partitioned into a 0.15 °C component due to a 2 W/m^2 SC change and a 0.22 °C component due to the increasing concentration of CO_2 and other greenhouse gases. A relatively large component of the variance in the global temperature, cloudiness, and radiation budget signals is due to interannual climatic system variability over time periods much shorter, e.g., 2–4 years, than a solar cycle, for which the solar luminosity experiences no comparable fluctuations (ENSO induced perturbations are clearly seen in this variability).

The 20 W/m² annual cycle of the global radiation budget is almost entirely explained by the 23 W/m² seasonal variation of insolation. The long-term global average ERB is equal to 6 W/m². Such an imbalance may be partially explained by an undersampling of the polar caps (1.3 W/m²) and by a 0.4% overestimate in the SC value (1.6 W/m²). The remaining part may be a real imbalance (the uncertainty of the absolute calibration is about 1%). A trend of about 0.2 W/m² per year is evident in annually averaged radiation budget values (the trend is completely driven by the OLR decrease; the albedo is stable).

The relationship between SAT (T_s), SC(S_o) and CO_2 concentration may be approximated by the following linear regression:

$$T_s = 0.26 + 0.076 S_o - 0.011 CO_2 \tag{5.3}$$

This relationship explains 32% of SAT variations and, thus, the 68% contribution belongs to the internal variability of the climatic system. The time dependence of aerosols injected into the stratosphere by the explosive 1982 eruption of El Chichón is found to be important, along with the global cloud amount, in describing the time dependence of the Earth's albedo during the period. As far as the Pinatubo eruption is concerned it may fully mask any greenhouse warming signal for at least the first half of the 1990s.

Ardanuy et al. (1992) have emphasized that reliable and well-characterized satellite data sets, ideally with lengths of 1 to 2 decades or more, are required to perform quantitative analyses of the relationship among different elements of the climatic system. The stability of instrument's calibration should be maintained to permit the analysis of interannual global variability at the 0.2% level. It should be added that various complementary conventional observations are also most important.

Numerical modeling results considered above with the aim of assessing the meteorological consequences of eruptions, have been preliminary, since they have been based on the prescribed aerosol content in the stratosphere and simplified AGC modeling. In this connection, Hunt (1977, 1978) performed a numerical modeling of the effect of volcanic aerosol on climate, with an account of realistically simulated diffusion of eruption products from their source in tropical latitudes (the 1883 Krakatau eruption has been considered, with the geographical coordinates 6° S, 105° W).

Numerical modeling was undertaken using a 3-D AGCM, which made it possible to assess both the direct effect of eruption products on the radiative and thermal regimes and the indirect effect on the temperature field through the atmospheric dynamics.

The 18-level hemispherical AGCM (the finite-difference analog to the system of primitive equations in stereographic projection) used in calculations is a further modification of the GFDL and NOAA AGCMs.

VOLCANIC ACTIVITY AND CLIMATE

An integration has been made for annual mean conditions without account of orography and land-ocean contrasts, but with an account of the hydrological cycle. For comparison purposes, the results of the previous numerical modeling have been used, as a reference to show the undisturbed atmosphere. The neglect of the thermal inertia of the ocean leads to an increased sensitivity of the model to the forcing, compared to real atmospheric conditions.

A consideration of radiative flux divergences due to shortwave and longwave radiation is based on prescribed climatological distributions of water vapor, carbon dioxide, ozone, clouds, and surface albedo. The initial source of aerosol is fixed as a ring located at a height of 23 km and 3° latitude wide, encircling the equator. Since the AGCM contains a wall at the equator, the aerosol diffusion is confined to the SH.

It is assumed that, reaching the lower level (0.85 km), the aerosol is immediately washed out from the atmosphere (this sink for aerosol is a single one). In accordance with the aerosol model suggested by Deirmendjian (1973) for the first phase after the Krakatau eruption, it was assumed that the aerosol particle size distribution is characterized by prevailing large (2 μm) particles. The initial total number of particles, determining their mixing ratio, is $8 \cdot 10^{23}$ particles.

The particles are assumed to be of silicate (purely scattering) with the refraction index 1.56, independent of wavelength within the shortwave radiation spectrum. This corresponds to the attenuation coefficient 2.9 km^{-1} at wavelength 0.45 μm, with the particles concentration 10^8 m^{-3}. The effect of stratospheric aerosol on the longwave radiation transfer is not taken into account, and in the case of shortwave radiation, it is determined only by backscattering, which for mean annual conditions, constitutes about 15% with respect to all scattered radiation.

The special feature of the radiation parameterization of the model is such that it does not permit one to take into exact account the effect of backscattering, since part of radiation scattered back is manifested as local heating (up to 10 °C at the upper level). This defect of the model can be considered, however, as a certain merit, since observations point a strong stratospheric heating after powerful eruptions.

Calculations of AGC were made with a time step of 10 min., but the effect of variations in the radiative flux divergence is taken into account every 24 hours. The "volcanic" experiment was started on day 254 of the control integration and continued for 150 days, when volcanic aerosol propagated over the globe. During 130 days, the control and volcanic experiments were performed simultaneously.

The results of the effects of eruptions were assessed by comparing the 10-day running means, starting from the first day of the volcanic experiment. Also, additional calculations were made during 14 days, with a diffusion of aerosol considered as a passive tracer (the interaction with the radiative factors is neglected).

A comparison of calculated values of the variability of hemispherical mean kinetic energy of the atmosphere has shown that substantial difference, compared to control calculations, occurred only during the second and third weeks when the concentration of aerosol in tropical latitudes was high. If the effect of radiative factors on the propagation of aerosol is not taken into account, then in this case the values of kinetic energy are close to control values.

The effect of volcanic aerosol on the wind field is weak and consists, as a rule, in a small attenuation of mean zonal winds and in some changes in the synoptic wind fields, which would be difficult to detect in conditions of the real atmosphere.

The temperature field changes substantially (though these changes have, apparently, been overestimated because of neglect of the thermal inertia of the ocean). There is a decrease of the mean hemispherical surface temperature by about 0.3 K and that of the the tropics by 0.7 K. In middle and high latitudes the effect of eruptions is masked by variations in local weather conditions.

Such a limited response of the mid and high latitude atmosphere is determined by the hemispheric (but not global) nature of the model and by comparatively short duration of the experiment (a consideration of global disturbances revealed a maximum of response to a high-latitude eruption). The temperature changes obtained outside the tropical band are, largely, caused by variations of meridional energy transport (especially of latent heat of condensation), but not by the direct effect of stratospheric dust loading on the solar radiation attenuation. The eruption does not practically affect the hydrological cycle. This conclusion will be, apparently, incorrect, if we consider the development of the processes for a longer time period.

The principal conclusion from the numerical experiment carried out by Hunt (1977) is that the effect of individual large eruptions on climate cannot be long. More serious consequences must take place in the case of a series of prolonged successive eruptions. Numerical experiments should be continued to assess the effect of volcanic eruptions on climate, with the use of a more realistic AGCM, which takes into account the oceanic circulation and the annual change.

The radiation parameterization of the model must be more precise, especially from the viewpoint of considering the effect of aerosol on the longwave radiation transport. The consideration of the hydrological cycle and of the surface characteristics must be more adequate. It is also important to consider the effects of different types of volcanoes located at various latitudes. Unfortunately, the observational data characterizing the products of volcanic eruptions, their propagation and transformation, are still scarce.

An important contribution to the application of 3-D global climate models for assessments of aerosol climate impacts has been made by Tanre et al. (1984). For this purpose the ECMWF spectral model with a low truncation (T21) was integrated twice for 3 months: with radiation computations with and without aerosols. Five aerosol types have been considered: continental, maritime, urban, volcanic, and stratospheric. The effect of the aerosols on the model's climate is just at the limit of the noise level for the dynamics but it is significant for temperature. The most significant effects are for both surface and stratospheric temperatures and to a lesser extent for the tropospheric temperature.

Consideration of mean zonal averages for July (differences in atmospheric temperature obtained with aerosols and without aerosols) reveals the following effects:

(i) Warming effect of aerosols over high latitudes in the lower levels. Antarctica is an exception because of the polar night.
(ii) Cooling effect near the surface over the desert regions and over latitudes without continents.
(iii) Warming of the troposphere almost everywhere, increasing to reach a maximum in the stratosphere (up to 3 K at 30° S).

The aerosol impact on land surface temperature is as follows:

(i) Saharan and continental aerosols have a cooling effect for the surface that reaches a maximum of 4 K over the Sahel.
(ii) Urban aerosols in such regions as northern Europe, northern United States, and northern China have a warming effect (because of their small single-scattering albedo).
(iii) The effect of aerosols near regions of high albedo (Greenland, Antarctic, northern Siberia, northern Canada) is to increase surface temperature, because of the increase of downward infrared radiation in relation to a higher atmospheric temperature.

Analysis of the evolution of computed climate made by Tanre et al. (1984) shows that 3 months might just have been enough to reach an equilibrium by the end of integration. The atmosphere is then warmer by about 0.5 K, the surface colder by 0.15 K (corresponding to 0.5 K on land) and the cloud cover has decreased by 0.02 under the influence of the aerosols. Other important values are a loss of about 2 W/m^2 for the whole surface-atmosphere system and a net increase of the atmospheric heating rate by 0.05 K/day when aerosols are introduced. Tanre et al. (1984) have pointed out that from all these numbers it seems reasonable to assume that, if we had an ocean/atmosphere coupled system (ocean/atmosphere interaction was not taken into account in the model) and could integrate it over larger periods, the average SAT would decrease between 0.5 K and 1 K under the effect of aerosols whereas the upper tropospheric temperature would increase by something less than 0.5 K with a maximum in the stratosphere. However, these assessments are only valid for July, which would mean annual values might be very different.

A number of new efforts to apply 3-D climate models have been undertaken recently to assess volcanic climatic impact. Rind et al. (1992) have investigated the effects of volcanic aerosols on the middle atmosphere and pointed out that such effects can be divided into three categories:

(i) The direct effect on the stratosphere itself
(ii) The direct effect on the troposphere, and the tropospheric changes that arise from the direct effect on the stratosphere; and
(iii) Feedback on the middle atmosphere.

Accordingly, three numerical experiments have been conducted with the use of the 9 level (with two stratospheric layers) GISS Global Climate Middle Atmosphere Model (GCMAM):

(i) Volcanic aerosols are put into the stratosphere (a visible sulfuric acid aerosol optical thickness is prescribed equal to 0.15 and is held constant in time and space) and the SST are not allowed to change; an emphasis of this experiment (STRAT) is on the direct effect of volcanic aerosols within the stratosphere.
(ii) The minimal SST change, that arise in the the first 3 years are analyzed. The purpose of the experiment TRANS is to investigate the impact of volcanic aerosols that remain in the stratosphere for only several years, as is normally the case (a transient volcanic influence).
(iii) Volcanic aerosols are incorporated and maintained in the stratosphere and the SSTs that occur after 50 years of integration are employed; the aim of the experiment (CLIM) is to allow full tropospheric cooling to be included (in this case global annual average SAT change is -4.7° C).

Computations made by Rind et al. (1992) have shown that within the first several years after an eruption (STRAT and TRANS), a period of similar influence to that of major eruptions in this century occurred: the warming of the lower stratosphere led to an increase in tropical static stability; the intensities of the Hadley and Ferrel cells decreased, midlatitude westerlies decreased and wave energy flux into the stratosphere at subtropical and lower midlatitudes increased. The additional wave energy flux convergence, along with the direct lower-stratospheric radiative heating due to aerosols, intensified the residual circulation in both hemispheres leading to some high-latitude warming and low-latitude cooling in the upper stratosphere and lower mesosphere. The dynamical changes are generally on the order of 10% and are similar to occurrences following major volcanic eruptions in the last 30 years.

On the longer time scale (CLIM) a strong hemispheric asymmetry arises. In the NH eddy energy decreases, as does the middle atmosphere residual circulation, and widespread stratospheric cooling results.

In contrast, in the SH, the large sea ice response to the global cooling substantially increased the latitudinal temperature gradient, leading to increased eddy energy, increased wave energy flux into the stratosphere, an amplified residual circulation, and some high latitude warming.

The important conclusion of the study by Rind et al. (1992) is that the middle atmosphere response to climate changes depends on both direct and indirect tropospheric effects. Similarly, the tropospheric changes are not simply the products of the direct climate perturbations: they depend as well on what happens to the stratosphere. Thus, climate change studies need to take into account a coupled troposphere-stratosphere system.

Pollack et al. (1993) have undertaken a new simulation of volcanic aerosol forcing on the basis of the GISS Model 2 with the three major objectives:

(i) To better understand the induced climatic changes by volcanic particles.
(ii) To assess similarities and differences in the response of the climatic system (aside from the sign differences) to forcing by volcanoes, greenhouse gases, and ice age conditions having in view, first of all, identification of "universal" and specific forcing-dependent responses. The aerosol optical thickness of 0.15 for a temporally and spatially constant amount of volcanic particles was chosen so as to produce approximately the same amount of forcing as results from doubling the present CO_2 concentration of the atmosphere and from boundary conditions associated with the peak of the last ice age 18,000 years ago.
(iii) To perform a limited assessment of parameterizations used in energy balance climate models.

The 9-level, from the surface to a pressure level of 10 hPa, GISS model 2 was run for 55 simulated years to reach a statistically steady state for the volcanic case with the use of a grid 8° latitude times 10° longitude. In cases of volcanic and CO_2 doubling impacts on sea surface temperature and sea ice are calculated, instead of being prescribed. For all three climate perturbation experiments, fractional cloud cover and cloud height were calculated although the radiative properties of clouds at a fixed level were held constant. Thus, cloud feedbacks are included in the simulations, although in a somewhat incomplete manner. The choice of the volcanic cloud composition and particle size distribution was made in accordance with the measured properties of the El Chichón and Mount Pinatubo volcanic clouds. The prescription of constant AOT for the whole 55 year period of integration means a possibility of closely spaced, multiple, sulfur-rich, volcanic eruptions. Such an assumption is not far from reality, because similar situations have taken place in the past: for instance, the 30-year period at the end of the nineteenth and beginning of the twentieth centuries when a sustained, significant AOT perturbation by volcanic particles occurred. The Little Ice Age was, probably, another example of prolonged AOT positive anomaly.

Pollack et al. (1993) have emphasized that, in certain ways, the climate response to the three different forcings is similar. Direct radiative forcing accounts for 30% and 25% of the total SAT change ΔT_s in the volcano and CO_2 doubling runs, respectively. Changes in atmospheric water vapor act as the most important feedback and are positive in all three cases. Albedo feedback is a significant, positive feedback at high latitudes for all three simulations, although the land ice feedback is prominent only in the 18 K run. Clouds produce a positive feedback in low latitudes and a negative feedback at high latitudes in all three altered runs. Remarkably, most of the cloud feedback at each latitude in these three experiments is cancelled by feedbacks of opposite sign due to the heat advection. The problem is, however, that the cloud parameterization in the model is very crude, which makes assessment of cloud feedbacks not reliable (cloud feedbacks remain poorly modeled and poorly understood in general). The importance of the advection raises questions concerning adequacy of energy balance climate models.

As Pollack et al. (1993) have noted, perhaps the largest surprise of the results considered is the obvious presence of asymmetries of several types. The climate response to a warm forcing, as exemplified by CO_2 doubling, and to a cool forcing, as exemplified by volcanic particles, are not mirror images of one another. Similarly, there are important differences in the climate response to two different cold forcing–volcanic particles and ice age–and there are important differences in the responses in the SH and NH. The sources of these asymmetries include differences in the latitudinal dependence of the primary radiative forcing; differences in the amount of land and ocean in the two hemispheres; the much greater expansion of sea ice for cold forcing than for warm forcing, the nonlinear dependence of the strength of turbulent mixing on the sign of temperature stratification; and the obviously important impact that extensive land ice sheets have on the 18 K simulations. Pollack et al. (1993) have stressed the importance of thorough model validation through comparisons with relevant observations. Major volcanic eruptions may serve as adequate "experiments" to accomplish such a validation.

5.3 Effects of Volcanic Eruptions on Atmospheric Ozone

After volcanic eruptions, which have caused unusually large increases in stratospheric gases and aerosol concentration, anomalies in both the total and the vertical ozone distribution have been noted (e.g., Angell, 1986) and attributed to the fine ash particles and/or to effects initiated by the volatile. It should be emphasized that some of the effects will contaminate the ozone measurements, others will cause real change in the ozone amount. Both these aspects have been discussed in detail earlier (Kondratyev, 1989). We have partly considered this problem in the Introduction and Chapter 3.

The effects initiated by particles can be of two types. One type that has received considerable attention already because of its direct climate implication is the perturbation in the radiative properties of the stratosphere by volcanic aerosols. The resulting changes in the radiation transmission and reflection characteristics of the lower stratosphere in the visible and near UV may cause some perturbation in dissociation rates of important stratospheric species including ozone.

As was already discussed in previous chapters, the solar radiation absorption by the volcanic aerosol can cause temperature increases of several degrees in lower stratospheric temperatures. Temperature changes will have a direct effect on minor constituents through temperature-dependent reaction rates. The effect of increasing temperature should lead to a decrease in ozone due to the inverse relationship between these two variables. However, the lower stratosphere is a region where dynamical time constants are significantly shorter than chemical time constants for ozone, implying that indirect effects of the temperature changes operating through dynamical perturbations may ultimately dominate the direct chemical effects.

A further complicating factor in the evaluation of the effects of particles from volcanic eruptions on stratospheric minor constituent chemistry is the possibility of heterogeneous reactions of the surfaces of these particles. Although heterogeneous reactions in the normal atmosphere seem to be negligible compared to gas phase chemistry, the possibility of significant effects following a large eruption is quite realistic.

The injection of gases such as H_2O, CO_2, SO_x, HCl, CO, HF, OCS, NH_3, into the stratosphere leads to a direct perturbation of minor constituent concentrations. In accordance with the presently available models considered by Stolarski and Cicerone (1974), the chlorine compounds, normally HCl, are contained in gaseous volcanic eruptions in variable amounts on the average 0.5% to 1.0% of the total gases emitted.

Volcanic clouds may affect total ozone observations in several ways, depending on the altitude of the injected material and the latitude of the volcano. First, the SO_2 in the clouds causes increased absorption at the short wavelengths of the Dobson AD pairs which produces measured total ozone values that could be up to 20% too high (Komhyr et al., 1984) depending on SO_2 concentrations after an eruption. Second, the optical properties of the volcanic aerosol, depending on their size distribution, may cause effects on the aerosol or dust term in the total ozone equation and affect calculated total ozone in either direction. Furthermore, the ozone depletion process may be attributed to the enhanced ozone photodissociation rates caused by the multiple scattering effects from the aerosol cloud and to the feedback effect of stratospheric heating and ozone mixing ratio that has been demonstrated in model calculations. And last but not least, depletion due to chlorine perturbations should be considered.

5.3.1 Influence on Ozone Retrievals

For Umkher observations, the sulfur dioxide in the volcanic cloud introduces excess ozone in those Umkher layers that contain the SO_2. However, the scattering effects of the aerosol are much larger and will cause derived ozone amounts in layer 9 (43–48 km) to be too small, while tropospheric ozone amounts may be too large. The size of the error increases with the optical depth. To understand the mechanism of the effect one should recall that the Umkher effect is produced by optical scattering and absorption of the solar UV radiation incident on the Earth's atmosphere. The Umkher method for deducing the gross features of the vertical ozone distribution requires measurements of the ratio of zenith-sky intensities of two wavelengths in the UV during the time when the solar zenith angle changes from 60° to 90°. As the solar zenith angle increases, the scattering layer appears at increasingly higher altitudes, and this produces the scattering effects that provide information on the vertical ozone distribution. The inversion algorithm for deducing the ozone profile from the Umkher measurements was developed for a molecular atmosphere only. Aerosol scattering and absorption will produce an additional effect that introduces an error in the calculated Umkher ozone profile (Dave et al., 1980; DeLuisi, 1969, 1979; DeLuisi et al., 1975, 1984).

In general, haze produces an apparent decrease in ozone concentrations above the region of the ozone maximum, and an apparent increase in ozone concentration in the region below the ozone maximum. The magnitude of the ozone error caused by haze depends on the amount and vertical distribution of aerosol concentrations and their optical scattering and extinction properties (DeLuisi, 1979). Compared to tropospheric aerosols, stratospheric aerosols are four to five times more effective in producing errors of the deduced ozone profiles (DeLuisi, 1979; Dave et al., 1980).

Using observational data from Aspendale (38° S) after the Agung eruption, DeLuisi (1979) has demonstrated and further explained the effect of volcanic aerosols on the Umkher ozone profile. The changes to the ozone profiles at Aspendale are so large that it is possible to deduce empirically the haze error effects to the Umkher ozone profile by comparing averages of ozone profiles observed nearly immediately after the Agung eruption with ozone profiles observed for several years during which the stratosphere was without serious aerosol contamination. It should be kept in mind, however, that the apparent ozone depletion there could have been complicated by the possible effect of massive atmospheric nuclear explosions during the previous years and, therefore, could not be attributed solely to the Agung eruption.

Shortly after the eruption of Agung, the ozone concentration in layer 9 underwent a sharp decline with respect to the normally expected concentration. A gradual recovery over a period of 2–3 years was seen to take place. A sharp decline also took place in layer 5 but the magnitude of the change in concentration, relative to the normally expected concentration, is much smaller than the change that is seen to occur in layer 9. There is considerable measurement noise in the long-term plots mainly because of insufficient data during some of the months, but the error signals caused by the Agung stratospheric dust clouds are clearly evidenced.

The reason why even small stratospheric aerosol enhancement could have an effect on the ozone calculations for the uppermost layers (30–50 km) is that near a solar angle of 90° the scattering layer for the long wavelength of the C-pair ($\lambda_1 = 311.4$, $\lambda_2 = 332.2$ nm) of the Dobson is near 20–25 km and the sun's direct rays must pass tangentially through the stratospheric dust layer, which is unusually dense at the same altitude. This can cause a substantial reduction in the radiation that reaches a point at the zenith of observation site *before* it is scattered downwards towards the surface and subsequently measured.

Using the Hunt (1977) three-dimensional simulation of the dispersion of the volcanic cloud of Krakatau centered at 23 km, and assuming that 1% of the erupted gas and ash carried Cl_x which spread around the whole, Stolarski and Butler (1979) calculated the possible ozone percent depletion at 30° latitude as follows:

After	10	30	70	150 days
Above 30 km	0.9	4.2	6.4	7.7
Above 15 km	1.1	4.8	6.2	6.6

Considering that more than 80% of the total atmospheric ozone is located above 15 km height the depletion should have been noticeable if measurements had been in 1883.

Turning to the Agung 1963 eruption, which shows a factor of 10 less ash than for Krakatau, the same assumption as above would yield about 0.5% total ozone depletion on hemispheric scale and this could be accommodated within the total ozone records of that time.

DeLuisi et al. (1984) considered standard Umkher ozone partial pressure (in hPa) in layers 5–9 observed during the time period May 1982 to December 1983, plotted as a function of cumulative Julian days with day 1 being January 1, 1982. The great anomaly during May–June 1982 and the slow recovery toward 1983 were obvious. The observations at Mauna Loa started about 1 month after the eruption of El Chichón and no Umkher ozone profile climatology exists at Mauna Loa so the current data set cannot be examined in the center of an unperturbed ozone profile record. However, if the ozone changes in individual months of 1982 are expressed in percentage of the ozone values of the same month in 1983 (that is, a period when the aerosol optical depth decreased from about 0.27 to 0.05) as done by Komhyr et al. (1984), it will appear that the combined effect of stratospheric and tropospheric aerosols caused ozone concentrations initially to be highly overestimated (up to 200%) in layers 1 to 3, and underestimated by up to 30–40% in layers 5 to 8. In layer 9, the ozone amount on May 1982 is underestimated by 115% causing fictitious negative ozone values to be recorded. By December 1982, however, the ozone underestimates in layers 5 to 9 were reduced between 5 and 15%, and remained essentially unchanged through March 1983, suggesting that the aerosol effect had substantially diminished toward the end of 1982 and that the early 1983 ozone concentrations were lower than those of early 1984.

The analysis of the errors caused by the El Chichón aerosols on the Mauna Loa Umkher observations by DeLuisi et al. (1984) brought them to the conclusion that the stratospheric aerosol layer coexistence with the ozone layer, with maximum concentrations for both between 25 and 28 km, suggest the possibility of a real-induced ozone depletion in layers 5–7 (31.2 – 3.9 hPa). This was reproduced in broad lines by model calculations made by Vupputuri and Blanchet (1984). They used a coupled one-dimensional radiative-convective photochemical model, which has been briefly discussed. The model determines the new temperature and trace constituents vertical distribution using a step marching method. The interaction between the El Chichón volcanic aerosols and the oxygen-hydrogen-nitrogen chemistry appears through the alteration of photodissociation rates of O_3 and NO_2.

The results showed in Figure 5.2.5 indicate ozone depletion between 15 and 50 km with a maximum of 1.5% or 5.5% at about 24 km depending on the assumed aerosols optical properties. The integrated effect of the El Chichón aerosol cloud according to this model calculation leads to up to 1.8% depletion of the total ozone which is expected to recover to background values within 2 to 3 years. Although one should consider the model calculations with the necessary precaution, due to the numerous assumptions made, the above results indicate the magnitude of the expected effects.

VOLCANIC ACTIVITY AND CLIMATE 335

An estimate of the effect of El Chichón aerosols on SBUV satellite measurements of ozone has been made by Mergenthaler (1985). He found that during the summer of 1982 the aerosols could have caused an error in the neighborhood of -25% in the infrared cumulative ozone between zero and 100 hPa, if their effect is neglected.

Since volcanic aerosols in some way will affect practically every type of ground or space based remote sensing inversions method for observing the vertical ozone distribution, it seems worthwhile to make stratospheric aerosol measurements on a more routine basis and to utilize the data for correcting the long-term ozone profiles as suggested earlier by DeLuisi (1979) particularly if one wishes to use the data for long-term trend analysis or to deduce possible solar, or other effects.

5.3.2 Volcanic Impact on Ozone Concentration and Relevant Climatic Implications

To monitor ozone changes before and after the Mount Pinatubo eruption Grant et al. (1992) obtained ozone profiles from ECC (electrochemical concentration cell) sondes at Brazzaville, Congo (4° S, 15° E), and Ascención Island (8° S, 14° W). Aerosol lidar sounding in the western Pacific (4°–6° N, 125° W) showed that most of the material injected into the stratosphere was located between 18 and 24 km with highest amounts at 24–25 km for the period 3–6 months after the eruption.

Grant et al. (1992) have discovered decreases in ozone at 16 to 29 km, with peak decreases as large as 20% found at 24 km. Integrated ozone content in the 16- to 28-km layer had decreased by 13–20 Dobson units in comparison with pre-eruption conditions. The altitude at which the most pronounced ozone decrease was found strongly correlated with peak aerosol loading determined by the lidar. A small increase of ozone was found above 28 km. The observed ozone variations are, apparently, due to a combined effect of heterogeneous chemistry, radiation, and dynamics.

In connection with the problem of ozone dynamics it is of interest to monitor changes of HCl or HF in the stratosphere after the eruption. Ground based spectroscopic observations made by Wallace and Livingston (1992) showed no such changes.

Observations made after the El Chichón and Pinatubo eruptions show, that, apart from the causes of TOC anomalies mentioned above, there are other potential mechanisms due to impacts of volcanic effluents. Now, we shall discuss this aspect of the problem.

To analyze a possible volcanic impact of the Mount Pinatubo eruption on TOC and vertical ozone distribution Hofmann et al. (1993a) have considered ozonesonde observations at Hilo (Hawaii, 20° N) before (1985–1990 time period) and after the eruption (1991–92). The ozone vertical profiles obtained since 1985 have been averaged over 2-km intervals from 14 to 30 km, from which TOC monthly averages have been determined as well as average ozone vertical profiles. The general nature of the ozone anomalies in 1991–92 can be summarized as lower than normal ozone below about 25 km (the largest deficits occur at 16–18 km in the spring; i.e., just above the tropopause where ozone is up to 50% below normal) and higher than normal ozone above. The net result was that TOC was somewhat lower than average and, during late 1992, was also as low as recorded in 1982, following the eruption of El Chichón.

Hofmann et al. (1993a) have pointed out that elevated temperatures in the region of the volcanic aerosol layer (heating of 2–3 °C in excess of the average in the layer 20–25 km appeared in the autumn of 1991) and upward motions of the aerosol layer were observed at Hilo following the eruption. Although the nature of the perturbed ozone profile may be the result of enhanced upward motion associated with volcanic aerosol particle heating, and the coupling of QBO effects (El Chichón eruption occurred during the westerly phase while Mount Pinatubo arose during the easterly phase of the QBO), the persistent nature of the perturbation, which was present more than a year after the eruption, is not easily explained. If a portion of the anomalous ozone in 1991–92 was indeed associated with the eruption of Pinatubo and its aftermath, then it is likely that these effects were predominantly ozone redistribution by dynamical effects and enhanced ozone destruction by homogeneous chemistry related to this dynamic redistribution.

Unprecedented low TOC values were observed during the winter of 1992–93 at all midlatitude stations in the USA. These values, which were considerably lower than those following the El Chichón eruption (Hofmann, 1993a; Hofmann et al., 1993b), have stimulated the development of a new viewpoint on post-eruption stratospheric changes. The principal ozone losses in the lower stratosphere (below about 24 km) took place in the same region occupied by the Mount Pinatubo aerosol cloud. Hofmann (1993a, 1983b) has pointed out, however, that although low ozone values were quite likely related to the Pinatubo eruption, explaining the long delay in this process, the fact the ozone was actually higher than normal values above 24 km is currently a topic of considerable debate. An important circumstance, mentioned by Hofmann et al. (1993a), is that trajectory analyses suggest, except at low latitudes in spring, air parcels on the days of ozone measurements generally arrived at Boulder (one of the points of surface observations) from higher latitudes, although seldom higher than 60° N. Hence, these air parcels may have been subjected to a heterogeneous process that is believed to be more effective under the low winter sunlight level of higher latitudes.

Dutton et al. (1993) have emphasized that the loss of heating in the stratosphere after the Mount Pinatubo eruption, revealed also by the MSU satellite data during 1992, could be due to volcanically induced ozone reduction (a similar but weaker phenomenon was observed after the 1982 El Chichón eruption). The stratospheric cooling contributed further to the tropospheric cooling. Thus, a quite different picture of a volcanic impact on the stratosphere arises; not just heating due to absorption by volcanic aerosol of solar radiation and upwelling longwave radiation, but also cooling because of the ozone reduction in the lower stratosphere with subsequent tropospheric cooling.

The geographical distribution of the TOC changes is rather complex. Komhyr et al. (1993) show that for 1979–1991 the trends derived from Dobson spectrophotometer observations over the USA were from -0.3% to -0.4% per year, but approached zero at Mauna Loa and American Samoa.

Chandra (1993) and Chandra et al. (1993) have undertaken an analysis of changes in stratospheric ozone and temperature due to the eruptions of Mount Pinatubo (June 15-16, 1991) on the basis of satellite TOC data from the Nimbus-7, TOMS and the NOAA II SBUV/2 spectrometers (the ozone anomalies inferred from the two instruments agree within 1%–2% in the presence of large volcanic clouds produced by Pinatubo). As Chandra (1993) has pointed out, within a few months after the eruptions, the TOC values decreased by 5%–6% in the tropics, 3%–4% in midlatitudes and 6%–9% at high latitudes in the NH.

Such TOC drops cannot be entirely explained, however, by volcanic impact only, because significant contributions to TOC changes were also made by the QBO and as such present a different dynamical scenario for the changes in stratospheric ozone (the Mount Pinatubo eruption took place during the easterly phase of the QBO compared with the El Chichón eruption, which happened during the westerly phase of the QBO). After the effects of QBO and interannual variability are taken into account, the decrease of TOC attributed to volcanic eruption at these latitudes may not be more than 2%–4% (this conclusion is in general agreement with a similar study of the El Chichón effects on the stratospheric ozone).

In fact, the most noticeable effect of the Mount Pinatubo eruption pointed out by Chandra (1993) is the breakdown of the phase relation between ozone and temperature (a similar situation was observed during the El Chichón period). This is attributed to additional heating in the lower stratosphere caused by volcanic aerosols.

The observational data show that there is no evidence of a TOC decrease in the SH where the aerosol cloud spread rapidly after the eruption. The empirical regression model, based on the 30-hPa zonal wind at Singapore, accounts for most of the observed characteristics of ozone changes both at the equator and at southern latitudes. The negative TOC anomalies at the equator and midlatitudes in the SH, to a large extent, are manifestations of the QBO as inferred from the 30-hPa zonal wind at Singapore. This model is not adequate, however, for the NH possibly because of relatively large interference from the planetary wave activity in this hemisphere. The principal conclusion is that the study considered does not support model predictions of massive ozone loss due to heterogeneous chemical processes, which have been discussed by Hofmann and Solomon (1985), Michelangeli et al. (1989) and others.

Kerr et al. (1993) have analyzed ozone measurement results from the Canadian total ozone network consisting of 12 stations (including 4 ozonesonde stations) that range from 44° N to 84° N as well as spanning 80° of longitude (thus, the stations are well located for studying NH ozone depletion to the north of 45° N). Comparison of the results, obtained during the first 4 months of 1993, with previous observations (pre-1980 averages that are taken to represent ozone climatology prior to the onset of induced ozone depletion), show the TOC has been about 11% to 17% below normal (the average value is $-14.1 \pm 1.9\%$). As Kerr et al. (1993) have emphasized, these low values were recorded, without exception, at all 9 of the Canadian midlatitude ground-based monitoring sites. Record low averages for the period from January to April were measured at three of the four midlatitude stations where measurements have been made since the 1960s.

Comparison of January to April 1993 ozonesonde profile measurements with those from earlier years indicate the ozone deficit is in the lower stratosphere between 40 to 200 hPa, the same altitudes where aerosols from the Mount Pinatubo volcanic eruption have been observed. The peak loss is 30% at 100 hPa (16 km).

The ozonesonde data for the troposphere do not support the recent reports of upward trends of tropospheric ozone. At all four stations the tropospheric ozone contents in 1993 are less than those of the previous years.

More and more evidence accumulates that demonstrates a significant impact of volcanic eruptions on the global ozone layer. D'Altorio et al. (1992) have discussed evidence for total ozone depletion during the 1991–92 winter months on the basis of the analysis of ground-based lidar (DIAL) monitoring of Mount Pinatubo aerosols and ozone at l'Aquila, Italy (42° N, 13° E). A 10-month record of aerosol and ozone measurements (August 1991–May 1992) has been compared to the monthly mean 1985–86 SBUV and SAGE II satellite ozone background data as well as to ozonesonde data for Hohenpeissenberg, (Germany: 48° N, 11° E) and S. Pietro Capofiume (Italy: 45° N, 11° E). The differential lidar DIAL (wavelengths 309 nm and 351 nm) permits retrieval of the vertical ozone profile at night for the 12- to 5-km altitude range with an accuracy within 2%–15%. The 2-year average of satellite data smoothed out the QBO signal but not the ozone trend related to the atmospheric composition changes (a 3%–4% decrease in 5 years occurred in northern midlatitude ozone during the winter season due to TOMS data).

The January-averaged DIAL ozone profile is in reasonable agreement with the ozonesonde data except for an average 14% underestimation in a 3-km layer centered at 20.5-km altitude (presumably due to aerosol interference). The ozone underestimation shown by DIAL and ozonesondes with respect to SAGE II and SBUV above 25 km may be partly attributed to the chemically driven decadal ozone trend. A comparison of DIAL and ozone data for columnar content in the 18–28-km layer with SBUV data shows a good fit up to the end of November 1991, while a sharp column difference is experienced during December 1991 and January 1992 as an 8%–10% loss of ozone. Although considerable caution is necessary (as well as further observations) as far as the interpretation of the data is concerned, it is quite realistic to consider the ozone loss as determined by the impact of the eruption. The following heterogeneous chemical reactions over sulfuric acid aerosol particles could contribute to the loss of ozone:

$$N_2O_5 + H_2O \Rightarrow 2HNO_3;$$
$$ClNO_3 + H_2O \Rightarrow HOCl + HNO_3$$

As D'Altorio et al. (1992) have pointed out, in the presence of volcanic aerosol, the total particle surface available for these reactions is so large that ClO can increase by about an order of magnitude. Mechanisms of volcanic impact are, however, more complicated than just mentioned.

The role of sulfur photochemistry in tropical ozone changes after the Mount Pinatubo eruption has been discussed by Bakki et al. (1993). Although stratospheric ozone losses following volcanic eruptions are generally attributed to the presence of sulfate aerosol, Bakki et al. (1993) have presented model calculations demonstrating that gas-phase sulfur chemistry may also have played an important part in the tropical ozone perturbations that followed the Pinatubo eruption.

This general conclusion is in accordance with observational data for August and September 1991 (ozonesonde measurements) that show the maximum O_3 reduction (18%–20%) was observed at an altitude around 24–25 km but, on the other hand, ozone concentrations had increased by 5%–10% above 28 km. TOMS data indicate local increases in the TOC of about 2% in July, but decreases of 1% in August and 3%–4% in September (zonally averaged TOMS O_3 changes do not exceed 2%–4%). Microwave Limb Sounder (MLS) data for September show TOC reductions in the tropical lower stratosphere, but not significant TOC increases around 30 km. Since the QBO signal has not been smoothed out observed TOC changes cannot be entirely attributed to the Mount Pinatubo eruption.

The calculation made by Bakki et al. (1993) shows that in the first month or so after the eruption, the large amount of SO_2 injected into the tropical atmosphere catalyzes midstratospheric ozone production.

Ozone production catalyzed by SO_2 is possible through the following sequence of reactions:

$$SO_2 + h\nu \Rightarrow SO + O. \quad (\lambda < 220 \text{ nm})$$
$$SO + O_2 \Rightarrow SO_2 + O,$$
$$2(O + O_2 + M \Rightarrow O_3 + M)$$
$$\text{Net: } 3O_2 \Rightarrow 2O_3.$$

Sulfur dioxide also has the potential to reduce transmission of solar radiation since it absorbs strongly between 180 and 235 nm, weakly in the 260- to 340-nm interval, and very weakly between 340 and 390 nm. Such a reduction leads to a decrease of the photolysis rate for key species such as O_2 and therefore results in reducing the rate of ozone production. The simulation modeling made by Bakki et al. (1993) with the use of a zonally averaged chemical-radiative-dynamical model (model integrations cover the period from June to September 1991) has demonstrated that the two effects mentioned cancel each other out at an altitude of about 25 km. After 1 or 2 months the SO_2 has been mostly oxidized to sulfate; the efficiency of these two mechanisms then becomes negligible (although ozone remains perturbed in the lower stratosphere because of its long photochemical lifetime in this region).

The changes in O_3 concentrations in July are predominantly confined to the latitude band of volcanic injections. Ozone concentration increases in the midstratosphere (the top region of the volcanic cloud) by up to 7%. By September, the initial SO_2 cloud has spread throughout the tropics. Ozone increase (2%) in the mid-stratosphere remains low with a reduction of up to 8%.

The rates of the competing processes mentioned impact the amount of SO_2 present in the stratosphere. As most of the SO_2 cloud is converted to sulfate by September the perturbations of the photochemical balance of O_3 are substantially reduced at this time compared with July. Ozone production now represents less than 10% of the ozone production in the midstratosphere, and oxygen photolysis is reduced only by about 10% in the lower stratosphere.

The model features show good agreement with initial ozone measurements following the eruption, including both the midlatitude switch from ozone loss to ozone gain, and the increase and subsequent decrease in the total ozone content.

There are, however, some discrepancies between calculations and observations, e.g., underestimated model TOC values (in comparison with TOMS data) and estimates of the altitude of the crossing over from the production to ozone distribution (~ 25 km model value and observed 28 km).

An important aspect of the problem is that volcanically induced ozone changes may lead to a substantial impact on the radiative forcing for the "surface-troposphere system" and, subsequently, on climate.

VOLCANIC ACTIVITY AND CLIMATE 341

Some recent numerical modeling results show that variations of TOC and its vertical profile may belong to significant climate-forming processes. As Schwarzkopf and Ramaswamy (1993) have pointed out, observations from satellite and ground-based instruments indicate that significant changes in atmospheric ozone concentrations have taken place during the 1980s. Decreases in TOC in middle and high latitudes have been measured by the TOMS instrument, while SAGE data indicate ozone depletion in the lower stratosphere at all latitudes, including the tropics. On the other hand, observations at several locations in the midlatitudes of the NH have shown significant decadal increases in tropospheric ozone. The earlier calculations led to the conclusion that the decadal changes in lower stratospheric ozone produce substantial negative RF of the surface-troposphere system at middle and high latitudes. The magnitude of ozone forcing is sensitive to both TOC loss and the altitude profile on the ozone depletion.

Schwarzkopf and Ramaswamy (1993) have undertaken new calculations of the sensitivity of RF to ozone changes with the use of SAGE I and II data on ozone vertical profiles above 17 km (below this level the presence of clouds complicates the remote sounding) as well as data on the increase of tropospheric ozone. The following parameters have been fixed: vertical temperature and humidity profiles, cloudiness, surface temperature and albedo. Three scenarios for annual-averaged profiles within the 17- to 50-km layer for decadal ozone losses in tropical (4.5° N) and middle (40.5° N) latitudes (similar in each latitude belt) have been chosen:

(i) No ozone depletion between 17 km (S1);
(ii) A linear decrease with altitude of the percentage of ozone depletion between 17 km and the troposphere, where the depletion becomes zero (S2);
(iii) The depletion percentage between the tropopause and 17 km is equal to the SAGE depletion percentage at 17 km (S3).

Additionally a "standard" profile (S4) has been used that employs a percentage ozone depletion adjusted so that the TOC is equal to that obtained using profile S1.

Tables 5.3.1 and 5.3.2 illustrate some of the results obtained by Schwarzkopf and Ramaswamy (1993).

These results show that the amount of stratospheric ozone RF during the 1980s depends principally on two factors: the TOC depletion and the altitude profile of the ozone change. In the tropics, where stratospheric ozone change profiles are available almost to the troposphere, the principal question is the quantity of lower stratospheric ozone loss. An important conclusion is that the magnitude of the decadal NH stratospheric ozone RF would increase significantly if ozone depletion has occurred in the tropics, as implied by SAGE observations.

Table 5.3.1. Decadal stratospheric column ozone change in DU, stratospheric ozone forcing in W/m^2, and stratospheric radiative forcing gradient (SRFG) in W/m^2/DU for "standard" tropical (January, 4.5°N) profiles. The SRFG derived here is valid over a stratospheric column ozone decrease of 0-10 DU. After Schwarzkopf and Ramaswamy (1993). [© by American Geophysical Union]

Profile	S1	S2	S3	S4
Column change	-7.34	-8.07	-8.79	-7.34
Forcing	-0.052	-0.071	-0.090	-0.052
SRFG	0.0071	0.0087	0.0102	0.0071

Table 5.3.2. Same as Table 5.3.1 except for midlatitude (January 40.5°N). The SRFG derived here is valid over a stratospheric TOC decrease of 0-30 DU. After Schwarzkopf and Ramaswamy (1993). [© by American Geophysical Union]

Profile	S1	S2	S3	S4
Column change	-10.38	-15.55	-20.72	-10.38
Forcing	-0.035	-0.084	-0.134	-0.078
SRFG	0.0034	0.0054	0.0065	0.0075

In midlatitudes, the stratospheric ozone forcing depends strongly on the altitude profile of decadal ozone change in the midlatitude lower stratosphere, especially in the layers nearest the tropopause (accurate measurements of ozone change in the vicinity of the tropopause are necessary to arrive at a precise estimate of the stratospheric ozone RF). If observed increases of tropospheric ozone in various regions of the NH are hemispherically representative, one can conclude that a substantial positive tropospheric ozone RF has occurred during the 1990s, of a magnitude possibly comparable to that produced by decadal stratospheric ozone decreases. Although the results considered require further confirmation, it is without doubt that both the magnitude and the sign of the decadal total ozone changes of TOC and vertical ozone profiles (both in the stratosphere and troposphere) belongs to significant climate-forming processes.

Since recent analysis of ozone data for middle and high latitudes of the NH has led to the conclusions about the trends of ozone drop (during the last few decades) in the lower stratosphere and the growth of tropospheric ozone (especially in the upper troposphere), Wang et al. (1993) have conducted calculations of an impact of such ozone change on the radiative forcing for the surface-troposphere system, including its short-wave and long-wave components. Monthly mean ozone profiles have been prescribed on the basis of ozonesonde observational data for seven stations in mid and high latitudes of the NH. Earlier calculations showed that observed lower stratospheric O_3 depletion during the 1980s may have induced a significant RF cooling of the surface-troposphere climate system in contrast to the positive RF due to an increase of GHGs during the same time period. Wang et al. (1993) have pointed out, however, that the calculations at mid and high latitudes of the NH did not include the effect of observed tropospheric O_3 increases. It is in this context that new calculations have been made to intercompare RF values for clear-sky conditions for the following time periods:

(i) 1971–1980 and 1981–1990 (Hohenpeissenberg, Payerne, Goose Bay and Resolute data);
(ii) 1981–1985 and 1986–1990 (Churchill, Edmonton and Tateno data).

Two approaches have been used to calculate RF due to lower stratospheric ozone depletion: for FT, and adjusted temperature according to FD heating assumptions.

Table 5.3.3 data illustrate comparative impacts on RF of ozone changes (including separate assessment for tropospheric ozone increase) and GHG increase (CO_2, CH_4, CFC-11, CFC-12, and NO_2 have been taken into account). For FT treatment, the total RF at Hohenpeissenberg is calculated to be 0.79 and 1.05 W/m^2, respectively for January and July, while the effect due to O_3 changes contributes 0.41 and 0.57 W/m^2, respectively. For FD treatment, the total RF for January and July is calculated to be 0.71 and 1.03 W/m^2. For both treatments it is quite clear that the tropospheric O_3 increase contributes substantially to the total radiative forcing. An important aspect of the results is that they are sensitive to stratospheric temperature responses associated with O_3 changes. Wang et al. (1993) have emphasized that the large sensitivity of the O_3 radiative forcing to lower stratospheric temperatures points out the need of studying the climatic implications of O_3 changes using general circulation models in which the interactions between radiation and dynamics are properly incorporated.

Significant changes of the total ozone content in the atmosphere after explosive volcanic eruptions has attracted serious attention to careful analyses of combined observational data from satellite, ozonesondes, and ground-based

Table 5.3.3. Comparison of changes in radiative forcing (Wm^{-2}) between O_3 changes and the combined O_3 changes and increases of greenhouse gases CO_2, CH_4, CFC-11, CFC-12 and N_2O. After Wang et al. (1993). [© by American Geophysical Union]

	O_3 changes		Combined O_3, and other GHGs			
	Total		Troposphere		Total	
	FT	FD	FT	FD	FT	FD
Hohenpeissenberg (47° N, 11° E)						
January	0.410	0.327	0.222	0.136	0.794	0.710
July	0.570	0.548	0.490	0.467	1.050	1.027
Payerne (47° N, 7° E)						
January	0.356	0.190	0.242	0.074	0.797	0.630
July	0.397	0.136	0.097	-0.167	0.879	0.617
Resolute (75° N, 95° W)						
Winter	-0.038	-0.085	0.012	-0.036	0.230	0.183
Summer	0.194	-0.009	0.052	-0.153	0.620	0.415
Tateno (36° N, 140° E)						
Winter	0.239	0.096	0.084	-0.060	0.491	0.348
Summer	0.304	0.287	0.295	0.279	0.585	0.569

instrumentation as well to the search of potential causes of ozone changes. Although similar changes took place during nonvolcanic periods, various attempts to explain post-eruption observations have understandably been concentrated on revealing volcanically induced mechanisms of TOC variations, especially in connection with possible enhancement of heterogeneous chemical reactions on surfaces of volcanic aerosol particles that lead to ozone destruction (e.g., Antarctic spring ozone minimum leading to ozone hole formation). In this context Schoeberl et al. (1993) have analyzed Nimbus 7 TOMS data of equatorial TOC change after the Mount Pinatubo eruption.

The data show a decrease of up to 6% over climatology within the belt ±12° latitude. Ozone losses began approximately a month following the eruption, consistent with the time required for the SO_2 to convert to sulfuric acid aerosol. Ozone values remained below climatology until December 1991.

VOLCANIC ACTIVITY AND CLIMATE 345

Following the eruption, retrieved TOC values increased from June 15 through the first week of July by about 10 DU. Subsequently, total ozone declined by about 20–30 DU reaching a minimum shortly before November, then levelling off. The net decrease is about 10% from the peak values during this period. This anomaly is, however, confined to the tropics. No anomalous minimum values are seen in the 12° N – 18° N latitude band, but the 12° S – 18° S band does not show a weak signal.

As Schoeberl et al. (1993) have pointed out, under usual conditions equatorial TOC follows the phase of the QBO in lower stratospheric equatorial winds. During the easterly phase, ozone amounts are lower while during the westerly phase the amounts are higher as a result of the secondary circulation associated with the QBO. Since Singapore 50 hPA winds switched to easterly in early 1991, it is possible that the tropical ozone decrease is simply due to the QBO. However, the ozone decrease following the Mount Pinatubo eruption amounts to a 5%–6% decrease over climatology of the easterly phase QBO years.

As has been noted above, the "heterogeneous chemistry" hypothesis has been developed in a number of studies (Prather, 1992; Deshler et al., 1993; and others). It is hardly probable, however, that this hypothesis explains the basic cause of ozone loss. As Schoeberl et al. (1993) have emphasized, the tropical lower stratosphere is poor in available chlorine as most of the chlorofluorocarbons emerging from the troposphere through the tropical Hadley circulation have yet to be photolyzed. A study by Kinne et al. (1992) seems more persuasive, suggesting a model of the shift of the volcanic aerosol layers by 1–2 km upward over several months due to additional heating. Thus, the aerosol layer carries low ozone air to higher latitudes. Additional chemical measurements in the vicinity of a tropical volcanic aerosol layer are necessary to solve the problem.

5.4 Conclusion

After the three powerful explosive volcanic eruptions during the time period of about one decade—Mount St. Helens (1980), El Chichón (1982), and Mount Pinatubo (1991)—the development of new ground-based, aircraft, balloon and satellite means of observation, as well as new numerical models, have brought about an increased level of understanding of what happens in the atmosphere under the impact of eruptions. This is especially true in the case of the Mount Pinatubo eruption when it has become possible to use a unique observing system and new 3-D global models of effluents transport and climate change. The problem is, however, far from being solved.

The principal conclusion is that perhaps the problem is much more complex that it was thought before. A persuasive proof of this simple truth is the discovery of interaction between volcanic aerosols and atmospheric (both stratospheric and tropospheric) ozone. It has become clear that not only is the impact of eruptions on ozone significant, but the effect of ozone changes on climatic consequences of eruptions. This means that an explanation of volcanic impact requires development of new global climate models with coupled dynamics, radiation, and photochemistry, as well as adequate simulation of interaction between the stratosphere and troposphere. For verification of new modeling results many more complete observations are necessary than those already accomplished. A dedicated international program of both observations and modeling is needed to resolve existing uncertainties.

REFERENCES

Angell, J.K., 1986: Environmental impact of volcanism. In *Norman D. Watson Symposium*. University of Rhode Island, 9-17.

Angell, J.K., 1989: Variations and trends in tropospheric and stratospheric global temperatures, 1958-1987, *J. Climate* **2**, 1404-1416.

Angell, J.K., and Korshover, J., 1975: Estimate of the global change in tropospheric temperature between 1958 and 1973. *Mon. Weather Rev.* **103**, 1007-1012.

Angell, J.K., and Korshover, J., 1977: Estimate of the global change in temperature, surface to 100 mbar between 1968 and 1973. *Mon. Weather Rev.* **105**, 375-388.

Angell, J.K., and Korshover, J., 1978: Comparison of stratospheric trends in temperature, ozone and water vapor in north temperature latitudes. *J. Appl. Met.* **17**, 1397-1401.

Angell, J.K., and Korshover, J., 1983a: Global temperature variations in the troposphere and the stratosphere, 1958-1982. *Mon. Weather Rev.* **111** (5), 901-921.

Angell, J.K., and Korshover, J., 1983b: Comparison of stratospheric warmings following Agung and Chichón. *Mon. Weather Rev.* **111** (10), 2129-2135.

Ardanuy, Ph.E., Kyle, H.L., and Hoyt, D., 1992: Global relationships among the Earth's radiation budget, cloudiness, volcanic aerosols, and surface temperature. *J. Climate* **5**, 1120-1139.

Bakki, S., Tuomi, R., and Pyle, J.A., 1993: Role of sulphur photochemistry in tropical ozone changes after the eruption of Mount Pinatubo. *Nature* **362**, 331-333.

Belle, T.L., and Abdullah, A., 1985: Detecting global climate change predicted by climate models. In *Third Conf. on Climate Variations and Symposium* on Contemporary Climate: 1850-2010. American Meteorological Society, Boston, Massachusetts, 89-90.

Bryson, R.A., and Goodman, B.M., 1980: Volcanic activity and climatic changes. *Science* **27**, 1041-1044.

Bryson, R.A., and Goodman, B.M., 1982: The climatic effect of explosive volcanic activity: Analysis of historical data. *Proc. Symp. on Atmospheric Effects and Potential Climate Impact of the 1980 Eruptions of Mt. St. Helens*, NASA Conference Publication 2240, 191-202.

Cess, R.D., Coakley, J.A. Jr., and Kolesnikov, P.M., 1981: Stratospheric volcanic aerosols: A model study of interactive influences upon solar radiation. *Tellus* **33** (5), 444-452.

Chandra, S., 1993: Changes in stratospheric ozone and temperature due to the eruption of the Mt. Pinatubo. *Geophys. Res. Lett.* **20** (1), 33-36.

Chandra, S., Jackman, Ch.K., Douglass, A.R., Fleming, E.L., and Contidine, D.B., 1993: Chlorine catalized destruction of ozone: Implications for ozone variability in the upper stratosphere. *Geophys. Res. Lett.* **20** (5), 351-354.

Charlock, T.P., 1983: The effect of volcanic aerosols on the thermal infrared budget of the lower atmosphere. *Fifth Conf. on Atmospheric Radiation.* Baltimore, Maryland, American Meteorological Society, Boston, Massachusetts, 350-353.

D'Altorio, A., Masci, F., Pitari, G., Visconti, G., Rizi, V., Cervino, M., and Giovanelly, G., 1992: Ground-based monitoring of Pinatubo aerosols and ozone at L'Aquila, Italy: 1. Evidence for ozone depletion during 1991/1992 winter months. *Il Nuovo Cimento* **16C** (1), 91-95.

Dave, J.V., DeLuisi, J.J., and Mateer, C.L., 1980: Results of a comprehensive theoretical examination of the optical effects of aerosols on the Umkehr measurements. Papers of WMO Tech. Conf. on Reg. and Global Observ. of Atmos. Poll. Relative to Climate. *Spec. Env. Rep.* **14**, 15-22.

Deirmendjian, D., 1973: On volcanic and other particulate turbidity anomalies. *Advances in Geophysics* **16**, 267-297.

DeLuisi, J.J., 1969: A study of the effect of haze upon Umkehr measurements. *J. Roy. Meteorol. Soc.* **95**, 181-187.

DeLuisi, J.J., 1979: Umkehr ozone profile errors caused by the presence of stratospheric aerosols. *J. Geophys. Res.* **84**, 1766-1770.

DeLuisi, J.J., Herman, B.M., Browning, S.R., and Sato, R K., 1975: Theoretically determined multi-scattering effects of dust on Umkehr observations. *J. Roy. Meteorol. Soc.* **101**, 325-331.

DeLuisi, J.J., Mateer, C.L., and Komhyr, W.D., 1984: Effects of El Chichón stratospheric aerosol cloud on Umkehr measurements at Mauna Loa, Hawaii. *Proc. Quadrennial Ozone Symp.* Halkidiki-Greece. 3-7 Sept. 1984.

Deshler, T., Johnson, B.J., and Rozier, W.R., 1993: Balloon-borne measurements of Pinatubo aerosol during 1991 and 1992 at 41° N: Vertical profiles, size distribution, and volatility. *Geophys. Res. Lett.* **20** (14), 1435-1438.

Dutton, E.G., Hofmann, D.J., Oltman, S.J., and Christy, J.R., 1993: Radiative transfer calculations of observed decreases in stratospheric temperatures and ozone beginning 1.5 years after the Eruption of Mt. Pinatubo. *American Geophysical Union Fall Meeting*, Dec 6-10, San Francisco. Paper # A12F-06. Abstract published in October 26, Supplement to *EOS*.

Ellsaesser, H.W., 1977: Effect of Mt. Agung on global temperature. *Mon. Weather Rev.* **105**, 1200.

Ellsaesser, H.W., 1983: Isolating the climatogenic effects of volcanoes. Preprint UCRL-89161 Lawrence Livermore National Laboratories, California.

Gerstell, M.F., Crisp, J., and Crisp, D., 1995: Radiative forcing of the stratosphere by SO_2 gas, silicate ash, and H_2SO_4 aerosol shortly after the 1982 eruption of El Chichón. *J. Climate* **8** (5), 1060-1070.

Gilliland, R., 1982: Solar, volcanic, and CO_2 forcing of recent climatic changes. *Climatic Change* **4**, 111-131.

Gilliland, R.L., and Schneider, S.H., 1984: Volcanic, CO_2 and solar forcing of Northern and Southern Hemisphere surface air temperatures. *Nature* **310** (5972), 38-41.

Grant, W.G., Fishman, J., Browell, E.V., et al., 1992: Observations of reduced ozone concentration in the tropical stratosphere after the eruption of Mt. Pinatubo. *Geophys. Res. Lett.* **19** (11), 1109-1112.

Hammer, C.U., 1977: Past volcanism revealed by Greenland ice sheet impurities. *Nature* **270**, 482-486.

Hansen, J.E., Wang, W.C., and Lacis, A.A., 1977: Mt. Agung eruption confirms climatic effects of global radiative perturbations of the atmosphere. Preprint, Goddard Inst. for Space Studies, New York, 4 pp.

Hansen, J.E., Wang, W.C., and Lacis, A.A., 1978: Mount Agung eruption provides test of a global climatic perturbation. *Science* **199** (4333), 1065-1068.

Hansen, J.E., Johnson, D., Lacis, A.A., Lebedeff, S., Lee, P., Rind, D., and Russel, G., 1981: Climate impact of increasing atmospheric carbon dioxide. *Science* **213**, 957-966.

Hansen, J.E., Lacis, A., Ruedy, R., and Sato, M., 1992: Potential climate impact of Mount Pinatubo aerosols. *Geophys. Res. Lett.* **19** (2), 215-218.

Harshvardhan, 1979: Perturbation of the zonal radiation balance by stratospheric aerosol layer. *J. Atmos. Sci.* **36** (7), 124-128.

Harshvardhan and Cess, R.D., 1975: Stratospheric aerosols: Effect upon atmospheric temperature and global climate. In *Coll. Abstr. Second Conf. on Atmos. Radiation*, Arlington, Virginia, 135-137.

Harshvardhan and Cess, R.D., 1976: Stratospheric aerosols: Eeffect upon atmospheric temperature and global climate. *Tellus* **28** (1), 1-10.
Hofmann, D.J., 1993a: Twenty years of balloon-borne tropospheric aerosol measurements at Laramie, Wyoming. *J. Geophys. Res.* **98**, 12753-12766.
Hofmann, D.J., 1993b: Comparing the volcanic signal in ozone depletion after El Chichón and Pinatubo eruptions. American Geophysical Union, AGU 1993, Annual Meeting, *EOS Suppl.* 114.
Hofmann, D.J., and Solomon, S., 1985: Ozone destruction through hetereogeneous chemistry. *Geophys. Res. Lett.* **94**, 5029-5041.
Hofmann, D.J., Oltman, S.J., Harwis, J.M., et al., 1993a: Ozonesonde measurements at Hilo, Hawaii following the eruption of Pinatubo. *Geophys. Res. Lett.* **20** (15), 1555-1557.
Hofmann, D.J., Oltman, S.J., Latrop, J.A., et al., 1993b: Unprecedented low ozone over the United States during 1992 and 1993: Dobson and Ozonesonde data. American Geophysical Union, AGU 1993, Annual Meeting, *EOS Suppl.* 138.
Hunt, B.G., 1977: A simulation of the possible consequences of a volcanic eruption on the general circulation of the atmosphere. *Mon. Weather Rev.* **105** (3), 247-260.
Hunt, B.G., 1978: A simulation of the possible consequences of a volcanic eruption on the general circulation of the atmosphere. In *Climate Change and Variability, South Perspectives*, Cambridge, e.a., 263-268.
Kent, G.S., and Yue, G.K., 1991: The modelling of CO_2 lidar backscattter from stratospheric aerosol. *J. Geophys. Res.* **96** (D3), 5279-5292.
Kerr, J.G., Wardle, D.I., and Tarasick, D.W., 1993: Record low ozone values over Canada in early 1993. *Geophys. Res. Lett.* **20** (13), 1975-1982.
Kinne, S., Toon, O.B., and Prather, M.J., 1992: Buffering of stratospheric circulation by changing amounts of tropical ozone: A Pinatubo case study. *Geophys. Res. Lett.* **19** (19), 1927-1930.
Komhyr, W.D., Oltman, S.J., Chopra, A.N., Leonard, R.K., García, T.E., and McFee, C., 1984: Results of Umkher, ozonesonde, total ozone and SO_2 observations in Hawaii following the eruption of El Chichón in 1982. *Proc. Ozone Symp., Halkidiki*.
Komhyr, W.D., Grass, R.D., Evans, R.D., et al., 1993: Ozone trends over the contiguous U.S.A., Hawaii, and Samoa, 1979-1993. American Geophysical Union, AGU 1993, Annual Meeting, *EOS Suppl.* **113**.
Kondratyev, K.Ya., 1989: *Global Ozone Dynamics*. Moscow, VINITI Publ., 212 pp. (in Russian).
Kondratyev, K.Ya., and Moskalenko, N.I., 1984: Radiative heat exchange in the atmosphere disturbed by volcanic eruptions. *Dokl. Akad. Nauk SSSR* **274** (4), 799-801 (in Russian).
Kondratyev, K.Ya., and Grassl, H., 1996: Global Climate Dynamics in the context of global change. Berlin e.a., Springer-Verlag (in press).

Lacis, A., Hansen, J., and Sato, M., 1992: Climate forcing by stratospheric aerosols. *Geophys. Res. Lett.* **15** (15), 1607-1610.

Lamb, H.H., 1970: Volcanic dust in the atmosphere, with a chronology and assessment of its meteorological significance. *Proc. Roy. Soc. London,* **A226**, 425-535.

Lamb, H.H., 1972: *Climate: Present, Past and Future, I. Fundamentals and Climate Now.* Methuen and Co., Ltd., London, 1972, pp. 614.

Lamb, H.H., 1977: *Climate: Present, Past and Future, II. Climatic History and Future.* London, Methuen, New York, Barnes and Noble, pp. 835.

Landsberg, H.E., and Albert, J.M., 1974: The summer of 1816 and volcanism. *Weatherwise* **27**, 63-66.

Lenoble, J., 1986: Detection of stratospheric aerosol characteristics in relation with climate impact. *Adv. Space Res.* **6** (10), 67-72.

Loginov, V.F., 1984: *Volcanic Eruptions and Climate.* Leningrad, Gidrometeoizdat (in Russian).

Loginov, V.F., Pivovarova, Z.I., and Kravchuk, E.G., 1983a: An assessment of the contribution of natural and anthropogenic factors to solar radiation variability on the Earth's surface. *Meteorol. Gidrol.* **8**, 55-60 (in Russian).

Loginov, V.F., Pivovarova, Z.I., and Kravchuk, E.G., 1983b: Variability of direct solar radiation and temperature in the Northern Hemisphere due to volcanic eruptions. *Izv. Vses. Geogr. Obschestva* **115** (5), 401-411 (in Russian).

Lough, J.M., and Pritts, H.C., 1987: An assessment of the possible effects of volcanic eruptions on North American climate using tree ring data: 1602 to 1900 AD. *Climate Change* **10** (5), 219-240.

MacCracken, M.C., 1976: Climate-model results of stratospheric perturbations. In Proc. *4th. Conf. on CIAP,* February 1975, DOT, Washington, D.C., 183-194.

MacCracken, M.C., and Potter, C.L., 1975: Comparative climatic impact of increased stratospheric aerosol loading and decreased solar constant in a zonal climate model., reprint UCRL-76132, WMO/IAMAP Symp. on Long-Term Climate Fluctuations. Aug. 17-21, University of East Anglia, England.

MacCracken, M.C., and Luther, F.M., 1983a: Radiative and climatic effects of the El Chichón eruption. Proc. 5th. Conf. Atmospheric Radiation, Oct. 31- Nov. 4, 1983, Baltimore, Maryland. American Meteorological Society, Boston, Massachusetts, 346-349.

MacCracken, M.C., and Luther, F.M., 1983b: Radiative and climatic effects of the El Chichón eruption. *Res. Activ. Atmos. Ocean Model* **5**, 6.28.

Mason, B.J., 1976: Towards the understanding and prediction of climatic variations. *Quart. J. Roy. Meteorol. Soc.* **102**, 473-498.

Mass, C., and Schneider, S.H., 1977: Statistical evidence of the influence of sunspots and volcanic dust on long-term temperature records. *J. Atmos. Sci.* **34** (12), 1995-2004.

McInturff, R.M., Miller, A.J., Angell, J.K., and Korshover, J., 1971: Possible effects on the stratosphere of the 1963 Mt. Agung volcanic eruption. *J. Atmos. Sci.* **28**, 1304-1307.

Mergenthaler, J.L., 1985: An estimate of the effect of El Chichón aerosol on SBUV profiling for summer 1982. *J. Geophys, Res.* **90**, D1, 2355-2359.

Michelangeli, D.V., Allen, M., and Yung, Y.L., 1989: El Chichón volcanic aerosols: Impact of radiative, thermal, and chemical perturbations. *J. Geophys. Res.* **94** (ND15), 18429-18443.

Newell, R.E., 1970: Stratospheric temperature change from the Mt. Agung volcanic eruption of 1963. *J. Atmos. Sci.* **27** (4), 077-978.

Newell, R.E., 1984: Volcanism and climate. In *1985 Yearbook of Science and Technology*, 206-225, New York: McGraw-Hill.

Oliver, R.C., 1976: On the response of hemispheric mean temperature to stratospheric dust: An empirical approach. *J. Appl. Meteorol.* **15**, 933-950.

Parker, D.E., 1984: The influence of the southern oscillation and volcanic eruption on temperature in the tropical troposphere. *Trop. Ocean-Atmos. Newslett.* **26**, 4-51.

Parker, D.E., and Brownscombe, J.L., 1983: Stratospheric warming following the El Chichón volcanic eruption. *Nature* **301** (5899), 406-408.

Pollack, J.B., and Ackerman, Th. P., 1983: Possible effects of El Chichón volcanic cloud on the radiation budget of the northern tropics. *Geophys. Res. Lett.* **10** (11), 1057-1060.

Pollack, J.B., Toon, O.B., Sagan, C., Summers, A., Baldwin, B., and Van Camp, A., 1976a: Stratospheric aerosols and climatic changes. *Nature* **263**, 551-555.

Pollack, J.B., Toon, O.B., Sagan, C., Summers, A., Baldwin, B., and Van Camp, A., 1976b: Volcanic explosions and climatic change: A theoretical assessment. *J. Geophys. Res.* **82**, 1071-1083.

Pollack, J.B., Toon, O.B., Danielsen, E.F., Hofmann, D.J., and Rosen, J.M., 1983: The El Chichón volcanic cloud: An introduction. *Geophys. Res. Lett.* **10** (11), 982-992.

Pollack, J.B., Rind, D., Lacis, A., Hansen, J.E., Sato, M., and Ruedy, R., 1993: GCM simulations of volcanic aerosol forcing, Part 1: Climate changes induced by steady-state perturbations. *J. Climate* **6**, 1719-1741.

Portman, D.A., and Gutzler, D.S., 1996: Explosive volcanic eruptions, the El Niño-southern oscillation, and U.S. Climate variability. *J. Climate* **9** (1), 17-33.

Prather, M., 1992: Catastrophic loss of anthropogenic ozone in dense volcanic clouds. *J. Geophys. Res.* **97**, 10187-10191.

Quiroz, R.S., 1983: The isolation of stratospheric temperature change due to the El Chichón eruption from nonvolcanic signals. *J. Geophys. Res.* **C88** (11), 6773-6780.
Rampino, M.R., and Self, S., 1982: Historic eruptions of Tambora (1815), Krakatao (1883), and Agung (1963)—their stratospheric aerosols and climatic impact. *Quart. Res.* **18** (2), 127-143.
Rampino, M.R., Self, S., and Stothers, R.B., 1988: Volcanic winters. *Ann. Rev. Earth and Planet. Sci.* **16**, 73-99.
Rind, D., Snozzo, R., Balachandran, N.K., and Prather, M.J., 1990: Climate change and the middle atmosphere, Part I: The doubled CO_2 climate. *J. Atmos. Sci.* **47**, 475-494.
Rind, D., Balachandran, N.K., and Snozzo, R., 1992: Climate change and the middle atmosphere, Part II: The impact of volcanic aerosols. *J. Climate*, **5**, 189-208.
Robock, A., 1978: Internally and externally caused climate change. *J. Atmos. Sci.* **35** (6), 1111-1122.
Robock, A., 1984: Climate model simulations of the effects of the El Chichón eruption. *Geofis. Int.* **23** (3), 403-414.
Robock, A., and Mao, J., 1995: The volcanic signal in surface temperature observations. *J. Climate* **8**, 1086-1103.
Schneider, S.H., 1983: Volcanic dust veils and climate: How clear is the connection?— An editorial. *Clim. Change* **5** (2), 111-113.
Schneider, S.H., 1984: Atmospheric double exposure. *Nat. Hist.* **93** (4), 100-101.
Schneider, S.H., and Mass, C., 1975: Volcanic dust, sunspots, and temperature trends. *Science* **190**, 741-746.
Schoeberl, M.R., Bhartia, P.K., Hilsenrath, E., and Torres, O., 1993: Tropical ozone loss following the eruption of Mt. Pinatubo. *Geophys. Res. Lett.* **20** (1), 29-32.
Schwarzkopf, M.D., and Ramaswamy, V., 1993: Radiative forcing due to ozone in the 1980's: Dependence on altitude of ozone change. *Geophys. Res. Lett.* **20** (2), 205-208.
Self, S., Rampino, M.R., and Barbera, J.J., 1981: The possible effects of large 19th and 20th century volcanic eruptions on zonal hemispheric surface temperatures. *J. Volcanol. Geotherm. Res.* **11**, 41-60.
Stolarski, R.S., and Cicerone, R.J., 1974: Stratospheric chlorine: A possible sink for ozone. *Can. J. Chem.* **52**, 1610-1615.
Stolarski, R.S., and Butler, D.M., 1979: Possible effects of volcanic eruptions on stratospheric minore constituents chemistry. *PAGEOPH* **117**, 486-497.
Stothers, R.B., and Rampino, M.R., 1983a: Historic volcanism, European dry fogs, and Greenland acid precipitation, 1500 BC to 1500 AD. *Science* **222** (4622), 411-412.

Stothers, R.B., and Rampino, M.R., 1983b: Volcanic eruptions in the Mediterranean before 630 AD from written and archaelogical sources. *J. Geophys. Res.* **B88** (8), 6357-6372.

Tanre, D., Geleyn, J.F., and Slingo, J., 1984: First results of the introduction of an adavanced aerosol-radiation interaction in the ECMWF low resolution global model. In *Aerosols and their Climatic Effects*. H.E. Gerber, and A. Deepak, (Eds.). A. Deepak Publ., Hampton, Virginia, 133-177.

Taylor, B.L., Gal-Chen, T., and Schneider, S.H., 1980: Volcanic eruptions and long-term temperature records: An empirical search for cause and effect. *Quart. J.R. Meteorol. Soc.* **106** (447), 175-200.

Vinnikov, K.Ya, and Groisman, P.Ya., 1981: Empirical analysis of CO_2 influence on variations in the mean annual surface air temperature of the Northern Hemisphere. *Meteorol. i gydrol.* **11**, 30-45 (in Russian).

Vonder Haar, T.H., and Ellis, J., 1975: Albedo of the cloud-free earth-atmosphere system. *Preprints Second Conf. Atmospheric Radiation, Arlington*, Amer. Meteor. Soc., Boston, Massachusetts, 107-110.

Vupputuri, R.K.R., and Blanchet, J-P., 1984: The possible effects of El Chichón eruption on atmospheric thermal structure and surface climate. *Geofis. Int.* **23** (3), 423-447.

Wallace, L., and Livingston, W., 1992: The effect of the Pinatubo cloud on hydrogen chloride and hydrogen fluoride. *Geophys. Res. Lett.* **19** (12), 1209.

Wang, W.-Ch., Zhuang, Yi.,-Ch., and Bojkov, R., 1993: Climate implications of observed changes in ozone vertical distributions in middle and high latitudes of the Northern Hemisphere. *Geophys. Res. Lett.* **20** (15), 1567-1570.

Yamamoto, R., 1981: Change of global climate during recent 100 years. In *Proc. Techn. Conf. on Climate — Asia and Western Pacific (Guanghou, China, Dec., 1980)*, WMO No. 578, World Meteorological Organization, Geneva, 360-375.

SUMMARY

In summary, it is worthwhile to discuss the present day situation in studies of weather and climate, and in this context, the place of the volcanoes and climate component.

In conditions of intensive industrial and technological progress human activity depends, in some respects, more and more on weather and climate. It refers, for example, to agriculture, energy production and transport. Also, the impact of man's activities on the environment and climate grows (for example, the effect of chlorofluorocarbons on the ozone layer and the climatic consequences of increased CO_2 and other minor gaseous components). Consequently, first priority must be given to the following three aspects:

1. Weather predictions, including the development of nowcastings, as well as synoptic forecasts, with emphasis on precipitation and particularly dangerous weather conditions.
2. The climate, the possibility to forecast climatic anomalies with a characteristic time scale of a month and longer; the climatic response to regional variations in the surface properties, for example, as a result of urbanization and deforestation.
3. Atmospheric chemistry: global chemical cycles, especially those that can affect the regional and global climate, problems of regional chemistry and atmospheric pollution.

The current observational means enable one to improve substantially the nowcasting 1 to 12 hours ahead. In this connection further effort is needed for a better understanding of the laws of various mesoscale processes. During the last two decades the reliability of the 48-hour forecasts of the fields of temperature and winds has largely been improved. The 24-hour forecasts of precipitation zones have become more reliable, mainly due to the development and application of high resolution models for limited geographical regions. However, predictions of precipitation amounts and of the time of their beginning and ceasing are less successful. Further theoretical studies and mesometeorological field experiments are needed here, aimed at improving

VOLCANIC ACTIVITY AND CLIMATE 355

techniques for precipitation forecasts 24 to 72 hours ahead. The emphasis must be placed on the development of the forecasting technique and methods of distributing information about the formation and evolution of local dangerous meteorological conditions with a characteristic time scale of less than 24 hours.

Apparently, the most serious errors in the forecasts of large-scale motions stem from inadequate analysis of long waves whose variability determines the location of anomalies in temperature and jet streams that stimulate the evolution of cyclones. The errors result from three causes: an inadequate description of the atmospheric state from data of the global observational network; imperfect techniques for an assimilation of observational data in numerical models; and the inadequacy of the physical and mathematical "constructions" of the models. The use of various databases will make it possible to determine whether a denser network may play a key role in raising the reliability and enlarging the range of the forecasts. The most important challenge in this context is the substantiation of an optimized global climate observing system based on the use of conventional and satellite observations.

Empirical data now exist that illustrate the potential predictability of climate changes a month and more ahead. This particularly refers to such persistent large-scale weather-forming features as blocking or stable trajectories of the cyclones. Therefore it may be possible in the near future to develop techniques to forecast the monthly and seasonal climatic anomalies. A major strategy should consist of combined empirical and numerical modeling studies aimed at the recognition of nonrandom and potentially predictable quantities of soil moisture or sea surface temperature. It must be borne in mind here that the climatic signals are most easily detected in certain regions of the globe.

An analysis of the responses of global and regional climates to regional-scale changes in the properties of the surface is very important. These changes take place under the influence of both natural conditions (ice and snow cover extent, desert surface) and anthropogenic activities (urbanization, deforestation), including heat and moisture releases by power plants. The local-scale consequences of these changes have been well studied, but they require more reliable quantitative estimates. The effects on regional and global climates continue to be studied inadequately, though rapid development of man's activities emphasizes the growing urgency of these problems. It is most important to intensify fundamental studies of the processes of the atmosphere-ocean interaction as well as to continue developing regional and global climate models. To accomplish it, the program of satellite observations must be enhanced (like an experiment on studies of the ERB) and more efficient computing facilities must be provided.

The strategy of climate studies should foresee the development in the following interdependent directions: the climatic database and methods for their analysis, studies of the key climatic features (for example ENSO, monsoons); analysis of significant climate-forming processes (especially the cloud-radiation interaction); a statistical analysis of the climatic data; the development of climate models; analysis of climate sensitivity to various factors; and estimations of the predictability of the climatic means and anomalies.

With respect to enhancing the volume and diversity of climate data, further development of satellite observations is of primary importance. Studies of the water cycle and of latent heat as factors of atmospheric energetics require more reliable data on evaporation and precipitation for every region of the globe. To understand the atmosphere-land surface interaction, data on soil moisture, albedo, snow and vegetation cover, and evapotranspiration are needed. A consideration of the ocean-atmosphere interaction is based on data on SST and the temperature of the mixed layer. Information on sea ice is very important, too. To determine a possible sun-induced climatic variability, continuous measurements are needed of the solar irradiance, UV radiation, and high-energy particles, using satellites, balloons and aircraft. Further success of the WCRP and its components (TOGA, WOCE, GEWEX, ISCCP, ISLSCP), as well as a realization of the International Geosphere-Biosphere Program will open up new horizons in this respect.

To plan an observational system in the interests of the climate problem is essential to having more reliable substantiation of observational data, based on further development of climate models. Of course, the process of planning will go step-by-step, as the observational system is enhanced and the process of numerical climate modeling develops. The first stage, now under realization, is based on the use of data from operational meteorological satellites and on a combination of conventional means of observation on the basis of the programs, such as GCOS, GOOS, GTOS and others.

Experience revealed particular urgency of studying some climatic phenomena, which are relatively independent components of the global climate change, including:

- The mode of circulation in the equatorial Pacific as a separate dynamic system in which interactions between large-scale fields of wind and precipitation, the atmospheric and ocean temperatures at the interface, and between the sea currents are manifested;
- The monsoon circulation (probably, it is connected with the equatorial mode or QBO in the tropical stratosphere), which is, apparently, the most important climatic subsystem and can seriously affect the climate of the midlatitude NH;
- The QBO, the blocking (the anticyclonic systems in the middle latitudes which can be persistent for many days to a week and sometimes cause catastrophic droughts); long planetary-scale waves.

A special unresolved problem is climatic impact of explosive volcanic eruptions.

From the viewpoint of understanding the climatic laws, the following physical processes are of the first priority:

- The cloud-radiation interaction as well as the effect of natural (first of all, volcanic) and anthropogenic aerosols on the radiative regime of the atmosphere, cloud formation, and climate change.
- The atmosphere-ocean interaction, the role of the ocean in the accumulation and transport of heat (contributions from the oceanic gyres, the wind-driven overturning of surface waters, and small-scale horizontal diffusion); the land-surface processes (the heat and moisture budgets at the surface level, a consideration of the soil moisture, albedo, snow cover and ice cover characteristics and evapotranspiration); the biospheric impact.
- The effect of CO_2 and other GHGs on the atmospheric radiative regime and climate.
- The solar-climatic interrelationships; in this context interaction between the stratosphere and troposphere is of special importance.

In view of the rudimentary state of development in atmospheric chemistry, studies of the GHG that either enter the atmosphere or leave it due to natural and anthropogenic processes, and the role of the GHGs in the physical processes that govern weather and climate, have become very important. Priority must be given here to the aspects of atmospheric chemistry that are directly relevant to the climate problem. This concerns, in particular, studies of the global cycles of the optically active minor components—greenhouse and aerosol-producing gases (firstly, water vapor, carbon dioxide, ozone, sulfur and nitrogen compounds) governing the atmospheric radiative regime, as well as the components affecting the optically active compounds (the hydroxyl radical, the oxides of nitrogen, hydrocarbons, and chlorofluorocarbons). Considerable progress in atmospheric chemistry can be reached only with an enhancement of the program of global measurements of the atmospheric chemical composition using aircraft, balloons and satellites. Laboratory studies of the rates of chemical reactions, studies of biogeochemical cycles, wash-out processes, and interactions between chemical and meteorological processes are very urgent.

Gas-to-particle reactions producing aerosol particles from gaseous compounds claim special attention. The first reactions of concern comprise a chain of processes of SO_2 oxidation into H_2SO_4 and the formation of sulfate aerosols. About half the mass and 90% of the surface of atmospheric aerosol particles (both liquid and solid) are products of gas-to-particle conversion. The fact that the surface area of all the particles exceeds that of the globe, illustrates possibilities of reactions on the surface of particles with gas components. Of special interest in this context are such powerful natural "experiments" as explosive volcanic eruptions that are accompanied by heterogeneous processes of gas-to-particle conversion leading to stratospheric sulfate aerosol formation. An important but still not adequately studied climate-forming role of volcanic stratospheric aerosol requires substantiation and accomplishment of relevant observing systems (see Table S1).

Table S1. Principal Observational Methods Used to Study Stratospheric Aerosols. After Kent and Yue (1991). [© by American Geophysical Union]

Technique	Notes
In situ Aerosol size, composition Wire impactor Particle size spectrometer	Aircraft ceiling limits sampling to lower stratosphere.
Quartz crystal microbalance Balloon-based particle counter	Numerous flights to 30 km by University of Wyoming; long-term database
Remote ground-based Lidar backscatter	Numerous stations, mainly in the NH; considerable data
Solar photometer optical depth	Long-term data set from Mauna Loa, Hawaii.
Airborne: lidar backscatter	Numerous flights by NASA-LaRC, particularly at times of volcanic activity and at high latitudes in winter
Satellite solar occultation	Considerable database from SAM II, SAGE I, and SAGE II.
Thermal emission and limb scattering	SME database exists from 1981.

Intensive studies in the field of regional atmospheric chemistry and pollution, which have been recently initiated, gain importance. An important component of pollution is acid rain due to industrial emissions (primarily, SO_2) that are converted into acid compounds harmful to forests, gardens, lakes and crops, and, to a lesser extent, construction. Simultaneously, the visibility worsens. To assess the extent of regional pollution extensive investigations are needed of the atmospheric chemical composition of key processes governing the atmospheric composition and the long-range transport of pollutants, as well as changes that they exhibit prior to scavenging. Such studies must be based on both laboratory and field measurements (surface and aircraft) and on numerical modeling.

Bearing in mind the importance and urgency of the various problems, the following order of their priority must be established: a better understanding of the weather-forming processes, and weather forecasts, with emphasis on precipitation in cyclones and processes that determine dangerous meteorological conditions; further studies of the climatic system and its variability on a time scale of seasons to decades including assessment of volcanic impact; studies of biogeochemical cycles and balances, as well as their connection with atmospheric processes.

Problems of the solar-climatic relationships and the impact on climate deserve serious attention, too. The results from studies obtained during the last two decades have given new evidence for numerous feedbacks between the various manifestations of solar activity, weather, and climate. For example, it is established that during the so-called Maunder Minimum (1645-1715), when the sun had almost no spots, the weather of Europe was much colder, compared to the norm. Similarly, earlier periods were found from analysis data of tree rings, and ice cores, and historical documents. The planned observations from satellites will enable one to accomplish a more adequate, reliable and continuous monitoring of the processes on the sun, which opens prospects for intensive searches for physical mechanisms of the sun-weather-climate relationships. In this context, the following studies claim serious attention: variations in the chemical composition of the upper atmospheric layers caused by variations in the electromagnetic and corpuscular emissions of the sun; the electrical interaction between different atmospheric layers due to solar activity, response of stratospheric and mesospheric dynamics to a heating of the ozone layer caused by varying UV radiation; and dynamic stratosphere-troposphere interactions, to determine a possible impact of the upper atmosphere warming on the tropospheric dynamics.

REFERENCE

Kent, G.S., and Yue, G.K., 1991: The modeling of CO_2 lidar backscatter from stratospheric aerosol. *J. Geophys. Res.* **96** (D3), 5279-5292.

GLOSSARY OF ABBREVIATIONS AND ACRONYMS

A	albedo
ABC	aerosol backscattering coefficient
ACS	anthropogenic climatic signal
AE	atmospheric emission
AEC	aerosol extinction coefficient
AGC	atmospheric general circulation
AGCM	atmospheric general circulation model
AGT	atmospheric greenhouse trapping
AOD	atmospheric optical depth
AOT	aerosol optical thickness
ARM	Atmospheric Radiation Measurement Program
ARMA	autoregressive moving average
ASR	absorbed solar radiation
ATC	Ångstrom turbidity coefficient
ATSR	along-the-track scanning radiometer
AVHRR	Advanced Very High Resolution Radiometer
BR	band ratio
BT	brightness temperature
CAC	Climate Analysis Center
CAENEX	Complete Atmospheric Energetics Experiment
CCM	Community Climate Model
CEL	Centre d'Essai des Landes et Biscarosse
CHAMMP	climate change monitoring and modeling
CI	complex indicator
CLIM	development of small satellites to study climate
CN	condensation nuclei
CNRS	Centre National de la Recherche Scientifique
COADS	Comprehensive Ocean-Atmosphere Data Set
DJF	December, January, February
DJFM	December, January, February, March
DOE	Department of Energy
DR	depolarization ratio

ABBREVIATIONS AND ACRONYMS

DS	dustsonde
DVI	dust veil index
ECMWF	European Centre for Medium Range Weather Forecasts
EF	eigenfunction
EN	El Niño
ENSO	El Niño/Southern Oscillation
EOF	empirical orthogonal functions
EOS	Earth Observation System
ERB	Earth's radiation budget
ERBE	Earth Radiation Budget Experiment
FD	fixed dynamic heating
FT	fixed air temperature
GAAREX	Global Atmospheric Aerosol Radiation Experiment
GCM	General Circulation Model
GCMAM	Global Climate Middle Atmosphere Model
GEWEX	Global Energy and Water Cycle Experiment
GFDL	Geophysical Fluid Dynamics Laboratory
GHG	greenhouse gases
GIRS	Geodynamic Laser Ranging System
GISS	Goddard Institute for Space Studies
GMT	Greenwich Meridional Time
GOES	Geostationary Observation Environmental Satellite
GOMR	Global Ozone Monitoring Radiometer
GPI	global precipitation index
GR	global solar radiation
GVaP	Global Water Vapor Project
GVI	global vegetation index
GWP	global warming potential
HIRIS	High-Resolution Imaging Spectrometer
IDV	interdecadal variability
IPCC	Intergovernmental Panel of Experts on Climate Change
IR	infrared radiation
ISCCP	International Satellite Cloud Climatology Project
ITCZ	Intertropical Convergence Zone
ITIR	Intermediate Thermal Infrared Radiometer
JJA	June, July, August

VOLCANIC ACTIVITY AND CLIMATE

LST	lower stratosphere temperature
LW	longwave radiation
MAM	March, April, May
MECCA	Model Evaluation Consortium for Climate Assessment
MFR	multichannel flux radiometer
MGC	minor gaseus components
MIC	multi-ion complexes
MISR	Multiangle Imaging Spectro-Radiometers
MODIS	Moderate Resolution Imaging Spectrometer
MLS	microwave limb sounder
MSD	mean-square-deviation
MSU	microwave sounding unit
N-CH	Nyamuragira-El Chichón
NAO	North-Atlantic Oscillation
NASA	National Aeronautics and Space Administration
NCAR	National Center for Atmospheric Research
NDVI	normalized difference vegetation index
NH	Northern Hemisphere
NOAA	National Oceanic and Atmospheric Administration
NRBS	non-Rayleigh backscatter
OHP	Observatory of Haute-Provence
OLR	outgoing longwave radiation
OSR	outgoing shortwave radiation
PNA	Pacific - North American
PSC	polar stratospheric clouds
PW	precipitable water
QBO	quasi-biennial oscillation
RB	Radiation Budget
RF	Radiative Forcing
RMSD	root mean-square difference
RSSN	relative sun spots number
SAGE	Stratospheric Aerosol and Gas Experiment
SAM	Stratospheric Aerosol Monitor
SAR	Synthetic Aperture Radar
SAT	surface air temperature

SBUV	solar backscattered ultraviolet radiation
SC	solar constant
SCE	snow cover extent
SD	standard deviation
SH	Southern Hemisphere
SICE	sea ice cover extent
SO	Southern Oscillation
SOI	Southern Oscillation Index
SON	September, October, November
SR	scattering ratios
SRB	surface radiation budget
SRFG	stratospheric radiative forcing gradient
SME	Solar Mesosphere Explorer
SSA	supersonic stratospheric aviation
SSM/I	Special Sensor Microwave/Imager
SSMR	channel 22 GHz
S/N	signal to noise
SST	sea surface temperature
STF	spectral transmission functions
SW	shortwave radiation
TES	Tropospheric Emission Spectrometer
TOC	total ozone content
TOGA	Tropical Ocean and Global Atmosphere
TOMS	Total Ozone Mapping Spectrometer
TOVS	Tiros Operational Vertical Sounder
TP	total precipitation
TT	tropospheric temperature
UKMO	United Kingdom Meteorological Office
UV	ultraviolet radiation
VEI	volcanic explosivity index
VF	volcanic forcing
VS	volcanic signal
WFOV	wide field of view
WV	water vapor

SUBJECT INDEX

Absorption
- aerosols 1, 235, 138, 257
- coefficients 139, 230, 231, 250
- continuum 139
- functions 142
- index 250
- molecular 256
- pressure-induced 139

Acid rains 21, 291
Actinometric observations 110, 189, 253, 254, 260, 262, 280
Aerological observations 235, 298
Aerosols (particles) 1
- absorption 138, 292, 337
- angular distribution 254
- anthropogenic 4, 5, 7, 11, 48, 290, 295
- atmospheric content 11, 260
- background 7, 176, 180, 198, 199, 200, 252, 295, 297, 309
- backscattering 11, 13, 175, 177, 189, 230, 232, 262, 264, 265, 296, 297
- chemical composition 2, 5, 6, 7, 8, 16, 21, 138, 179
- climatic impacts 8, 12, 117
- coagulation 182, 199, 204, 206, 209, 211, 212, 228, 253, 263
- complex refractive index 4, 8, 14, 18, 21, 198, 253, 255-257, 260, 261, 312, 175, 249-251, 325, 265, 309, 311
- imaginary 198, 250, 251, 255-257, 260
- colloidal system 22
- cooling 5, 58, 250, 255
- concentration 127, 170, 179, 181, 183, 189, 200, 204, 206, 208, 209, 214, 226, 249, 250, 294
- condensation 211, 214, 262, 263
- condensation nuclei 1, 13, 21, 49, 87, 125, 182-184, 187, 197, 199, 204, 209, 210, 225-227
- counter 196
- continental 327
- daily variations 7
- density 169, 254
- deposition
    - dry 3
    - wet 3

SUBJECT INDEX

Aerosol(s) particles
- desert 9
- diffusion 186, 189, 211
- disintegration 1
- dispersion 214, 250, 263
- emissions 3, 12
- evaporation 203
- evolution 5
- extinction coefficient 203, 264, 265
- formation 118, 205
    - nucleation 128, 131
- gas-to-particle conversion 1, 4, 6, 87, 136, 168, 169, 189, 197, 199, 203, 204, 206, 208, 210, 211, 222, 226 252, 253, 138, 292, 308
- generation 6
- gravitational settling 197, 206, 208, 228
- global distribution 132, 198, 203, 214, 217, 220, 263
- growth 205, 206, 208, 210, 211
- heterogeneous chemistry 2, 4, 7, 21, 195, 196, 215, 219, 335, 336, 338, 339, 345
- homogeneous 336
- ion clusters 209
- layer 178, 195, 204, 223, 231, 256, 257, 312, 336
    - multilayer 166
- local pollution 251
- maritime 327
- man's impact 1, 7
- mass 131, 132, 169, 176, 196, 200, 205-207, 209, 215, 216, 223, 226, 253, 254
- mineralogical composition 179
- mixing ratio 180, 181
- models 15, 16, 203, 214, 233, 234, 237, 256
- modeling properties 4, 6, 16, 183, 185, 218
- natural 5
- nucleation 186, 187, 197, 214, 227
- number density 140, 181, 196, 199, 204, 205, 209, 226, 251, 263, 265, 266
- optical depth 9, 14, 15, 19, 20, 96, 98, 104, 108, 118, 120, 200-203, 322, 323, 332, 358
- optical properties 2, 5, 7, 9, 16, 250
- optical thickness 89, 111, 112, 114, 137, 138, 168, 169, 171, 175, 176, 189, 194, 197, 204, 215-218, 231, 232, 236, 249, 253, 259, 252-255, 261, 262, 265, 267, 268, 271, 294-296, 306, 312, 313, 315, 321, 322, 323, 329, 330
    - mapping 218

- organic components 6
- oxidation 263
- particle formation 16, 167, 182
- particle removal 6
- phase function 15, 138, 250, 253, 255, 255, 294
    -asymmetry 249, 254, 294
- physical properties 5
    -size 249
    -shape 5, 249, 250, 255
- radiative characteristics 13
- Saharan 327
- scattering 13, 197
- sedimentation 182, 186, 196, 253, 263, 211
- silicate 181, 196, 200, 223, 293, 306, 312, 325
- size distribution 5, 14, 16, 115, 139, 179, 180, 182, 183, 187, 189, 199, 200, 204, 214, 226, 276, 298, 250, 253 255, 257, 259, 261, 263, 325, 330
    - bimodal 131, 173, 198, 204, 210, 222, 226 , 227, 232
    - gamma 138, 140
    - Junge 254, 261
    - lognormal 170, 182, 209, 253, 263
    - modified gamma-distribution 254, 294
    - monomodal 222
    - multimodal 140, 167, 185
- stratospheric 7, 10, 14, 15, 87, 131, 133, 170, 171, 179, 201, 204, 206, 229, 236, 249-252, 254, 255, 262, 265, 267, 294, 295, 306, 308, 316, 325, 327, 332-334
- sulfate 2, 5, 10, 59, 71, 133, 136, 177, 178, 181, 185, 199, 200, 214, 223, 226, 227, 252, 306, 308, 340
- sulfuric acid 111, 118, 175, 178, 179, 182, 187, 189, 196, 197, 199, 206, 207, 209, 219, 222, 227, 228, 292-295, 304, 339, 344
    - droplets 179, 197, 203, 209
    - layer 196
    - vapor 179, 187, 197, 204, 205, 209
    - water solution 206
- sulfur dioxide 177-179, 185, 187, 197, 199, 263, 265
- transport 7, 206, 227
- tropospheric 7, 332, 333
- types 6, 357
- urban 257, 327
- vertical profile 172, 225, 227, 264, 334
- volcanic 138, 208, 218, 223, 227, 233, 235, 251, 252, 263, 297, 315, 316, 318, 327, 336

SUBJECT INDEX

Aircraft observations 16, 49, 86, 89, 107, 110, 115, 117, 118, 123, 127, 132, 167-169, 177, 179, 180, 189, 196, 199, 200, 213, 215, 221, 250, 251, 262, 268, 292, 345
- Fourier spectrometer 224, 228
- lidar sounding 175, 196, 202, 206, 223, 251, 252, 262
- photometer 196

Albedo 10, 68, 219, 252, 296, 311, 330, 341
- cloud 11, 20
- feedback 311
- Earth 3, 150
- planetary 136, 145, 148, 150, 151, 169, 202, 309, 310, 320, 321
- single scattering 14, 15, 148, 151, 198, 294, 296, 326
- surface 10, 55, 149, 154, 293, 315, 325
- surface-atmosphere system 222, 294, 296, 312
- mean global 154, 251, 294
- tropospheric 57

Angstrom turbidity coefficient 171, 176

Anthropogenic climate signal 56, 59, 60
- activities 353
- variations 43

Arctic haze 2, 22

Arid regions 2

Ash 1, 172, 178, 181, 185--188, 199, 173, 224, 293, 308, 333
- clouds 167-169, 292
- chemical composition 198
- ejection 167, 168, 217
- layer 188
- particles 178, 179, 222, 182, 331
- plume 175
- silicate 200, 292
- size distribution 176

Atmosphere
- Earth 134, 136
  - initial 145, 147, 150
  - evolution 144
- Venus 134

Atmospheric chemistry 353

Atmospheric emission 55

Atmospheric pollution 356

Atmospheric transmission 280
- transparency 176, 252-254, 259, 260, 262, 293
  - infrared 167, 201
  - spectral 200, 202, 204

VOLCANIC ACTIVITY AND CLIMATE 369

Atmospheric turbidity 171, 259
Backscattersonde 228
Balloon observations 6, 117, 122, 131, 176, 204, 206, 210, 211, 213, 225, 227, 345
    - photoelectric counter 225
Biomass burning 4
Bioproductivity 4
    - oceanic 4
Brewer spectrometer 178
Carbon dioxide (CO_2) signal 30, 175, 201, 287, 279, 280, 292, 331
    - absorption bands 140, 256
    - concentration 58, 59, 66, 67, 69, 73, 75, 79, 96, 113, 114, 234, 314, 322, 324
    - distribution 325
    - doubling 9, 12, 35, 42, 43, 60, 61, 62, 63, 70, 71, 79, 80, 297, 329, 330
        increased 297, 353
    - radiative disturbances 10
Canonic correlation technique 75
Cascade impactor measurements 126, 226
Chlorine 178, 185, 224, 332, 238, 345
Chlorhydric acid 187, 224, 331, 335
Chlorofluorocarbons (CFC) 67, 75, 129, 266, 353
Circulation cells
    - Ferrel 328
    - Hadley 73, 309, 313, 328, 345
    - Walker 73
Climate 1, 168, 189, 220, 344
    -anthropogenic global change warming 65, 322
    - cloud feedback 355
    - deforestation 353
    - fluctuations 37
    - forecasting 355
    - impact 166, 233, 235, 238
    - models 10
    - numerical modeling 2, 38, 43, 66, 73, 74, 313
    - sensitivity 9
    - signal 216
    - stratospheric 50
    - variability 36, 291
        - interannual 38
        - interdecadal 37
        - long-period 65
        - low frequency 36

- natural 1
- warming 32

Climate change 233, 329
- anthropogenic 297, 322
- global 213, 322
- interdecadal 37
- interannual 38
- long-term 39
- volcanically induced 278, 297

Clouds
- aerosol 225
- amount 309
- condensation nuclei 173
- cover 329
- distribution 325
- formation 2, 87
- height 329
- microphysical properties 7, 13
- mother-of-pearl 266
- multimodal size distribution 140
- noctilucent 266
- polar stratospheric 21
- radiative characteristics 13
- total amount 323

Coefficient
- absorption 139
- attenuation 15, 139, 265, 266
- backscattering 177, 189, 203, 204, 222, 223
- extinction 14, 169, 170, 264, 217, 267, 294
- scattering 15
- transmission 251

Comprehensive Ocean-Atmosphere Data Set (COADS) 30, 47
Cooling 290
- global scale 291, 329

Condensation 326
- anthropogenic 13
- latent heat 326

Conveyor belt 40
- Atlantic 40

Correlation Spectrometer (COSPEC) 126, 171, 214, 223
Cyclones 23
- blocking 355
- evolution 355

- orographically induced 74
- trajectories 355

Delphi experiment 56, 57
Delta-Eddington approximation 140, 315
Dendroclimatic data 280
Dimethylsulfide 1, 3, 4, 5, 94
- biogenic emissions 4
- natural 3
- oceanic emissions 3, 4
- oxidation 5

Dobson spectrophotometer 178, 196, 337
Dry-fogs 291
Droughts 72, 95
Dust 1, 172, 332
- aerosols 151
- cloud 166
- ejection 166
- load 9, 180, 309, 311
- mineral 6
- outbreaks 9, 218
- road 7
- Saharan 9, 218
- storms 151
- stratospheric 326
- Veil Index 279, 280
- volcanic 95, 96, 307

Dustsondes 181, 182, 226, 266, 268
Earth
- orbital parameters 278
- Observation System (EOS) 56, 220
- Radiation Budget (ERB) 9, 50, 51, 54, 55, 69, 116, 118, 120, 218, 219, 255, 295, 322-324

Earth Radiation Budget Experiment (ERBE) 1
Eddy energy 329
Electrochemical concentration cell (ECC) 335
El Niño/Southern Oscillation (ENSO) 211, 212, 219, 234, 237, 279, 284, 323
- climatic impact 16, 18, 19, 32, 39, 41, 47, 50, 51, 52, 55, 64, 73, 114, 115, 286, 290, 298, 305-307

Empirical orthogonal functions (EOF) 50, 51, 52
Environment 1
- anthropogenic loading 1

Equilibrium
- convective 308
- radiative 308

Evaporation 309-311
Field experiments
- CAENEX 2
- GAAREX 2
Fluorhydric acid 335
Gas ejections 166
Gas phase chemistry 331, 339
General atmospheric circulation 36, 37, 40, 73, 184
Geopotential anomalies
- Pacific North American 53
Glaciological volcanic index
Global change 1
Global Climate Observing System 305
Global Vegetation Index (GVI) 87
Greenhouse climate signal 37 56
Greenhouse effect 9, 39, 61, 63, 113, 136, 292, 294, 295, 297
- CO_2 induced 61
Greenhouse gases 12, 41, 43, 51, 58, 59, 67, 71, 76, 77, 88, 237, 322, 343, 344, 357
Greenhouse warming 15, 38, 39, 67, 237, 324
Haze 332, 333
Heat
- advection 330
- balance 9, 311
- diabatic 311
- divergence 9, 257, 308, 314
- exchange 255
- latent 311
- turbulent 310
-transport 315
- horizontal 315
- vertical 315
Henyey-Greenstein function 294
Humidity 341
- profile 341
- relative 16, 315
- specific 65
Hurricanes 70, 71
- activity 40, 41
Hydrocarbon compounds 7
Hydrological cycle 41, 325, 311, 326
- processes 8
Hydroxyl radicals 4, 7,

Ice
- ages 278, 329
- core 291
    - acidity 93, 100, 279, 280
    - analysis 97, 118
    - data 87, 279
- concentration 54
- crystals 173
- distribution 74
- cover
    - extent 54
- thickness 42, 54
- sea 329

Industrial emissions 4, 7
- sulfur dioxide 358

Industrial revolution 7, 237
Insolation 33, 180, 229, 294, 296, 309
- extra-atmospheric 33
- variations, 33

Interferometer (Fourier) spectrometer 215
Intertropical convergence zone (ITCZ) 51, 313, 314, 322
IPCC Report 67, 71
Knollenberg photoelectric counter 196, 199
Kuwait 225
- field fires 225

La Niña 47
Lidar observations 107, 108, 166, 169-171, 175, 181, 189, 196, 215, 223, 229-230, 235, 358, 251, 261, 263, 264, 338, 339
- depolarization ratio 223, 224
- Raman 232
- Rayleigh-Mie 230
- scattering ratio 169, 267-269

Lindzen effect 63
Little Ice Age 94, 292
Linke's turbidity factor 194
Long-wave radiation 8, 15, 293, 316, 322, 325
- upward flux 8, 316
- absorption 62, 136, 250, 293, 295, 316

Magma
- ejection 166

Mars 134, 135, 142
- aerosols 151

- albedo
    - planetary 151, 155
    - single scattering 151
    - spectral 154
    - surface 154
- atmospheric circulation 151
- atmospheric greenhouse effect 152, 153. 155
- atmospheric composition 151
- CO_2, 151
- climatic impacts 151
- dust
    - clouds 155
    - storms 151
- general circulation 152
- long-wave radiation absorption 154, 155
- Mariner 151
- optical thickness 151, 152
- pressure 154
- rain 155
- relative humidity 154
- temperature 154
    - vertical profile 152-153
- radiative-convective equilibrium 154
- radiative cooling 151
- sensible heat exchange 151
- subterranean water 155
- sulfur mixing ratio 153
- SW radiation absorption 151, 154, 155
- thermal emissions 154
- Viking probe 155
- volcanic activity 151, 155
    - eruptions 154

Mauna Loa 134, 174, 196, 215, 228, 252-255, 259, 280, 312, 334, 337
Maunder minimum 359
Meridional circulation 311
    - energy transfer 311
Mie theory 232, 249, 257
Minor gaseous components 217, 352
Models 343
    - Aerosol Global Circulation Model (AGCM) 60, 70, 77, 324, 325
    - Autoregressive Moving Average Model (ARMA) 35
    - Coupled Climate Sea-Ice (CCSI) 42

- Climate Change Monitoring and Modeling Program (CHAMMP) 6 66, 69
- Colorado State University Regional Atmospheric Model 75
- Coupled Climate Sea-Ice Model (CCSI) 42
- Energy balance climate models 329, 330
- ECMWF 36, 45, 46, 327
- GISS Global Climate Middle Atmosphere Model (GCMAM) 236, 237, 328, 329
- Geophysical Fluid Dynamics Laboratory (GFDL) 64, 324
- Model Evaluation Consortium for Climate Assessment (MECCA Phase I) 66
- MOGUNTIA 3
- NCAR Community Climate Model 44, 64, 214
- Photochemical models 79, 315, 334
- United Kingdom Meteorological Office (UKMO) 64, 69, 71, 73, 74
- Zero dimensional empirical models 307
- Zero dimensional energy balance models 307
- Zonal averaged-chemical-radiative-dynamical model 340
- 1-D energy balance models 307
- 2-D zonal model 307, 308
- 1-D Time-dependent physico-chemical model 185
- 1-D radiative-convective climate 132, 138, 307
- 1-D interactive-radiative convective climate model 315
- 3-D general circulation models 307

Monsoons 73
- circulation 316

Net long-wave radiation 310, 311, 323
Nitrogen oxides 59, 181, 215, 233, 266
Normalized difference vegetation index (NDVI) 55
North-Atlantic Oscillation (NAO) 37
North Polar Vortex 228
Nuclear explosions 280
Numerical modeling 3, 10, 12, 29, 39, 41, 59, 62, 68, 71, 72, 138, 297, 307, 311, 313, 325, 341
Ocean 257, 322
- thermal inertia 325, 326

Orographic effects 73, 325
Outgoing long-wave radiation (OLR) 10, 19, 48, 51, 55, 58, 60, 219, 252, 291, 292, 322-324
Outgoing shortwave radiation 218, 219, 252, 337
Oxidized nitrogen 169
Oxyhydril (OH) 185, 187, 196

Oxygen 340
Oxygen-hydrogen-nitrogen chemistry 334
Ozone 75, 178, 185, 188, 194, 197, 224, 233, 239, 256, 258, 266, 292, 310, 315, 316, 331, 338, 343, 346
- absorption bands 201
- catalysis 186
- climatology 338
- concentration 225, 236, 266, 332, 334, 335, 340, 341
- distribution 325
- depletion 50, 78, 79, 129, 215, 236, 319, 321, 331-334, 337-340, 341, 342
- dynamics 238, 335
- hole 79, 80, 238, 344
- layer 181, 213, 238, 353, 293, 308, 334, 338
- measurements 331, 338, 340
- mixing ratio 315, 316, 332
- molecules
    - disintegration 177
- observations (ozonesonde) 188, 336, 338, 339, 343
- photodissociation 239, 332, 334
- total content (TOC) 39, 55, 63, 78, 79, 128, 211, 215, 216, 238, 239, 319, 332, 335, 336-342, 343-345
- tropospheric 67, 338, 339, 341-343, 346
- vertical profile 76, 315-317, 336, 338, 339, 341, 343

Pacific-North American anomalies 53
Paleoclimate 168, 291
- ice-core acidity 93, 94
- mid Holocene 71
- tree-rings 93
Photochemical processes 293
- Photodissociation rate 315
- Photolysis rate 340
- reactions 315
Photometer 200
- Eppley 196
- solar 176
- Volz 171
Precipitation (rainfall) 33, 65, 72, 73, 75, 181, 306, 309, 313, 353
- acidity 2
- index 47
- meridional distribution 313
Pyrheliometer 180, 196, 259, 260

Quartz crystal micro-balance cascade impactor 177, 196
Quasi-biennial oscillation (QBO) 105, 108, 188, 189, 235, 236, 259, 289, 290, 304, 336-339, 345, 356
 - droughts 356
Radiation budget 9, 14, 15, 105, 108, 168, 249, 294, 323
Radiative
 - cooling 202, 343
 - flux divergence 292, 312, 315, 316, 325
 - forcing 215, 330, 341-344
 - processes 293
 - regime 252, 256, 312, 319, 321
 - warming (heating) 168, 202, 256, 257, 328
Radiometer (multichannel) 215
Radiosonde network data 196, 280, 235, 236
 - observations 50, 204, 211
Rayleigh scattering 189, 223, 264, 265
Research vessels observations 230, 231
Saharan dust outbreaks 9
Sahel region 41, 72
Salinity 33
 - great salinity anomaly 38
Satellite observations 16, 44, 56, 67, 107, 131, 189, 196, 206, 213, 216, 235, 267
 - AVHRR 49, 55, 56, 215, 216, 218, 222, 267
 - GOES 48, 172, 196
 - Microwave sounding unit (MSU) 18, 19, 44, 45, 46, 47, 50, 53, 55, 3376
 - Nimbus-3 266
 - Nimbus-6 50
 - Nimbus-7 170, 196, 199, 215, 216, 265, 337, 344
   -7 THIR/TOMS 50, 322
 - NOAA-2 219, 337
 - NOAA-6 168-171, 267, 289
 - SAGE-1 169, 170, 215, 216, 222, 239, 264
 - SAGE 2 203, 217, 239, 266, 338, 339, 341
 - SAM-2 223, 265-267
 - SBUV 219, 335, 337-339
 - SSM/I 51
 - Solar Mesosphere Explorer (SME) 197
 - TOMS 187, 215, 216, 219, 220, 224, 337-341, 344
 - TOVS 49
Sea ice cover
 - extent 53, 54

Short-wave radiation 8, 15, 292, 322, 325, 343
- absorption 1, 10, 136, 138, 292, 297, 310, 292, 340
- backscattering 229, 232, 251, 252, 254, 261, 325
Smithsonian Astro-Physical Observatory 259
Solar constant 309, 310, 324
Solar radiation 170, 171, 175, 176, 180, 189, 194, 229, 293, 321, 337, 340
- absorption 196, 215, 255, 261, 201, 235, 304, 331
- characteristics 252
- direct 189, 194, 200, 215, 222, 224, 252-254, 321
- diffuse 171, 175, 176, 189, 222
- global 171, 184, 185, 189, 215, 222, 251-254, 261, 312, 321
- reflected 49, 293, 294
- scattered 176, 218, 219, 310, 321, 325
- spectral distribution 194, 256
- UV- absorption 331, 332
- visible 202, 331
Spectral Transmission Function (STF) 138, 139, 140, 142, 143, 146
Stratospheric Aerosol and Gas Experiment (SAGE) 128
- solar occultation 358
Stratospheric heating 169, 316, 325
- cooling 311
Smoke particles 6, 7, 226
Snow 280
- cover extent 51, 53, 311
- interannual variability 51, 53
Soils 42
Solar
- activity 122, 233, 234, 279
- aureole 176
- climatic impact 42, 359
- sun spots 42, 43
- 11-year cycle 42, 43
Solar constant 42, 62, 256, 294, 297, 322
- variations 311
Solar radiation 1, 111, 265, 311
- absorption 50, 88, 110
- attenuation 18, 104, 164
- diffuse (sky) 104
- diffuse-to-direct ratio 260, 261
- direct 104, 105, 251, 259, 260, 313
- global 9, 10, 18, 55, 251, 260, 311, 313, 324
- photometer 358

- reflected 20
- scattering 251, 313, 332
- transmission 265
- total 105

Solfataric activity 187
Southern Oscillation Index (SOI) 16, 73, 291
Standard atmosphere 316
Stratosphere 197, 216, 328, 329, 340
- aerosols 293
- arctic 89
- clouds 89
- circulation 269
- cooling 14, 234, 337
- dust loading 281
- heterogeneous reactions 89
- ions 121
- layer 289
- neutralization 122
- plumes 265
- radiative heating 292
- silicate ash 198
- warming 211, 215, 234-236, 281, 289-291, 316, 319, 321, 332
- water vapor 89

Sulfuric acid 131, 169, 215, 217, 293
- concentration 214, 263
- droplets 132, 210, 211, 226, 232, 237
- growth 211
- shell 312
- vapor 210, 211, 214
- water solution 226, 227

Sulfur compound emissions 91, 92
- volcanic 168
- yield 9, 91

Sulfur dioxide 200, 203, 207, 209, 228, 292, 293, 331, 332, 339, 340, 344
- absorption spectrum 137, 219, 228
- concentration 331
- distribution 225
- emissions 10, 13, 168, 169, 171, 219, 224, 225
- photochemistry 119, 120, 339
- mass 218
- mixing ratio 200, 207
- oxidation rate 119, 207

- stratospheric 21, 337, 339, 341-343, 346
- total content (TSC) 215-217

Sun photometers
- aircraftborne 267
- balloonborne 267
- ground-based 267
- lidars 267

Superimposed epoch method 285, 287, 288, 306

Surface cooling 321

Tectonic plates: American, Cocos and Caribbean 187

Temperature 236, 334, 337
- anomalies 34, 216, 281, 301, 302
- brightness 44, 4
- gradient 173
- global 286, 289, 291, 299, 308
- sea surface (SST) 9, 51, 55, 58, 64, 70, 72, 73, 74, 101, 114, 211, 212, 218, 253, 254, 280, 286, 290, 298, 301, 313, 328, 329
- stratospheric 19, 105, 106, 234-236, 308, 312
- surface 9, 10, 29, 30, 59, 168, 184, 185, 189, 203, 278-283, 288, 291, 293, 309-311, 326
- surface air (SAT) 10, 53, 63, 67, 70, 75, 107, 212, 219, 233-237, 237, 260, 280, 297, 298, 301, 305, 306, 308, 322-324, 328, 330
- volcanic eruptions 101, 314
- trends 31, 110, 112, 113
- tropospheric 19, 96, 281, 299, 304, 327, 328
- variations 113, 285
    - interannual 19
- vertical profile 9, 14, 56, 312, 315-318,
- zonal mean 37

Thermal emission 1, 197, 236, 314

Thermal infrared radiation 320

Tropical Ocean and Global Atmosphere (TOGA) 63

Troposphere 211
- cooling 169, 212, 318, 337

Umkehr observations 332-334

Urban heat islands 29

Urbanization 357

Visibility 16, 171, 176, 180, 188

Volcanic eruptions 14, 42, 86, 95, 295, 331
- Agung-Awu 17, 87, 95, 96, 98, 100, 103, 106, 108, 109, 111, 114, 118, 120, 129, 175, 211, 212, 235, 236, 251-253, 255, 259-261, 280, 281, 283, 287, 289-291 293, 304, 305, 311, 321, 332, 333

- Alaid 204, 208
- Amargura 387
- Andesitic 213
- Ash 89, 98, 110, 111, 118, 125, 129, 312
- Bezymiannaya 111
- Cerro Hudson 89, 236
- climatic effects 115
- column height 89
- Cosiguina 87, 286
- dust 102, 180
- Dust Veil Index 86, 110, 116
- El Chichón 14, 17, 50, 87, 91, 95, 96, 103, 104, 106, 108, 111, 117, 118, 120, 121, 129, 131, 132, 133, 166, 168, 169, 181, 187, 190, 194-200, 203, 204, 208-213, 216-218, 223, 226, 229, 230, 233, 235-237, 250-253, 259, 260, 262, 267-269, 271, 289-293, 298, 299, 301, 311, 314-317, 319-322, 324, 330, 334-337, 345
- explosive 87, 111, 166
- explosivity index 86
- Fuego 106, 129, 131, 132, 174, 175, 183, 211, 252, 255, 259-263
- glaciological index 86
- Krakatau 87, 92, 95, 96, 99, 102, 103, 109, 117, 184, 212, 253, 287, 299, 324, 325, 333
- Katmai 17, 93, 95, 96, 104, 108, 109, 168, 212, 217, 299
- Kilauea 174
- Ksudach 17, 168
- Laki crater 94, 97, 109
- Lamington 260
- mass eruption rate (intensity) 82, 99
- magma ejection 91, 98, 103, 118, 120, 127, 130, 166
- Mount St. Helens 95, 96, 97, 107, 108, 109, 117, 130, 131, 132, 178, 181, 183, 185, 186, 189, 204, 208, 209, 250, 251, 254, 255, 262, 266, 267, 271, 345
- Mt. Pelée 17, 78, 111, 212, 299
- Mount Pinatubo 18, 19, 39, 49, 87, 95, 109, 117, 166, 168, 170-172, 174-176, 213-218, 222-225, 227-231, 233, 235-237, 298, 324, 330, 335, 337-339, 344, 345
- Nevado del Ruiz 91, 203, 217
- Nyamuragira 18, 19, 199, 201, 262, 290
    - andesyte pyroclasts 188
- Pacaya 126, 129, 174
- Pagan 262
- Paluweh 260

- phreatic 174
- plinian 89, 121, 167
- Puyehue 260
- Quizapu 260
- St. Augustine 125, 130, 177
- Santa María 17, 109, 212, 299
- Santiaguito 126, 129, 174
- Sierra Negra 255, 262
- Soufriere 17, 111, 212, 255, 263, 265, 299
- Spurr 260
- stratospheric veil 99
- Tambora 86, 93, 98, 99, 100, 101, 109, 110, 286
- tephra 97
- Toba 90
- total eruption mass (magnitude) 89
- transport 90
- Ulawun 262

Volatile compounds (CO_2, HCl, HF) 181, 182, 198, 206, 226, 227

Volcanic cloud 185, 219, 331, 333
- thermal emission 167
- trajectory 169, 171
- transport 251

Volcanic explosivity index (VEI) 217, 234

Volcanic hazards 220

Volcanic plume 124, 169, 171-173, 175, 177, 198, 221, 229, 230, 293
- height 306
- water vapor 130

Volcanic signal 19, 29, 92, 211, 236, 287, 299, 300, 305, 306

Water vapor 65, 185, 209, 308, 330, 331
- absorption 197, 256
- concentration 169, 315, 319
- condensation 196
- content 49, 51, 260, 309, 310, 311
- distribution 325
- dynamics 225
- production 353

Weather 3

Wind field
- zonal 216, 231, 326, 338
- synoptic 326

Wire impactor 196